Praise for *Genomics in the Cloud*

This book captures the essence of what's been learned about bringing genomics to the cloud. And it lays out an accessible path for newcomers to join this exciting and important ecosystem.

—*Eric S. Lander, Founding Director,*
The Broad Institute of MIT and Harvard

This book is a fantastic introduction to modern genome analysis using state-of-the-art tools and practices. It covers everything a reader needs to get their own analyses running in an open, repeatable way. This is the quintessential primer on the GATK and cloud-based analysis with Terra.

—*Jonathan Smith, Principal Software Engineer,*
The Broad Institute of MIT and Harvard

This is a great primer about reproducible bioinformatics in the cloud. Geraldine and Brian are at the forefront of this field, so we are learning from the best. And for those who have yet to work with Terra, look no further for an excellent introduction to it!

—*Jessica Maia, Data Scientist, BD*

Transferring from physics to cancer research as I did, I learned genomics, sequencing, statistics piecemeal. I could have used a book like this back then, because no matter how much time you've spent in the field or if it's your first contact, there's something new to learn and an appreciation for the bigger picture to be gained.

—*Aaron Chevalier, PhD Candidate, Boston University*

Genomics in the Cloud
Using Docker, GATK, and WDL in Terra

Geraldine A. Van der Auwera and Brian D. O'Connor

Beijing · Boston · Farnham · Sebastopol · Tokyo

Genomics in the Cloud

by Geraldine A. Van der Auwera and Brian D. O'Connor

Published by O'Reilly Media, Inc., 1005 Gravenstein Highway North, Sebastopol, CA 95472.

O'Reilly books may be purchased for educational, business, or sales promotional use. Online editions are also available for most titles (*http://oreilly.com*). For more information, contact our corporate/institutional sales department: 800-998-9938 or *corporate@oreilly.com*.

Acquisitions Editor: Rebecca Novack
Development Editor: Michele Cronin
Production Editor: Katherine Tozer
Copyeditor: Octal Publishing, LLC
Proofreader: Sharon Wilkey

Indexer: Ellen Troutman-Zaig
Interior Designer: David Futato
Cover Designer: Karen Montgomery
Illustrator: Rebecca Demarest

April 2020:   First Edition

Revision History for the First Edition
2020-04-02:   First Release
2020-09-11:   Second Release

See *http://oreilly.com/catalog/errata.csp?isbn=9781491975190* for release details.

978-1-491-97519-0

[LSI]

Table of Contents

Foreword

I migrated from mathematics into the field of genomics in 1985—roughly a year before the field officially came into existence. The word *genomics* was coined in 1986, which also saw the first public debate, at the Cold Spring Harbor Laboratory, about the notion of mounting a Human Genome Project.

It's hard to imagine how much has changed since then. Computers hardly figured in biomedicine—the initial design for the Whitehead Institute for Biomedical Research, founded in the early 1980s, included no provision for a computer. Large amounts of data were seen as a nuisance, not an asset—in a Nature article reporting on the Human Genome Project debate, the journal's biology editor wrote, "If the skill and ingenuity of modern biology are already stretched to interpret sequences of known importance, such as those of the DMD and CGD genes, what possible use could be made of more sequences?"

Despite such doubts, biologists eventually decided to press on—launching the Human Genome Project, their first major data gathering effort, in 1990. One of the important motivations was the prospect of deploying systematic methods—rather than guesswork—to discover the genes responsible for human diseases. In 1980, a brilliant biologist, David Botstein, had conceived how to find the location of genes for rare monogenic diseases by tracing their inheritance in families relative to a genetic map of DNA variants across the human genome. Realizing the full power of the idea, though, would require mapping—and eventually sequencing—the entire human genome.

The Human Genome Project was an extraordinary collaboration that spanned six countries and twenty institutions, took thirteen years, and cost $3 billion. When the dust settled, the world had the three billion nucleotide-long DNA sequence of a single human genome.

With this project completed, many biologists thought that business would return to usual. But what happened next was even more remarkable. Over the next 15 years, biology became an information science—in which the generation of massive amounts of data reshaped the field. For example:

- Genetic mapping in families revealed the genes responsible for more than 5,000 serious rare monogenic disorders.

- New kinds of genetic mapping in populations led to the discovery of ~100,000 robust associations of specific genetic regions with common diseases and traits.

- Genetic analysis of thousands of tumors uncovered hundreds of new genes in which mutations propelled cancer.

Remarkably, the cost of sequencing a human genome fell by a factor of five million—from $3 billion to $600—and the cost is likely to reach $100 in the coming years. More than one million genomes have been sequenced so far. Overall, genomic data of all kinds is doubling roughly every eight months.

None of this would have been possible without the development of powerful new computational methods and tools to work with the many new types of data that were being generated. A good example is the Genome Analysis Toolkit, developed by colleagues at the Broad Institute, which you'll read a lot more about in this book.

Today, life sciences are in the midst of new data explosions. Many countries are undertaking systematic efforts to collect genomic and medical data into national biobanks, which will give researchers the ability to probe even further into the genetics of both common and rare diseases and traits. It will be especially important to ensure that the world's full genetic diversity is represented in these large-scale efforts—not just people of European descent.

Because of the amazing technological progress in recent years, we can now read out not just the DNA blueprint, but how this blueprint is read out as RNA in individual cells. Methods have been developed to read out gene expression at the single-cell level, with an initial analysis of 18 cells soon leading to analyses of more than 18 million cells. This work has given rise to an international Human Cell Atlas project, involving more than 60 countries around the world. These datasets are beginning to make it possible to use computational methods, including modern machine learning, to systematically infer the underlying circuitry of cells.

As the biological applications burgeon, though, we are often held back by systemic limitations in how we access and share data. Most of the world's biomedical data has traditionally been held in silos—accessible only through servers from which each authorized researcher or group must download their own copies to their own institution's computing infrastructure. From a purely technical standpoint, this is unsustainable. Instead of bringing data to researchers, we need systems that allow researchers to operate on the data where it resides. We also need more transparent models for managing custody of the data, as well as efficient ways to assess, enforce and audit who can access the data and for what purpose. We should aim to abide by these four principles: (1) copying data should not be the default mode of sharing data; (2) security and auditing should be baked in and enterprise-grade; (3) large-scale analysis should be accessible to all research groups; and (4) computational resources should be elastic, so that they can be scaled up or down as needed.

Cloud computing has emerged as the leading solution for the technical aspect of these challenges. In practice, though, it creates new obstacles that require creative solutions.

At the Broad Institute, we started moving to the cloud four years ago, to cope with the rising tide of genomic data. We cut our teeth by converting our genomic data-processing operation from a traditional on-premises system to one that runs on the cloud from the moment the data is generated in our genome sequence platform. That move required rethinking every aspect of the process and building entirely new systems from scratch to handle the terabytes of data that come streaming off sequencing machines every day. But that was just the beginning. Once the data was up on the cloud, we hit the next obstacle: the available cloud services, in their current state, can be daunting to use for life sciences researchers without advanced training. So, we teamed up with partners to develop a software and analysis platform, Terra.

Other such platforms also have emerged as the move to the cloud has picked up steam in biomedical research. Today we are working with many other groups to build a federated data ecosystem of interconnected components that offer complementary services and capabilities. We expect these platforms will help facilitate the kind of open collaboration that is needed to bring together data, tools, and expertise spanning multiple domains and disciplines. We also want to lower the technical thresholds for individual researchers to participate in the cloud-based ecosystem, especially those with fewer IT resources at their disposal.

By all accounts, the transition of genomics to the cloud is still in its early phases. At the Broad Institute, we've learned many hard lessons on our own journey to the cloud, and we're learning more every day. In a time of such disruptive change, it's essential that groups share their experiences with each other.

That's why I'm so excited that the incomparable Geraldine Van der Auwera, longtime advocate for the research community at the Broad Institute, and Brian O'Connor, an ardent campaigner for software and data interoperability at UCSC, have written this book. The book captures the essence of what we have learned so far, and lays out an accessible path for newcomers to join the genomics cloud ecosystem.

— Eric S. Lander, Founding Director,
The Broad Institute of MIT and Harvard

Preface

If cloud technology is the future of biomedical science, then for genomics, the future is already here.

Genomics is the first biomedical discipline to move en masse to the cloud. Perhaps inevitably so, given that it was the first to experience explosive growth in data generation, leading to rapidly escalating compute and storage requirement issues that a cloud infrastructure is ideally positioned to address. Major genomic datasets and their derived resources are now available in the cloud, and many tools like the industry-leading Genome Analysis Toolkit (GATK) produced by the Broad Institute are now offered in forms optimized to run efficiently on a cloud infrastructure. As a result, many researchers making use of genomic data and related analysis tools are now or will soon be confronted with the need to learn to use cloud resources, which can represent a huge challenge to many. Meanwhile, many informatics and bioinformatics support staff are being pulled in to help researchers to achieve this transition, sometimes with only minimal or no training relevant to the science of genomics. Taken together, these two populations form a continuum of people who need to get on the same page and work together to solve the challenges they face.

Purpose, Scope, and Intended Audience of This Book

With this book, we aim to provide a hands-on orientation tour of major tools, mechanisms, and processes involved in performing genomic analysis in the cloud that can serve as a middle ground for the majority of people on this spectrum. We try to assume as little prior knowledge as possible, and we provide two primer-style chapters, one focused on genomics and one on technology, to ensure that everyone has a firm grounding in the fundamental concepts we rely on from both domains. In addition, we deliberately chose a particular open source technology stack—GATK, Workflow Description Language (WDL), Terra, Docker, and Google Cloud Platform—that provides end-to-end functionality and is backed by robust user support systems in order to guarantee a successful educational experience.

To be clear, this book is not intended to be comprehensive, either in terms of tooling options or the scientific scope of genomic analyses. Our operational definition of genomics, centered on variant discovery and immediately related analyses, is intentionally narrow; and for every step of the processes we describe, there often exist several, if not many, alternative tools that you could substitute for those we chose to showcase. However, we designed the topics and exercises presented here to provide patterns and takeaways that are largely transferable and extensible to other tools and analyses in order to maximize their long-term value to readers. In addition, we plan to release a series of companion blog posts and other online materials that will show complementary approaches using different platforms and technologies; see the book's GitHub repository (*https://oreil.ly/genomics-repo*) and its companion website (*https://oreil.ly/genomics-blog*).

What You Will Learn from This Book

The very idea of doing genomics in the cloud might seem intimidating on first approach, especially if you're new to either one or both, but it's not as complicated as you might think. Throughout this book, we walk you through all of the important pieces of the puzzle, step by step. You'll have the opportunity to run genomic analyses involving the GATK, selected for their broad appeal and interesting computational approaches. You'll do so first through the "bare" services provided by the Google Cloud Platform (GCP) and then on Terra, a scalable platform for biomedical research codeveloped by the Broad Institute and Verily, an Alphabet company, on top of GCP.

By the end of the book, you should expect to have learned or achieved the following:

- Fundamentals of computational infrastructure and processes
- Fundamentals of genomics including biological underpinnings, formats, and conventions
- Beginner- to intermediate-level hands-on usage of the core technology stack:
  - GATK, WDL, Terra, Docker, and Google Cloud
  - GATK Best Practices for variant discovery as formulated by the GATK development team at the Broad Institute, covering germline short variants, somatic short variants, and somatic copy-number alterations
  - Reading, authoring, and interpreting analysis workflows, first in a sandbox environment and then at scale through several modes of execution (from a standalone command-line package to a fully managed system)
  - Managing data and workflow execution in a workspace environment
  - Performing interactive analysis using Jupyter Notebooks

— Tying it all together: achieving computational reproducibility in publications through the use of cloud data storage, synthetic data generation, portable workflows, and containerized tools

- Secondary goals
  — Increased familiarity with computational concepts such as scaling and optimization approaches
  — Practical experience with several bioinformatics command-line packages, common commands, and file formats

What Computational Experience Is Needed for the Exercises?

For the exercises in Chapter 4 through Chapter 10, we assume that you are already somewhat familiar with command-line fundamentals, including the basics of navigating directories and interacting with text files in a Bash shell; composing and running simple commands; and the concepts of environment variables, path, and working directory. For Chapter 8 through Chapter 11 and Chapter 13, we assume that you are familiar with the concept of writing scripts, though we do not require you to have practical experience doing so. For Chapter 12 and Chapter 14, we assume that you have heard of the programming languages R and Python, and you will find it easier to understand the more complex examples if you have some familiarity with their syntax, though it is not required.

If at any point during the exercises you feel out of your depth in terms of the computational tooling and terminology, we recommend that you check out the lessons provided by the Software Carpentry (*https://software-carpentry.org*) organization, which are specifically designed for research scientists who have not had formal computational training. The lessons on the Unix shell (*https://oreil.ly/bnGo3*) can be particularly helpful if you don't have any prior command-line experience. They also have sets of lessons on Python (*https://oreil.ly/j73Ht*) and on R (*https://oreil.ly/400VG*) as well as other topics relevant to the book like version control with Git (*https://oreil.ly/85cEo*). These lessons are all open source and developed by volunteers in the community who understand the everyday challenges faced by researchers, so they're a truly fantastic resource.

Conventions Used in This Book

The following typographical conventions are used in this book:

Italic
Indicates new terms, URLs, email addresses, filenames, file extensions, table names and components, and workflows.

`Constant width`
> Used for program listings, as well as within paragraphs to refer to program elements such as variable or function names, databases, data types, environment variables, statements, and keywords.

`Constant width bold`
> Shows text that should be typed literally by the user.

`Constant width italic`
> Shows text that should be replaced with user-supplied values or by values determined by context.

$ *before code*
> Indicates a command run in the VM shell

before code
> Indicates a command run in the docker container

 This element signifies a note.

Using Code Examples

Supplemental material (code examples, exercises, full-size color figures, etc.) is available for download on GitHub (*https://oreil.ly/genomics-repo*).

This book is here to help you get your job done. In general, if example code is offered with this book, you may use it in your programs and documentation. You do not need to contact us for permission unless you're reproducing a significant portion of the code. For example, writing a program that uses several chunks of code from this book does not require permission. Selling or distributing examples from O'Reilly books does require permission. Answering a question by citing this book and quoting example code does not require permission. Incorporating a significant amount of example code from this book into your product's documentation does require permission.

We appreciate, but generally do not require, attribution. An attribution usually includes the title, author, publisher, and ISBN. For example: "*Genomics in the Cloud* by Geraldine A. Van der Auwera and Brian D. O'Connor (O'Reilly). Copyright 2020 The Broad Institute, Inc. and Brian O'Connor, 978-1-491-97519-0."

If you feel your use of code examples falls outside fair use or the permission given above, feel free to contact us at *permissions@oreilly.com*.

O'Reilly Online Learning

 For more than 40 years, *O'Reilly Media* has provided technology and business training, knowledge, and insight to help companies succeed.

Our unique network of experts and innovators share their knowledge and expertise through books, articles, and our online learning platform. O'Reilly's online learning platform gives you on-demand access to live training courses, in-depth learning paths, interactive coding environments, and a vast collection of text and video from O'Reilly and 200+ other publishers. For more information, visit *http://oreilly.com*.

How to Contact Us

Please address comments and questions concerning this book to the publisher:

O'Reilly Media, Inc.
1005 Gravenstein Highway North
Sebastopol, CA 95472
800-998-9938 (in the United States or Canada)
707-829-0515 (international or local)
707-829-0104 (fax)

We have a web page for this book, where we list errata, examples, and any additional information. You can access this page at *https://oreil.ly/genomics-cloud*.

Email *bookquestions@oreilly.com* to comment or ask technical questions about this book.

To learn more about our books, courses, and news, visit *http://www.oreilly.com*.

Find us on Facebook: *http://facebook.com/oreilly*

Follow us on Twitter: *http://twitter.com/oreillymedia*

Watch us on YouTube: *http://www.youtube.com/oreillymedia*

Acknowledgments

We would like to thank our countless colleagues at the Broad Institute and at the University of California, Santa Cruz (UCSC), who contributed in so many ways to making this book a reality.

We are hugely indebted to all the past and present members of the frontline support and education teams in the Data Sciences Platform at the Broad Institute who developed and maintain the original educational materials and resources on which we based many of the hands-on exercises presented in this book. Within the education team led by Robert Majovski, we'd like to highlight the work of Soo Hee Lee, whose thoroughness and exacting attention to detail produced some of the deepest resources available about GATK tools; Allie Hajian and Anton Kovalsky, who are tasked with the Herculean feat of documenting how to use Terra even as it wriggles and evolves from underneath them; and Kate Noblett, who wrote much of the original WDL documentation and now coordinates GATK, WDL, and Terra workshops with an iron hand. Within the frontline support team led by Tiffany Miller, we'd like to highlight the work of Beri Shifaw, who maintains the gatk-workflows pipelines on GitHub and in Dockstore as well as the featured workspaces in Terra; and Bhanu Gandham, who has so enthusiastically taken on the responsibility of obsessing about the well-being of the GATK user community. Other contributing members from these two teams, past and present, include Derek Caetano-Anolles, Sushma Chaluvadi, Sheila Chandran, Elizabeth Kiernan, David Kling, Ron Levine and Adelaide Rhodes.

We also recognize and appreciate the growing role played by the Broad DSP Field Engineering team led by Alexander Baumann in this arena. Star among the stars, Yvonne Blanco swooped in from the User Experience team to improve key diagrams and illustrations with her impeccable design mojo.

We are eternally grateful to the many members of the GATK development team who have provided critical input to educational resources and lent their expertise in GATK workshops across the globe. There are too many of them to enumerate here, but within that team, we would like to highlight the invaluable support of Eric Banks, Laura Gauthier, Yossi Farjoun, and Lee Lichtenstein; the seemingly endless patience of David Benjamin and Sam Lee; the unflappable aplomb of David Roazen and jovial fatalism of Louis Bergelson; the quiet expertise of Mark "Duplicates" Fleharty and the cheerful expertise of Megan Shand. Special shout-out also to Chris Norman for his work on the Barclay library, which powers the GATK documentation system.

On a more personal level, Geraldine would like to thank Mauricio Carneiro and Mark De Pristo, past member and founder of the original GATK team, respectively, for taking a chance and hiring a confused microbiologist all those years ago.

Speaking of too many to count, we could not begin to name everyone involved in the development of the chapters on WDL, Cromwell, and Terra, but we'd like to put in a special mention for Adrian "Notebooks Guy" Sharma, William Disman, Ruchi Munshi, and Kyle Vernest, who all contributed helpful insights and put up with our constant badgering about issues we hoped to see addressed before the book came out. On that note, we owe a big thank you to Chris Llanwarne and Adam Nichols for patching womtool just in time for Chapter 9 to make a lot more sense than it would have otherwise. And speaking of badgering, our deepest apologies go to Eric Karofsky and Jerôme Chadel from the User Experience team, who had to endure a constant barrage of questions about what elements of the Terra interface would change next and on what timeline. We're deeply grateful to Matthieu J. Miossec for collaborating with us to develop the project we present in Chapter 14.

Within the UCSC GI, we want to thank the Computational Genomics Platform (CGP) team whose members work on a variety of projects that leverage Terra and other cloud-based analysis ecosystem components we present in this book. Contributors include Jesse Brennan, Amar Jandu, Natan Lao, Melaina Legaspi, Geryl Pelayo, Charles Reid, Hannes Schmidt, and Daniel Sotirhos. Within CGP, the Lighthouse Point team—Michael Baumann (now at the Broad Institute), Lon Blauvelt, Brian Hannafious, and Ash O'Farrell, led by Beth Sheets—deserves special recognition for their role in writing excellent research tutorials that helped inspire sections of the book.

We also want to thank the Dockstore teams at both UCSC and the Ontario Institute for Cancer Research (OICR) for their feedback on this effort and support building a platform for workflow sharing that contributes to the Terra ecosystem. Charles Overbeck leads the technical team at UCSC, and we are grateful for contributions by Louise Cabansay, Abraham Chavez, Andy Chen, Trevor Heathorn, Nneka Olunwa, Kevin Osborn, Natalie Perez, Walter Shands, Emily Soth, Cricket Sloan, and David Steinberg. Denis Yuen leads the technical team at OICR with Lincoln Stein as the PI and contributions from Ryan Bautista, Kitty Cao, Andy Chen, Vincent Chung, Andrew Duncan, Victor Liu, Gary Luu, Shreya Radesh, and Jennifer Wu.

None of this would have been possible without the support of our respective leadership teams. At the Broad Institute, we would like to thank Eric Lander, Lee McGuire, and the Data Sciences Platform leaders, particularly Anthony Philippakis, Eric Banks again, and Danielle Ciofani for keeping the faith that this book would eventually materialize. At UCSC, we thank the Genomics Institute (GI) leadership including Benedict Paten and the institute director, David Haussler, for their support along with Greta Martin, whose organizational skills are unrivaled, and Nadine Gassner, who keeps us funded so that we can work on cool projects.

We are forever grateful to the reviewers who took the time to read through early draft versions in order to help us identify what didn't work reliably and to understand what could be improved. The book you see before you is very different from what we originally gave them to evaluate, for the better. In this category, we salute Titus Brown, Aaron Chevalier, Jeff Gentry, Sean Horgan, Lynn Langit, Lee Lichtenstein, Jessica Maia, David Mohs, Andrew Moschetti, Anubhav Shelat, and Jonn Smith.

We are also incredibly grateful to the editorial team at O'Reilly who performed the truly magical feat of turning our manuscript—a loose conglomerate of Google Docs —into an actual book. In particular, we thank our development editor, Michele Cronin, for shepherding us from early drafts to the finished product. It took a lot of cajoling and a few stern reminders about deadlines to get us there.

Last but most definitely not least, we would like to thank our loved ones for their patience and support during the more than two years that it took us to produce this book. Geraldine hopes that her lovely wife, Jessica, and daughters, Gabrielle and Melanie, will be suitably impressed and somehow forget her many late nights, obsessive behavior, and general inability to complete any home-improvement projects during that time period. Meanwhile, Brian thanks his partner Dhawal for his infinite patience, understanding, and encouragement to finish the book, along with his mom (Patty) and dad (Jim) for providing the occasional and appreciated push to "get it done!"

## CHAPTER 1

Introduction

We live in a time of great opportunity: technological advances are making it possible to generate incredibly detailed and comprehensive data about everything, from the sequence of our entire genomes to the patterns of gene expression of individual cells. Not only can we generate this type of data, we can generate a *lot of it*.

Over the past 10 years, we've seen a stunning growth in the amount of sequencing data produced worldwide, enabled by a huge reduction in cost of short read sequencing, a technology that we explore in Chapter 2 (Figure 1-1). Recently developed and up-and-coming technologies like long-read sequencing and single-cell transcriptomics promise a future filled with similar transformative drops in costs and greater access to 'omics experimental designs than ever before.

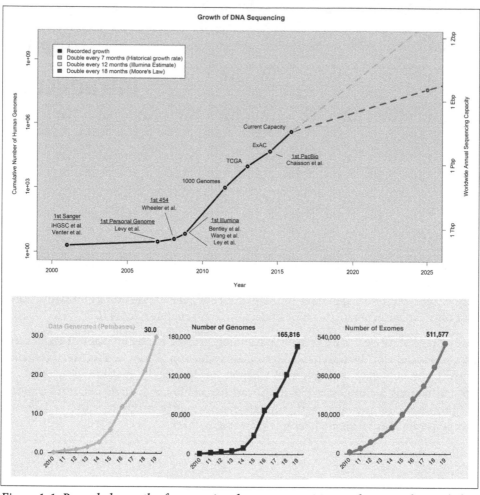

Figure 1-1. Recorded growth of sequencing datasets up to 2015 and projected growth for the next decade (top); growth in data production at the Broad Institute (bottom).[1]

---

1 Stephens ZD, et al. "Big Data: Astronomical or Genomical?" PLoS Biol 13(7): e1002195 (2015). *https://doi.org/10.1371/journal.pbio.1002195.*

The Promises and Challenges of Big Data in Biology and Life Sciences

Anyone, from individual labs to large-scale institutions, will soon be able to generate enormous amounts of data. As of this writing, projects that are considered large include whole genome sequences for hundreds of thousands of genomes. Over the next decade, we can expect to see projects sequencing millions of genomes and transcriptomes, complemented (and complicated) by a wide variety of new data types such as advanced cell imaging and proteomics. The promise is that the copious amounts of data and variety of new data types will allow researchers to get closer to answering some of the most difficult—if vexingly simple to ask—questions in biology. For example, how many cell types exist in the human body? What genetic variants cause disease? How do cancers arise, and can we predict them earlier? Because research is by its nature a team sport, we'll want to broadly share much of the data on the horizon, we'll want to share our algorithms to analyze this data, and we'll want to share the findings with the wider world.

Infrastructure Challenges

The dual opportunities of falling costs and expanded experimental designs available to researchers come with their own set of challenges. It's not easy being on the bleeding edge, and each new technology comes with its own complications. How do you accurately read single bases correctly as they whip on by through a nanopore? How do you image live cells in 3D without frying them? How do you compare one lab's single-cell expression data with another lab's while correcting for differences from batch effects? These are just a few examples in a very long list of technical challenges we face when developing or optimizing a new experimental design.

But the difficulty doesn't stop with the data generation; if anything, that's only the beginning. When the experiments are done and you have the data in hand, you have a whole new world of complexity to reckon with. Indeed, one of the most challenging aspects of 'omics research is determining how to deal with the data after it's generated. When your imaging study produces a terabyte of data per experiment, where do you store the images to make them accessible? When your whole genome sequencing study produces a complex mixture of clinical and phenotypic data along with the sequence data, how do you organize this data to make it findable, both within your own group and to the wider research community when you publish it? When you need to update your methodology to use the latest version of the analysis software on more than 100,000 samples, how will you scale up your analysis? How can you ensure that your analytical techniques will work properly in different environments, across different platforms and organizations? And how can you make sure your methods can be reproduced by life scientists who have little to no formal training in computing?

In this book, we show you how to use the *public cloud*—computational services made available on demand over the internet—to address some of these fundamental infrastructure challenges. But first, let's talk about why we think the cloud is a particularly appealing solution and identify some of the limitations that might apply.

This is not meant to be an exhaustive inventory of all available options; much like the landscape of experimental designs is highly varied, there are many ways to use the cloud in research. Instead, this book focuses on taking advantage of widely used tools and methods that include the Best Practice genomics workflows provided by the Genome Analysis Toolkit (*https://oreil.ly/6J8y5*) (GATK) (Figure 1-2), implemented using the Workflow Description Language (*https://openwdl.org*) (WDL), which can be run on any of the platform types commonly used in research computing. We show you how to use them in Google Cloud Platform (GCP) (*https://cloud.google.com*), first through GCP's own services, and then in the Terra (*https://terra.bio*) platform operated on top of GCP by the Broad Institute and Verily.

Figure 1-2. GATK provides a series of Best Practices to process sequence data for a variety of experimental designs.

By starting with this end-to-end stack, you will acquire fundamental skills that will allow you to take advantage of the many other options out there for workflow languages, analytical tools, platforms, and the cloud.

Toward a Cloud-Based Ecosystem for Data Sharing and Analysis

From Figure 1-1, you can see that data has already been growing faster than the venerable *Moore's law* can keep up, and as we discussed earlier, new experimental designs that generate massive amounts of data are popping up like mushrooms in the night. This deluge of data is in many ways the prime mover that is motivating the push to migrate scientific computing to the cloud. However, it's important to understand that in its current form, the public cloud is mainly a collection of low-level infrastructure components, and for most purposes, it's what we build on top of these components that will truly help researchers manage their work and answer the scientific questions they are investigating. This is all part of a larger shift that takes advantage of cloud-hosted data, compute, and portable algorithm implementations, along with platforms that make them easier to work with, standards for getting platforms to communicate with one another, and a set of conceptual principles that make science open to everyone.

Cloud-Hosted Data and Compute

The first giant challenge of this dawning big data era of biology lies in how we make these large datasets available to the research community. The traditional approach, depicted in Figure 1-3, involves centralized repositories from which interested researchers must download copies of the data to analyze on their institution's local computational installations. However, this approach of "bringing the data to the people" is already quite wasteful (everyone pays to store copies of the same data) and cannot possibly scale in the face of the massive growth (both in dataset numbers and size) that we are expecting, which is measured in *petabytes* (PB; 1,000 terabytes).

In the next five years, for example, we estimate that the US National Institutes of Health (NIH) and other organizations will be hosting in excess of 50 PB of genomic data that needs to be accessible to the research community. There is simply going to be too much data for any single researcher to spend time downloading and too much for every research institution to host locally for their researchers. Likewise, the computational requirements for analyzing genomic, imaging, and other experimental designs is really significant. Not everyone has a compute cluster ready to go with thousands of CPUs at the ready.

The solution that has become obvious in recent years is to flip the script and "bring the people to the data." Instead of depositing data into storage-only silos, we host it in widely accessible repositories that are directly connected to compute resources; thus, anyone with access can run analyses on the data where it resides, without transferring any of it, as depicted in Figure 1-3. Even though these requirements (wide access and compute colocated with storage) could be fulfilled through a variety of technological solutions, the most readily available is a public cloud infrastructure. We go into the details of what that entails in Chapter 3 as part of the technology primer; for now, simply imagine that a cloud is like any institution's high-performance computing (HPC) installation except that it is generally a whole lot larger, a lot more flexible with respect to configuration options, and anyone can rent time on the equipment.

Popular choices for the cloud include Amazon Web Services (AWS) (*https://aws.amazon.com*), GCP, and Microsoft Azure (*https://azure.microsoft.com*). Each provides the basics of compute and storage but also more advanced services; for example, the Pipelines API on GCP, which we use when running analysis at scale in Chapter 10.

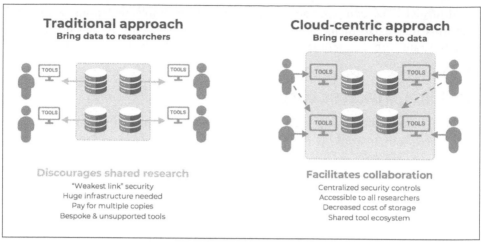

Figure 1-3. Inverting the model for data sharing.

Unlike the traditional HPC clusters for which you tend to script your analysis in a way that's very much dependent on the environment, the model presented in Figure 1-3 really encourages thinking about the portability of analysis approaches. With multiple clouds vying for market share, each storing and providing access to multiple datasets, researchers are going to want to apply their algorithms to the data wherever it resides. Highly portable workflow languages that can run in different systems on different clouds have, therefore, become popular over the past several years, including WDL (which we use in this book and explore more in Chapter 8), Common Workflow Language (*https://www.commonwl.org*) (CWL), and Nextflow (*https://www.nextflow.io*).

Platforms for Research in the Life Sciences

The downside of the move to the cloud is that it adds a whole new layer of complexity (or possibly several) to the already nontrivial world of research computing. Although some researchers might already have sufficient training or personal affinity to figure out how to use cloud services effectively in their work, they are undoubtedly the minority. The much larger majority of the biomedical research community is generally not equipped adequately to deal with the "bare" services provided by public cloud vendors, so there is a clear and urgent need for the development of platforms and interfaces tailored to the needs of researchers that abstract away the operational details and allow those researchers to focus on the science.

Several popular platforms present easy-to-use web interfaces focused on providing researchers a point-and-click means of utilizing cloud storage and compute. For example, Terra (which we explore in Chapter 11 and continue to use through the book), Seven Bridges (*https://www.sevenbridges.com*), DNAnexus (*https://*

www.dnanexus.com), and DNAstack (*https://dnastack.com*) all provide these sophisticated platforms to researchers over the web.

These and similar platforms can have different user interfaces and focus on different functionality, but at their core they provide a workspace environment to users. This is a place where researchers can bring together their data, metadata, and analytical workflows, sharing with their collaborators along the way. The workspace metaphor then allows researchers to run analysis—for example, on Terra this could be a batch workflow in WDL or Jupyter Notebook for interactive analysis—without ever having to dive into the underlying cloud details. We look at this in action in Chapters 11, 12, and 13. The takeaway is that these platforms allow researchers to take advantage of the cloud's power and scale without having to deal with the underlying complexity.

Federal Cloud Initiatives Supporting Biomedical Research in the US

Funding agencies such as the National Cancer Institute (NCI) in the United States have been leading the way of supporting the migration of biomedical research to public cloud infrastructure. One of the earliest examples of this shift to the cloud is the Cancer Genomics Cloud (CGC) pilot program funded by the NCI. That program supported the development of cloud infrastructure to host data from The Cancer Genome Atlas (TCGA), with appropriate security and access controls as well as colocated analysis tools.

The NCI granted awards to three applicants rather than a single winner: the commercial company Seven Bridges; the Institute for Systems Biology (ISB), a nonprofit research organization located in Seattle, WA; and the Broad Institute of MIT and Harvard, also a nonprofit research organization, located in Cambridge, MA. Seven Bridges offers the ability to run on either GCP or AWS, whereas the ISB and Broad Institute built their analysis platforms on top of Google Cloud. You can read more about the program and its evolution on the NCI blog (*https://oreil.ly/CrfE1*).

The NCI Cloud Pilots program has played a seminal role in enabling infrastructure development groups to get the ball rolling. For all three awardee groups and their collaborators, it has been a long road of identifying requirements, building out prototypes, and experimenting with technologies and interfaces involved in threading that finest of needles: empowering researchers with minimal computational background to get the most from massive computational resources, securely and efficiently. Yet today all three original pilot projects are operating stable services that are supporting thousands of users on a monthly basis, in many cases enabling important research that would not have been readily possible otherwise.

In this book, you will have the opportunity to use the Terra Community Workbench, a subset of the Terra platform developed by the Broad Institute and Verily based on our experiences with FireCloud (the Broad Institute's cloud pilot project). The Terra Community Workbench provides access to powerful data management and analysis capabilities for biomedical researchers. We use the name *Terra* throughout this book to refer to both the Community Workbench and the underlying capabilities of the Terra Platform that also support applications like the *All of Us* Researcher Workbench and Single Cell Portal.

Moving forward, the NIH, which is the largest biomedical research agency in the world, has committed to expanding the range of data and services available through public cloud infrastructure. In fact, the NIH is currently supporting multiple programs on the theme of hosting federally funded datasets and tools on the cloud. Likewise, other countries and funders have embraced using clouds for researchers as well, such as efforts like ELIXIR (*https://elixir-europe.org*) in Europe.

Standardization and Reuse of Infrastructure

So it sounds like multiple clouds are available to researchers, multiple groups have built platforms on top of these clouds, and they all solve similar problems of colocating data and compute in places that researchers can easily access. The other side of this coin is that we need these distinct data repositories and platforms to be interoperable across organizations. Indeed, one of the big hopes of moving data and analysis to the cloud is that it will break down the traditional silos that have in the past made it difficult to collaborate and apply analyses across multiple datasets. Imagine being able to pull in petabytes of data into a single cross-cutting analysis without ever having to worry about where the files reside, how to transfer them, and how to store them. Now here's some good news: that dream of a mechanism for federated data analysis is already a reality and is continuing to rapidly improve!

Key to this vision of using data regardless of platform and cloud are standards. Organizations such as the Global Alliance for Genomics and Health (*https://www.ga4gh.org*) (GA4GH) have pioneered harmonizing the way platforms communicate with one another. These standards range from file formats like CRAMs, BAMs, and VCFs (which you will see used throughout this book), to application programming interfaces (APIs) that connect storage, compute, discovery, and user identity between platforms. It might seem boring or dry to talk about APIs and file formats, but the reality is that we want the cloud platforms to support common APIs to allow researchers to break down the barriers between cloud platforms and use data regardless of location.

Software architecture, vision sharing, and component reuse, in addition to standards, are other key drivers for interoperability. For the past few years, five US organizations involved in developing cloud infrastructure with the support of NIH agencies and programs have been collaborating to develop interoperable infrastructure components under the shared vision of a Data Biosphere (*https://www.databiosphere.org*). Technology leaders at the five partner organizations—Vanderbilt University in Nashville, TN; the University of California, Santa Cruz (UCSC); the University of Chicago; the Broad Institute; and Verily, an Alphabet company—articulated this shared vision of an open ecosystem in a blog post on Medium (*https://oreil.ly/hsG1B*), published in October 2017. The Data Biosphere emphasizes four key pillars: it should be community driven, standards based, modular, and open source. Beyond the manifesto, which we do encourage you to read in full, the partners have integrated these principles into the components and services that each has been building and operating.

Taken together, community-based standards development in GA4GH and system architecture vision and software component sharing in Data Biosphere have moved us collectively forward. The result of these collaborative efforts is that today, you can log on to the Broad Institute's Terra platform, quickly import data from multiple repositories hosted by the University of Chicago, the Broad Institute, and others into a private workspace on Terra, import a workflow from the Dockstore (*https://dock store.org*) methods repository, and execute your analysis securely on Google Cloud with a few clicks, as illustrated in Figure 1-4.

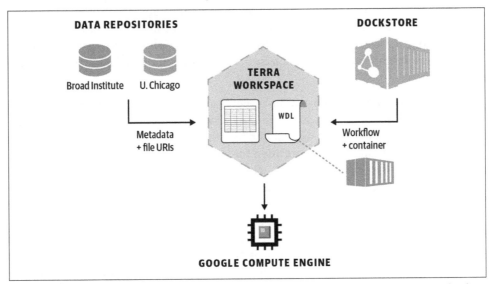

Figure 1-4. Data Biosphere principles in action: federated data analysis across multiple datasets in Terra using a workflow imported from Dockstore and executed in GCP.

To be clear, the full vision of a Data Biosphere ecosystem is far from realized. There are still major hurdles to overcome; some are, boringly, purely technical, but others are rooted in the practices and incentives that drive individuals, communities, and organizations. For example, there is an outstanding need for greater standardization in the way data properties are formally described in metadata, which affects searchability across datasets as well as the feasibility of federated data analysis. To put this in concrete terms, it's much more difficult to apply a joint analysis to samples coming from different datasets if the equivalent data files are identified differently in the metadata—you begin to need to provide a "translation" of how the pieces of data match up to one another across datasets (`input_bam` in one, `bam` in another, `aligned_reads` in a third). To solve this, we need to have the relevant research communities come together to hash out common standards. Technology can then be used to enforce the chosen conventions, but someone (or ideally several someones) needs to step up and formulate those in the first place.

As another example of a human-driven rather than technology-driven stumbling block, biomedical research would clearly benefit from having mechanisms in place to run federated analyses seamlessly across infrastructure platforms; for example, cloud to cloud (Google Cloud and AWS), cloud to on-premises (Google Cloud and your institution's local HPC cluster), and any multiplatform combination you can imagine on that theme. There is some technical complexity there, in particular around identity management and secure authentication, but one important obstacle is that this concept does not always align with the business model of commercial cloud vendors and software providers. More generally, many organizations need to be involved in developing and operating such an ecosystem, which brings a slew of complications ranging from the legal domain (data-use agreements, authority to operate, and privacy laws in various nations) to the technical (interoperability of infrastructure, data harmonization).

Nevertheless, significant progress has been made over the past several years, and we're getting closer to the vision of the Data Biosphere. Many groups and organizations are actively cooperating on building interoperable cloud infrastructure components despite being in direct competition for various grant programs, which suggests that this vision has a vibrant future. The shared goal to build platforms that can exchange data and compute with one another—allowing researchers to find, mix, and match data across systems and compute in the environment of their choosing—is becoming a reality. Terra as a platform is at the forefront of this trend and is an integral part to providing access to a wide range of research datasets from projects at the NCI (*https://www.cancer.gov*), National Human Genome Research Institute (NHGRI) (*https://www.genome.gov*), National Heart, Lung, and Blood Institute (NHLBI) (*https://www.nhlbi.nih.gov*), the Human Cell Atlas (*https://www.humancellatlas.org*), and Project Baseline by Verily (*https://www.projectbaseline.com*), to name just a few. This is possible because these projects are adopting GA4GH APIs and common architectural principles of the Data Biosphere, making them compatible with Terra and other platforms that embrace these standards and design philosophies.

Being FAIR

So far, we've covered a lot of ground in this chapter, starting with the phenomenal growth of data in the life sciences and how that's stressing the older model of data download and pushing researchers toward a better model that uses the cloud for storage and compute. We also took a look at what the community is doing to standardize the way data and compute is made accessible on the cloud, and how the philosophy of the Data Biosphere is shaping the way platforms work together to make themselves accessible to researchers.

The benefits of this model are clear for platform builders who don't want to reinvent the wheel and are motivated to reuse APIs, components, and architectural design wherever possible. But, from a researcher's perspective, how do these standards from GA4GH and architecture from Data Biosphere translate to improvements in their research?

Taken together, these standards and architectural principles as applied in platforms like Terra enable researchers to make their research more FAIR: findable, accessible, interoperable, and reusable.[2] We dive into this in more detail in Chapter 14. But for now, it's useful to think that the work described up to this point by platform builders is all in an effort to make their systems, tools, and data more FAIR for researchers. Likewise, by embracing the cloud, writing portable workflows in languages like WDL, running analysis in Terra, and sharing workflows on Dockstore, researchers can make their own work more FAIR. This allows other researchers to find and access analytical techniques, interoperate, run the analysis in different places, and ultimately reuse tools as a stepping-stone to novel discovery. Throughout the book, we come back to the FAIR principles from the perspective of both platform builders and researchers.

Wrap-Up and Next Steps

Now that we've given you a background on some of the central motivations for why genomics as a discipline is moving to the cloud, let's recapitulate how this book aims to help you get started in this brave new world, as outlined in the Preface. We designed it as a journey that takes you through a progression of technical topics, with an end goal of addressing the aforementioned infrastructure challenges, ultimately showing you how to get your work done on the cloud and make it FAIR to boot.

Remember, there are many different ways to approach these challenges, using different solutions, and we're focusing on just one particular approach. Yet, we hope the following chapters will give you robust foundations to build on in your own work:

Chapter 2 and Chapter 3
    We explore the fundamentals of biology and cloud computing.

---

2 The FAIR Guiding Principles for scientific data management and stewardship (*https://oreil.ly/JyTlX*) by Mark D. Wilkinson et al. is the original publication of this set of principles.

Chapter 5 through Chapter 7

We dive into the GATK toolkit and the current Best Practices pipelines for germline and somatic variant discovery.

Chapter 8 and Chapter 9

We describe how to automate your analysis and make it portable with workflows written in WDL.

Chapter 10 and Chapter 11

We begin scaling up analysis first in Google Cloud, then in Terra.

Chapter 12

We complement workflow-based analysis with interactive analysis using Jupyter in Terra.

Chapter 13 and Chapter 14

We show you how to make your own workspaces in Terra and bring together everything you learned, to show you how to make a fully FAIR paper.

By the end of the book, we want you to have a good understanding of the current best practices for genomics data analysis, feel comfortable using WDL to express your analytical processes, be able to use Terra for both workflow-based and interactive analysis at scale, and share your work with your collaborators.

Let's get started!

## CHAPTER 2

Genomics in a Nutshell: A Primer for Newcomers to the Field

Now that we've given you the broad context of the book in Chapter 1, it's time to dive into the more specific scientific context of genomics. In this chapter, we run through a brief primer on key biological concepts and related practical information that you need to know in order to benefit fully from working through the exercises in this book. We review the fundamentals of genetics and build up to a definition of genomics that will frame the scientific scope of this book. With that frame in place, we then review the main types of genomics variation, with a focus on how they manifest in the genome. Finally, we go over the key processes, tools, and file formats involved in generating and processing high-throughput sequencing data for genomic analysis.

Depending on your background, this might serve as a quick refresher, or it might be your first introduction to many of these topics. The first couple of sections are aimed at readers who have not had much formal training in biology, genetics, and related life sciences. If you are a life-sciences student or professional, feel free to skim through until you encounter something that you don't already know. In later sections, we progressively introduce more genomics-specific concepts and tools that should be useful to all but the most seasoned practitioners.

Introduction to Genomics

So what is genomics? The short answer, "it's the science of genomes," doesn't really help if you don't already know what a genome is and why it matters. So before we get into the gory details of "how to do genomics," let's take a step back and make sure we all agree on what we mean by genomics. First, let's go over the fundamental concepts involved, including the basics of what a gene is and how it relates to DNA. Then, we progressively deepen our exploration to cover key types of data and analyses within the specific defined scope. Keep in mind that this will not be an exhaustive compilation of all things that could possibly be grouped under the term of genomics, given that (a) this is not meant to be a scientific textbook, and (b) the field is still evolving very rapidly, with so many exciting new methodologies and technologies being actively developed that it would be premature to attempt to codify them in a book like this one.

The Gene as a Discrete Unit of Inheritance (Sort Of)

Let's start at the historical beginning—the gene. It's a term that is almost universally recognized but not necessarily well understood, so it's worth going into its origins and unrolling its evolution (so to speak).

The earliest mention of a *gene* was based on concepts formulated by key 19th-century thinkers such as Charles "Origin of Species" Darwin and Gregor "Pea Enthusiast" Mendel, at a time when the advent of molecular biology was still well more than a century away. They needed a way to explain the patterns of partial and combinatorial trait inheritance that they had observed in various organisms, so they postulated the existence of microscopic indivisible particles that are transmitted from parents to progeny and constitute the determinants of physical traits. These hypothetical particles were eventually dubbed *genes* by Wilhelm Johannsen in the early 20th century.

A fistful of decades later, Rosalind Franklin, Francis Crick, and James Watson elucidated the now famous *double-helix* structure of DNA: two complementary strands, each formed by a backbone of repeated units of a molecule called *deoxyribose* (the D in DNA), which carry the actual information content in the form of another type of molecule called *nucleic acids* (the NA in DNA). The four nucleic acids in DNA, more generally called *bases*, are famously referred to by their initials A, C, G, and T, which stand for *adenine*, *cytosine*, *guanine*, and *thymine*, respectively. The reason DNA works as a double helix is that the molecular structure of the bases makes the two strands pair up: if you put an A base across from a T base, their shapes match up, and they attract each other like magnets. The same thing happens with G and C, except they attract each other even more strongly than A and T, because G and C form three bonds, whereas A and T form only two bonds. The implications of this simple biochemical complementarity are enormous; sadly, we don't have time to go into all of that, but we leave you with Crick and Watson's classic understatement in their original paper on the structure of DNA:

> It has not escaped our notice that the specific pairing we have postulated immediately suggests a possible copying mechanism for the genetic material.

Combine the concept of genes and the reality of DNA, and it turns out that the units we call genes correspond to fairly short regions that we delineate within immensely long strings of DNA. Those long strings of DNA are what make up our chromosomes, which most people recognize as the X-shaped squiggles in their high school biology textbooks. As illustrated in Figure 2-1, that X shape is a condensed form of the chromosome in which the DNA is tightly wound.

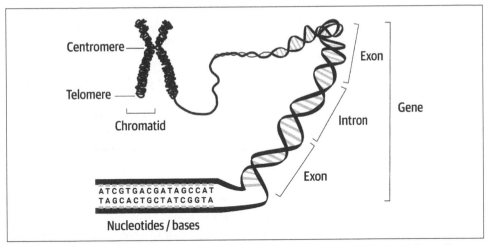

Figure 2-1. The chromosome (shown here in the form of two sister chromatids, each composed of one incredibly long molecule of double-stranded DNA) on which we delineate genes composed of exons and introns.

Like many other living creatures, human beings are *diploid*, which means that you have two copies of each chromosome: one inherited from each biological parent. As a result, you have two copies of every gene that is present on both chromosomes, which amounts to most of your genes. In a few cases, some genes might be present on one chromosome and absent from the other, but we won't go into those here. For a given gene, your two copies are called *alleles*, and they're typically not strictly identical in sequence to each other. If you produce a biological child, you pass on to them one chromosome from each of your 23 pairs and, accordingly, one of your two alleles for every gene. This concept of allele will be very important in the context of variant discovery, in which it will apply to individual variants instead of the overall gene.

Although the X-shaped form is their most common representation in popular media, our chromosomes don't spend much time looking like that. Most of the time they are in a rod-shaped form called a chromatid. When the single *chromatid* is copied ahead of a cell division event, the two identical chromatids are connected by a central region called the *centromere*, resulting in the popular X shape. Be careful not to confuse sister chromatids, which belong to the same chromosome, with the chromosome copies that you inherit from each biological parent.

If we look more closely at the span of a region of DNA that is defined as a gene, we see that the gene itself is split into smaller regions called *exons* and *introns*. When a particular gene is called into action, the exons are the parts that are going to contribute to its functionality, whereas the introns will be ignored. If you have a difficult time not mixing them up, try to remember that *ex*ons are the *ex*citing parts, and *in*trons are the bits *in* between.

All this suggests that genes are not really like the discrete particles that our forerunners originally envisioned. If anything, chromosomes are closer to being discrete units of inheritance, at least in the physical sense, given that each corresponds to an unbroken molecule of DNA. Indeed, many traits caused by different genes are inherited together because when you inherit one copy of a chromosome, there is a physical linkage between all the genes that it carries. Then again, the reality of chromosomal inheritance is more complex than you might expect. Just before the cell division that produces eggs and sperm cells, the parental chromosome copies pair up and swap segments with each other in a process called *chromosomal recombination*, which shuffles the combinations of physically linked traits that will be transmitted to the progeny. Heads-up: biology is messy that way.

Infinite Diversity in Infinite Combinations

Diploidy (the state of being diploid) might sound like the norm when you're human (or even an animal in general), but many creatures have more interesting chromosomal lifestyles. For example, microorganisms such as bacteria are generally *haploid*, meaning they have only one copy of their chromosome, but they can have an array of optional mini-chromosomes called *plasmids*. Fungi can be haploid or diploid at different stages of their life cycles.

As for plants—well, plants are the most open-minded of all, with abundant examples of *polyploid* plants such as bananas and some apples (*triploid*, 3 copies), potatoes (*tetraploid*, 4 copies), bread wheat (*hexaploid*, 6 copies), strawberries (*octaploid*, 8 copies), and some hybrids of sugar cane taking top prize (*dodecaploid*, 12 copies).

Now that we have a handle on the basic terminology and mechanisms of genetic inheritance, let's look into what genes actually do.

The Central Dogma of Biology: DNA to RNA to Protein

This central dogma is the next level in our understanding of what genes are and how they work. The idea is that the "blueprint" information encoded in genes in the form of DNA is transcribed into a closely related intermediate form called *ribonucleic acid*, or RNA, which our cells then use as a guide to string together amino acids into proteins. The main differences between DNA and RNA are that RNA has one different nucleotide (uracil instead of thymine, but it still sticks to adenine) and it's a lot less stable biochemically. You can think of DNA as the master blueprint, and RNA as working copies of sections of the master blueprint. Those working copies are what the cell will use to build proteins. These proteins range in function from structural components (e.g., muscle fiber) to metabolic agents (e.g., digestive enzymes) and ultimately act as both the bricks and bricklayers of our bodies.

The process as a whole is called *gene expression* (see Figure 2-2). It makes a ton of sense and explains a lot—yet this is not the complete picture. As a description of *a* gene expression process, it would have been true, but for several decades the theory went on to state that this flow of information was strictly unidirectional and comprehensive. Yep, you guessed it: just like the gene concept, this turned out to be somewhat wrong. We have since discovered various ways in which life violates this longheld dogma: for example, through DNA-to-RNA-only as well as RNA-only pathways, which modify gene expression in a variety of fun and baffling ways. Pathways also exist that lead to protein synthesis without the guidance of an RNA template, called *nonribosomal protein synthesis*, which some bacteria use to produce small molecules that play various roles such as intercellular signaling and microbial warfare (i.e., antibiotics). Again, keep in mind that reality is more complicated than the cartoon version, though the cartoon version is still useful to convey the most common relationship between the major players: DNA, RNA, amino acids, and proteins.

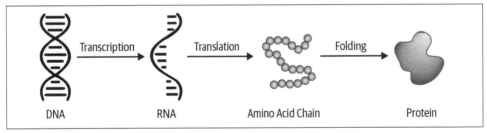

Figure 2-2. The central dogma of biology: DNA leads to RNA; RNA leads to amino acids; amino acids lead to protein.

So how does that relationship work in practice? How do you go from DNA sequence to the amino acids that make up the proteins? First, the transcription of DNA to RNA uses a neat biochemical trick: as we mentioned earlier, the nucleotides are complementary to one another—A sticks to T/U, and C sticks to G—thus, if you have a string of DNA that reads TACTTGATC, the cell can generate a "messenger" RNA string by placing the appropriate (i.e., sticky) RNA base opposite each DNA base, to produce AUGAACUAG. Then, it's time for the genetic code to come into play: each set of three letters constitutes a *codon* that corresponds to an amino acid, as shown in Figure 2-3. A cellular tool called a *ribosome* reads each codon and lines up the corresponding amino acid to add it to the chain, which will eventually become a protein. Our minimalist example sequence therefore codes for a very short amino acid chain:

just methionine (AUG) followed by asparagine (AAC). The third codon (UAG) simply signals the ribosome to stop extending the chain.

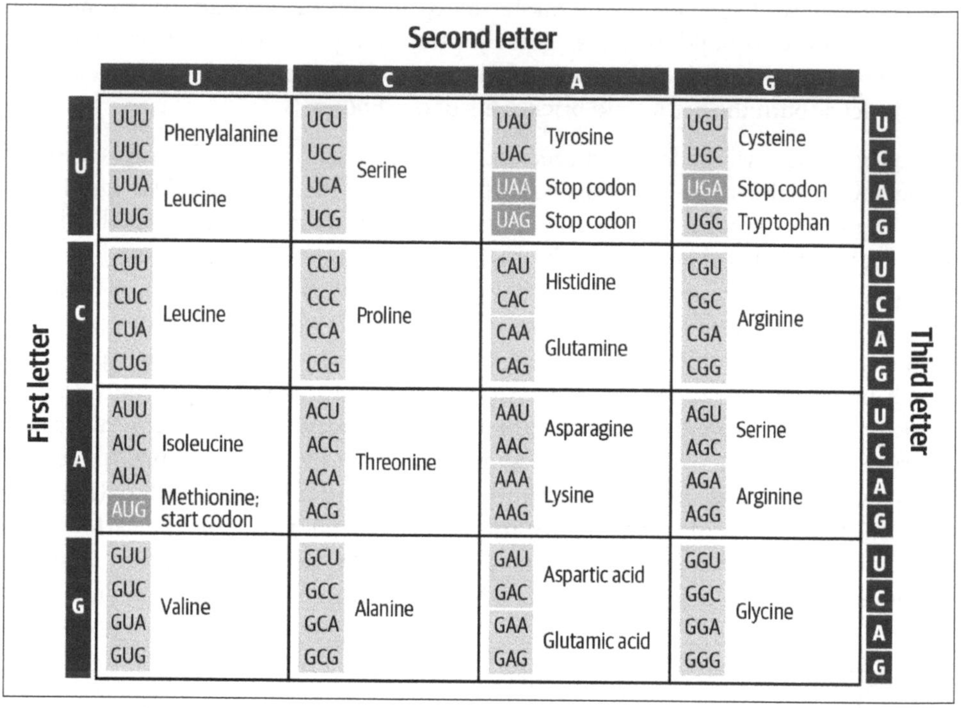

Second letter

First letter		U	C	A	G		Third letter
U	UUU, UUC	Phenylalanine	UCU, UCC, UCA, UCG — Serine	UAU, UAC — Tyrosine	UGU, UGC — Cysteine	U, C	
	UUA, UUG	Leucine		UAA — Stop codon; UAG — Stop codon	UGA — Stop codon; UGG — Tryptophan	A, G	
C	CUU, CUC, CUA, CUG	Leucine	CCU, CCC, CCA, CCG — Proline	CAU, CAC — Histidine; CAA, CAG — Glutamine	CGU, CGC, CGA, CGG — Arginine	U, C, A, G	
A	AUU, AUC, AUA	Isoleucine	ACU, ACC, ACA, ACG — Threonine	AAU, AAC — Asparagine; AAA, AAG — Lysine	AGU, AGC — Serine; AGA, AGG — Arginine	U, C, A, G	
	AUG	Methionine; start codon					
G	GUU, GUC, GUA, GUG	Valine	GCU, GCC, GCA, GCG — Alanine	GAU, GAC — Aspartic acid; GAA, GAG — Glutamic acid	GGU, GGC, GGA, GGG — Glycine	U, C, A, G	

Figure 2-3. The genetic code connects three-letter codons in a messenger RNA sequence to specific amino acids.

We could explain a lot more about how all of this works, as it's really quite fascinating, so we encourage you to look it up to learn more. For now, what we want to emphasize is the dependence of this mechanism on the DNA sequence: changing the sequence of the DNA can lead to changes in the amino acid sequence of the protein, which can affect its function. So let's talk about genetic changes.

The Origins and Consequences of DNA Mutations

DNA is pretty stable, but sometimes it breaks (e.g., because of exposure to radiation or chemicals) and needs to be repaired. The enzymes that are responsible for repairing these breaks occasionally make mistakes and incorporate the wrong nucleic acid (e.g., a T instead of an A). The same applies for the enzymes responsible for making copies of DNA during cell division; they make mistakes, and sometimes those mistakes persist and are transmitted to progeny. Between these and other mechanisms of mutation that we won't go into here, the DNA that each generation inherits from their parents diverges a little more over time. The overwhelming majority of these mutations do not have any functional effects, in part because some redundancy is present in the genetic code: you can change some letters without changing the meaning of the "word" that they spell out.

However, as illustrated in Figure 2-4, some mutations do have effects that cause an alteration in gene function, gene expression, or related processes. These effects might be positive; for example, improving the efficiency of an enzyme that degrades toxins and thereby protects your body from damage. Others will be negative; for example, reducing that same enzyme's efficiency. Some changes might have dual effects depending on context; for example, one well-known example of a mutation contributes to sickle-cell disease but protects against malaria. Finally, some changes can have functional effects that are not obviously associated with a direct positive or negative outcome, like hair or eye color.

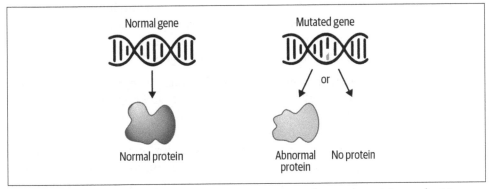

Figure 2-4. A mutation in the DNA sequence can cause the gene's protein product to function abnormally or disable its production entirely.

Again, this topic is vastly more complex than we show here; our goal is mainly to illustrate why we care so much about determining the exact sequence of people's genomes.

Genomics as an Inventory of Variation in and Among Genomes

Now that we've gone over the fundamentals, we can step back and take another stab at defining genomics. Let's say the relationship between DNA and genes can be roughly summarized as "genes are made of DNA, but your DNA overall comprises both genes and nongenes." This is an awkward formulation, so we're going to use the term *genome* to refer to "your DNA overall." But if genome means "all of your DNA including (but not limited to) your genes," what should *genomics* encompass as "the science of genomes"?

The US-based NHGRI website declares (*https://oreil.ly/Fm7Tb*):

> Genetics refers to the study of genes and the way that certain traits or conditions are passed down from one generation to another. Genomics describes the study of all of a person's genes (the genome).

The problem with that definition (beyond the rather cavalier omission of "all the DNA that is not part of genes") is that the minute you begin looking for something specific in the genome, it becomes a question of genetics. For example, out of all the human genome, what are the genes involved in the development of type 2 diabetes? Is that genomics or genetics? Technically you need to look at all of the genes first in order to narrow down your search to the genes that will be meaningful to your research. So the "some genes" versus "all the genes" distinction is a rather fuzzy one that is not entirely helpful, at least not in the context of our book. We therefore define genomics more narrowly as the discipline tasked with identifying the contents of genomes, whether at the level of individuals, families, or populations. In this view, the goal of genomics is to produce the information that feeds into the downstream genetic analyses that will ultimately yield biological insights.

This definition includes the process of identifying which genomic contents are specific to individuals, which are shared among individuals and populations, and which are different among individuals and populations. A few related disciplines complement this subdivision, including *transcriptomics*, which is the study of gene expression at the genome scale. The term genomics is sometimes used as a catch-all for these disciplines, though it has now become more common to use the more generic term *omics* for that purpose. We won't go into the details of those either, except to note the points of connection when the opportunity arises.

We can use *genome* in the individual sense (as in "my personal genome") and in the collective sense (as in "the human genome").

The Challenge of Genomic Scale, by the Numbers

Approximately three billion base pairs are in a human genome, or three *gigabases* (GB), representing an enormous amount of biochemically encoded informational content. The technology we use to decipher this content generates hundreds of millions of very short sequences for a single genome. Specifically, about 40 million sequences of ~200 bases, which amounts to a 100 GB file per genome. If you do the math, you'll realize that this dataset is redundant by a factor of 30 times! This is intentional, because the data is full of errors—some random, some systematic—so we need the redundancy to have enough confidence in our findings. Within that mass of data that is redundant but full of errors, we need to identify the four to five million small differences that are real (relative to a reference framework, which we discuss in the next section). That information can then be used to perform the genetic analysis that will determine which differences actually matter to a particular research question. Oh, and that's for just one person; but we might want to apply this to cohorts of hundreds, thousands, tens of thousands of people, or more. (Spoiler alert: definitely more.)

Given the scale involved, we're going to need some very robust procedures for analyzing this data. But before we begin tackling the computational methods, let's take a deeper look at the topic of genomic variation to define more precisely what we're going to be looking for.

Genomic Variation

Genomic variation can take multiple forms and can arise at different times in the life cycles of organisms and populations. In this section, we review the way variants are described and classified. We begin by discussing the concept of a reference genome that is used as a common framework for cataloging variants. Then, we go over the three major classes of variants that are defined on the basis of the physical change that is involved: short sequence variants, copy-number variants, and structural variants. Finally, we discuss the difference between germline variants and somatic alterations, and where *de novo* mutations fit into the picture.

The Reference Genome as Common Framework

When it comes to comparing variation between genomes, whether it's between cells or tissues of the same individual, or between individuals, families, or entire populations, almost all analyses rely on using a common genome reference sequence.

Why? Let's look at a similar, if simpler problem. We have three modern-day sentences that we know evolved from a common ancestor:

1. The quick brown f*a*x jumped over the lazy dog*e*.
2. The quick _ fox jumped over th*eir* lazy dog*e*.
3. The quick brown fox jump*s* over the lazy *brown* dog.

We'd like to inventory their differences in a way that is not biased toward any single one of them, and is robust to the possibility of adding new mutant sentences as we encounter them. So we create a synthetic hybrid that encapsulates what they have most in common, yielding this:

The quick brown fox jumped over the lazy doge.

We can use this as a common reference coordinate system against which we can plot what is different (if not necessarily unique) in each mutant:

1. Fourth word, o → a substitution; ninth word ends with *e*

2. Third word missing; seventh word ends with *ir*; ninth word ends with *e*

3. Fifth word ed → s substitution; duplication of the third word located after the eighth word; ninth word does not end with *e*

It's obviously not a perfect method, and what it gives us is not the true ancestral sentence: we suspect that's not how *dog* was originally spelled, and we're unsure of the original tense (*jumps* versus *jumped*). But it enables us to distinguish what is commonly observed in the population we have access to from what is divergent.

The more sentences we can involve in the initial formulation of this reference and the more representative the sampling, the more appropriate it will be for describing the variations we encounter in the future.

Similarly, we use a reference genome in order to chart sequence variation in individuals against a common standard rather than attempting to chart differences between genome sequences relative to one another. That enables us to identify what subset of variations in sequence are commonly observed versus unique to particular samples, individuals, or populations.

Reference Genome Format: FASTA

The standard format currently used for reference genome is FASTA (*https://oreil.ly/ RbeV3*), which is generally pronounced "fast-ay" in English (rather than "fast-ah"). FASTA is a fairly simple text format in which all segments of contiguous sequence in a genome, referred to as *contigs*, can be contained in a single file. Each contig record in a FASTA file consists of a description line that starts with an angle bracket and the name of the contig, plus some optional metadata, and one or more lines of sequence, as follows:

```
>my_contig some_metadata
ATTCGCGAGGATATAAGGATATAGGATA
GGATATTCGCGCCCTATATAGAGATTCG
```

Genome references in FASTA format are generally accompanied by two accessory files: a dictionary file (*.dict*) and an index file (*.fai*), which allow analysis tools to access specific positions in the reference sequence without having to load the entire file into memory. Note that FASTA can also be used beyond genome references to store other kinds of DNA, RNA, or protein sequences.

So whose genome do we use as a common framework? In the simplest case, any individual genome can be used as a reference genome. However, our analysis will produce better results if we can use a reference genome that is representative of the widest group of individuals that we might want to study.

That leads to a complicated problem: for our reference to be representative of many individuals, we need to evaluate the most commonly observed sequence across available individual genomes, which we call a *haplotype*, for each segment of the genome reference. Then, we can use that information to compose a synthetic hybrid sequence. The result is a sequence that is never actually observed wholesale in any particular individual genome.

Researchers are trying to address these limitations using more advanced representations of genomes, such as graphs for which nodes represent different possible sequences, and edges between them allow you to trace a given genome. However, for the purposes of our analysis presented in this book, we use a high-quality reference genome derived from the efforts of the scientific community starting with the Human Genome Project (HGP).

It's important to understand that the effectiveness of the common reference is limited by the diversity of the original samples used in its development. When the HGP was declared complete in 2003, the genome sequence that the HGP consortium published was, in fact, a synthetic reference based on a very small number of people, all from European background. This limited its effectiveness as a reference for non-European populations because, as we learned later, some of these populations have substantial enough differences that cannot be adequately mapped back to the original Eurocentric reference.

Successive "versions" of the human genome reference, commonly called *assemblies* or *builds*, have been published since then. In addition to delivering higher quality through purely technological advances, these updates to the common genome reference have also increased its representativeness by including more data from historically underrepresented populations. It is worth emphasizing that these efforts to compensate for historical bias in the selection of participants in genomic studies are very necessary. Indeed, lack of representation in research data leads to real inequalities in clinical outcomes because it affects our ability to identify meaningful variation in an individual's genome sequence if that person belongs to an underrepresented population.

In humans, the most significant improvements to date have been made in the representation of so-called *alternate haplotypes*, regions that are sometimes dramatically different in different populations. The latest build of the human reference genome, officially named *GRCh38* (for Genome Research Consortium human build 38) but commonly nicknamed *Hg38* (for human genome build 38), greatly expanded the repertoire of alternate (ALT) contigs representing the alternate haplotypes. These have a significant impact on our power to detect and analyze genomic variation that is specific to populations carrying alternate haplotypes.

The Reference Genome Version Conundrum

Although the gradual improvements to the human reference genome are absolutely beneficial and necessary, the existence of these different builds is a common source of confusion and errors for many researchers. To make matters worse, there existed for some time two parallel streams of reference genome evolution, named *hg** versus *b**, published by different groups, with different naming conventions, some differences in sequences, and inclusion of different noncanonical contigs. Many bioinformatics tools fail to enforce consistent use of a specific reference, which allows the unwary user to switch reference genomes halfway through a project without realizing that their comparisons suddenly become worthless, because, for example, all the positions are silently shifted by some coordinate index.

In our book, we use the older b37 reference in some exercises, and the newer hg38 reference in others. This is in part for the convenience of using previously existing materials, but also because the older reference is still widely used; thus we feel it's important to be aware of their coexistence and practice dealing with different references.

As a final plot twist, note that all current standard reference genome sequences are *haploid*, meaning they represent only one copy of each chromosome (or contig). The most immediate consequence is that in *diploid* organisms such as humans, which have two copies of each autosome (i.e., any chromosome that is not a sex chromosome, X or Y), the choice of standard representation of sites that are most often observed in *heterozygous* state (manifesting two different alleles; e.g., A/T) is largely arbitrary. This is obviously even worse in *polyploid* organisms, such as many plants including wheat and strawberries, which have higher numbers of chromosome copies. Although it is possible to represent reference genomes using graph-based representations (*https://oreil.ly/5BiFo*), which would address this problem, few genome analysis tools are able to handle such representations at this time, and it will likely take years before graph-based genomes gain traction in the field.

Physical Classification of Variants

Now that we have a framework for describing variation, let's talk about the kinds of variation we'll be considering. We distinguish three major classes of variants based on the physical change they represent, which are illustrated in Figure 2-5. *Short sequence variants* (single-base changes and small insertions or deletions) are small changes in DNA sequence compared to a reference genome; *copy-number variants* are relative changes in the amount of a particular fragment of DNA; and *structural variants* are changes in the location or orientation of a particular fragment of DNA compared to a reference genome.

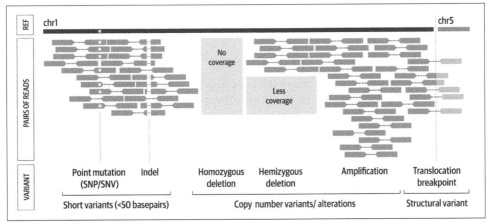

Figure 2-5. *The major types of variant classified by physical changes to the DNA.*

Single-nucleotide variants

Earlier, we examined the hybrid sentence: "The quick brown fox jumped over the lazy doge." Within the list of three sentences, the first sentence used *fax* instead of *fox*. Here we might say there was a single-letter substitution in the word fox, where the *o* was substituted with an *a* because the "reference" says *fox*.

When we apply this to DNA sequence—for example, when we encounter a G nucleotide instead of the A we expect based on the reference—we'll call it a *single-nucleotide variant* (Figure 2-6). Some variation exists in the exact terminology used to refer to these variants, and appropriately enough, the difference is a single-letter substitution: depending on context, we'll call them either *single-nucleotide polymorphism* (SNP; pronounced "snip"), which is most appropriate in population genetics, or *single-nucleotide variant* (SNV; usually spelled out but sometimes pronounced "sniv"), which is more general in meaning yet is mostly used in cancer genomics.

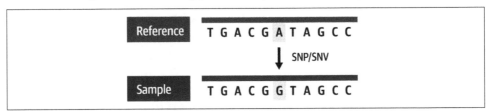

Figure 2-6. *A single-nucleotide variant.*

SNVs are the most common type of variation encountered in nature because the various mechanisms that cause them are fairly commonplace and their consequences are often minimal. Biological functions tend to be robust to such small changes because, as we saw earlier in our overview, the genetic code has multiple codons that code for the same amino acid, so single base changes don't always lead to changes in the protein.

Insertions and deletions

In the second sentence of our example, "The quick _ fox jump*s* over the*ir* lazy dog*e*," we see the two primary types of small insertions and deletions: we've lost the word *brown*, and one instance of *the* has become *their* through the insertion of two letters. We also have one complex substitution in the replacement of *jumped* by *jumps*, which is interesting because without additional information we can't say whether the change happened in a single event, *ed* becoming *s*, or whether the *ed* was lost first and the *s* was tacked on later. Or, the original ancestor of both *jumped* and *jumps* could have been *jump*, and both forms evolved later through independent insertions! That latter scenario seems more likely because it involves a simpler chain of events, but we can't know for sure without seeing more data.

Applied to DNA sequence, *indels* are insertions or deletions of one or more bases compared to the reference sequence, as shown in Figure 2-7. Interestingly, multiple studies have reported that insertions occur more commonly than deletions (at least in humans). This observation has not been fully explained but could be related to naturally occurring mobile genetic elements, a kind of molecular parasite that is fascinating but sadly out of scope for this book.

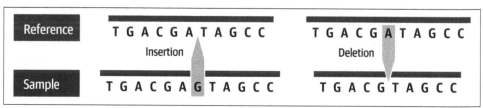

Figure 2-7. Indels can be insertions (left) or deletions (right).

 The term *indel* is a portmanteau created by combining *in*sertion and *del*etion.

Most indels observed in nature are fairly short, measuring less than 10 bases in length, but there is no maximum length of sequence than can be inserted or deleted. In practice, indels longer than a few hundred bases are generally classified as copy-number or structural variants, which we discuss next. Compared to single-nucleotide changes, indels are more likely to disrupt biological function, because they are more likely to cause phase shifts, where adding or removing bases shifts the three-letter reading frame used by the ribosome to read the codons. This can completely change the meaning of the sequence downstream of the indel. They can also cause protein folding changes when inserted into coding regions, for example.

Copy-number variants

With *copy-number variation* (Figure 2-8), we're moving into a more complex territory because the game is no longer just a matter of playing "spot the difference." Previously, the question was "What do I see at this location?" Now it shifts to "How many copies do I see of any given region?" For example, the third sentence in our evolved text example, "The quick brown fox jumps over the lazy *brown* dog" has two copies of the word *brown*.

Figure 2-8. Example of copy-number variant caused by a duplication.

In the biological context, if you imagine that each word is a gene that produces a protein, this means that we could produce twice as much "Brown" protein from this sentence's DNA compared to the other two. That could be a good thing if the protein in question has, for instance, a protective effect against exposure to dangerous chemicals. But it could also be a bad thing if the amount of that protein present in the cell affects the production of another protein: increasing one could upset the balance of the other and cause a deleterious chain reaction that disrupts cellular processes.

Another case is the alteration of *allelic ratio*, the balance between two forms of the same gene. Imagine that you inherited a defective copy of a particular gene, which is supposed to repair damaged DNA, from your father, but it's OK because you inherited a functional copy from your mother. That single functional copy is sufficient to protect you, because it makes enough protein to repair any damage to your DNA. Great. Now imagine that you have a for-now benign tumor in which one part suffers a copy-number event: it loses the functional copy from your mom, and now all you have is your dad's broken copy, which is unable to produce the DNA repair protein. Those cells are no longer able to repair any DNA breaks (which can happen randomly) and begin to accumulate mutations. Eventually, that part of the tumor starts proliferating and becomes cancerous. Not so great.

These are called *dosage effects* and illustrate the biological complexity of genetic pathways that regulate cell metabolism and health. The DNA copy number can affect biological function in other ways, and in all cases the degree of severity of the effects depends on a lot of factors.

As an example, *trisomies* are genetic disorders in which patients have an extra copy of an entire chromosome. The most well-known is Down syndrome, which is caused by an extra copy of chromosome 21. It is associated with multiple intellectual and physical disabilities, but the people who carry them can live healthy and productive lives if they are supported appropriately. Trisomies affecting sex chromosomes (X and Y) tend to have fewer negative consequences; for example, most individuals with triple X syndrome experience no medical issues as a result. In contrast, several trisomies affecting other chromosomes are tragically incompatible with life and always result in either miscarriage or early infant death.

Structural variants

This last class of variants is a superset that includes copy-number variants and large indels in addition to various copy-neutral structural alterations such as translocations and inversions, as shown in Figure 2-9.

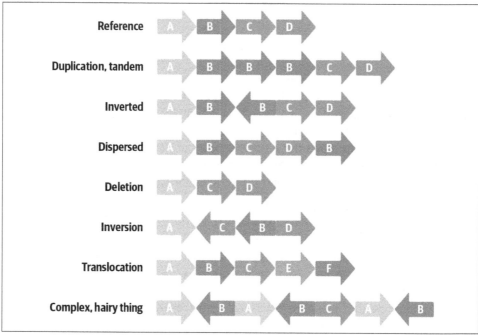

Figure 2-9. Examples of structural variants.

As you can see from the examples in Figure 2-9, *structural variants* can be rather complex because they cover a range of physical transformations of the genetic material. Some can span large distances, even affecting multiple chromosomes in some cases. We won't cover structural variation in this book because this is still an area of very active development and the relevant methods are far from settled.

Germline Variants Versus Somatic Alterations

The fundamental biological distinction between germline variants and somatic altera-tions has major consequences for the way we approach variant discovery, depending on the specific purpose of the analysis. Wrapping your head around this can be a bit tricky because it's tied to concepts of inheritance and embryo development. Let's start from the definitions of the terms *germline* and *somatic*.

Germline

The term *germline* is a combination of two related aspects of biological inheritance. The *germ* part refers to the concept of germination of a seed, and doesn't have any-thing to do directly with microbes, although it's the same idea that causes us to call microbes *germs*. The *line* part refers to the fact that all cells that eventually turn into either eggs or sperm cells come from a single *cell line*, a population of cells with a common ancestor that get differentiated early in the development of the embryo.

In fact, the term *germline* is often used to refer to that entire population of cells as *the germline* of an individual. So in principle, your germline variants are variants that were present in your biological parents' germline cells, *typically* the egg and/or sperm cell that produced you, before fertilization happened. As a result, your germline var-iants are expected to be present in every cell in your body. In practice, variants that arise very shortly after fertilization will be difficult or impossible to distinguish from germline variants.

Researchers are interested in germline variants for a variety of reasons: diagnosing diseases with a genetic component, predicting risk of developing disorders, under-standing the origin of physical traits, mapping the origins and evolution of human populations—and that's just on the human side! Germline variation within popula-tions of organisms is also of great interest in agriculture, livestock improvement, eco-logical research, and even epidemiological monitoring of microbial pathogens.

Somatic

The term *somatic* is derived from the Greek word *soma*, meaning *body*. It's used as an exclusionary term to refer to all cells that are not part of the germline, as illustrated in Figure 2-10. Somatic alterations are produced by mutation events in individual cells during the course of your life (including during embryo development, prior to birth), and are present only in subsets of cells in your body.

These mutations can be caused by a variety of factors, from genetic predispositions to environmental exposure to mutagens such as radiation, cigarette smoke, and other harmful chemicals. A great many of these mutations arise in your body but stay con-fined to one or a few cells and cause no noticeable effects. In some cases, the muta-tions can cause domino effects within the affected cells, damaging their ability to regulate growth and causing the formation of a cell tumor. Many tumors are thankfully benign; however, some become malignant and cause negative chain reactions, fueling the occurrence of many additional somatic mutations and resulting in metabolic diseases, which we ultimately group under the label of *cancer*.

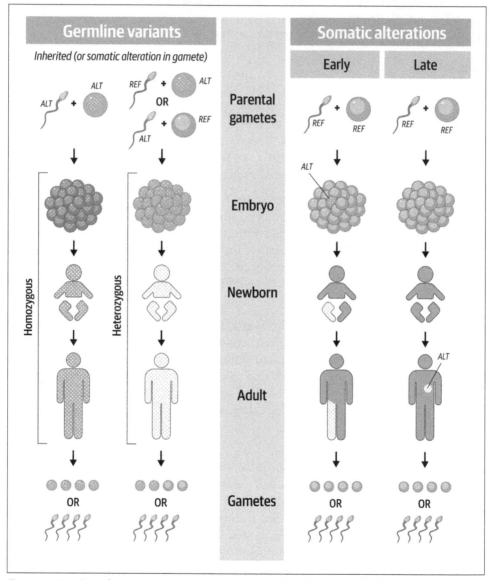

Figure 2-10. Germline variants are present in all cells of the body (left) while somatic alterations are present only in a subset of cells (right).

That being said, even though cancer research is one of the major applications of somatic variation analysis, the effects of somatic variation are not limited to tumor growth and cancer. They also include various developmental disorders such as Down syndrome, which, as we alluded to earlier, is caused by an erroneous duplication of chromosome 21; in other words, a form of CNV. Although Down syndrome can be inherited through the germline, it is most commonly observed as the result of an abnormal cell division event that occurs early on during the development of the embryo. Most of the time this happens so early that virtually all of the individual's cells are affected.

This is where the line between germline variation and somatic alteration becomes a bit blurry. As we've just discussed, you can have a child born with a variant that arose very early on during their development. As far as its origin is concerned, it's a somatic alteration; however, in the analysis of your child's genome, it will manifest as a germline variant because it's present in all of their cells. Along similar lines, you can get somatic mutations within your germline cells during the course of your life; for example, if something goes wrong during the cell division that produces an egg or sperm cell. The resulting variants are clearly somatic in origin, but if you go on to have a child who develops from one of the affected cells, that child will carry the new variants in all cells in their body. And so those variants will be germline variants as far as the child is concerned; doubly so, given that in this case they'll have inherited them from you. In both cases, we qualify those variants as *de novo variants* from the Latin *de novo* meaning "newly arisen," because in the child they appear to have come out of nowhere. In this book, we treat early somatic mutations as germline mutations.

Variant, Polymorphism, Alteration, and Mutation

The terminology in the field of variant discovery is not completely settled (especially on the somatic side of the fence), so you might encounter inconsistencies in the literature. In this book, we apply the usages that we have encountered most frequently and believe to be most established, as summarized here:

- The term *variant* is the most generic term that can be substituted for all the others.
- For short variants, we use the well-established SNP for germline single-base variants, and SNV for somatic single-base variants. The term *polymorphism* refers to a concept from population genetics and relies on the assumption that these are naturally occurring variants that are present in a population.
- For copy number variation, we use CNV (*V* for variant) for germline variants and CNA (*A* for alteration) for somatic variants, for which the term *alteration* emphasizes that these variants result from recent modification events.
- For structural variation, we use SV (*V* for variant) for both germline and somatic variants.

- So far, we have not encountered any usage of single-nucleotide alteration or *structural alteration* for somatic single-base and structural variants even though that would be the most logical specialized term to use.

- Finally, we note that many use the term *mutations* to refer generically to somatic variants, but we restrict our use of that term to refer only to the modification events themselves.

High-Throughput Sequencing Data Generation

Now that we know what we're looking for, let's talk about how we're going to generate the data that will allow us to identify and characterize variants: *high-throughput sequencing*.

The technology we use to sequence DNA has evolved dramatically since the pioneering work of Allan Maxam, Walter Gilbert, and Fred Sanger. A lot of ink, electronic or otherwise, has been spent on classifying the successive waves, or generations, of sequencing techniques (*https://oreil.ly/159S1*). The most widely used technology today is *short read* sequencing-by-synthesis popularized by Illumina, which produces very large numbers of *reads*; that is strings of DNA sequence (originally approximately 30 bases, now going up to roughly 250 bases). This is often called *next-generation sequencing* (NGS) in contrast to the original *Sanger sequencing*. However, that term has arguably become obsolete with the development of newer "third-generation" technologies that are based on different biochemical mechanisms and typically generate longer reads.

So far, these new technologies are not a direct replacement for Illumina short read sequencing, which still dominates the field for resequencing applications, but some such as PacBio and Oxford Nanopore have become well established for applications like *de novo* genome assembly that prioritize read length. Nanopore, in particular, is also increasingly popular for point-of-care and fieldwork applications because of the availability of a pocket-size version, the MinION, which can be used with minimal training and connects to a laptop through a USB port. Although the MinION device cannot produce the kind of throughput required for routine analysis of human and other large genomes, it is well suited for applications like microbial detection for which smaller yields are sufficient.

Exciting developments have also occurred in the field of single-cell sequencing, in which microfluidics devices are used to isolate and sequence the contents of individual cells. These are generally focused on RNA with the goal of identifying biological phenomena such as differences in gene expression patterns between tissues. Here, we focus on classic Illumina-style short read DNA sequencing for the sake of simplicity, but we encourage you to check out the latest developments in the field.

From Biological Sample to Huge Pile of Read Data

So how do we sequence someone's DNA using short read technology? First we need to collect biological material (most commonly blood or a cheek swab) and extract the DNA. Then, we "prepare a DNA library," which involves shredding whole DNA into fairly small fragments and then preparing these fragments for the sequencing process. At this stage, we need to choose whether to sequence the whole genome or just a subset. When we talk, for example, about *whole genome sequencing* versus *exome sequencing*, what differs isn't the *sequencing technology*, but the *library preparation*: the way the DNA is prepared for sequencing. We go over the main types of libraries and their implications for analysis in the next section. For now, let's assume that we can retrieve all of the DNA fragments that we want to sequence; we then add molecular tags for tracking and adapter sequences that will make the fragments stick in the correct place in the sequencing flowcell, as illustrated in Figure 2-11.

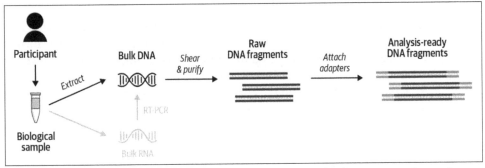

Figure 2-11. Library preparation process for bulk DNA (top); alternative pathway for bulk RNA (bottom).

The *flowcell* is a sort of glass slide with fluidic channels in which the sequencing chemistry takes place when the appropriate reagents are pumped through it in cycles. During that process, which you can learn more about in Illumina's online documentation (*https://www.illumina.com*), the sequencing machine uses lasers and cameras to detect the light emitted by individual nucleotide binding reactions. Each nucleotide corresponds to a specific wavelength, so by capturing the images produced from each reaction cycle, the machine can read out strings of bases called *reads* from each fragment of DNA that is stuck to the flowcell, as illustrated in Figure 2-12.

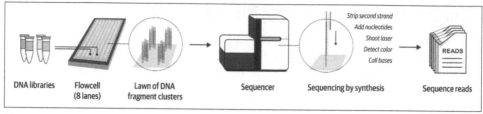

Figure 2-12. Overview of Illumina short read sequencing.

Typically, the reactions are set up to produce one read from each end of each DNA fragment, and the resulting data remains associated as read pairs. As you'll see later in this chapter, having pairs of reads per fragment rather than single reads gives us information that can be useful in several ways.

High-throughput sequencing data formats

The data generated by the Illumina sequencer is initially stored on the machine in a format called Basecall (BCL), but you don't normally interact with that format directly. Most sequencers are configured to output FASTQ (*https://oreil.ly/NbFcD*), a straightforward if voluminous text file format that holds the name of each read (often called *queryname*), its sequence and the corresponding string of base quality scores in Phred-scaled form, as illustrated in Figure 2-13.

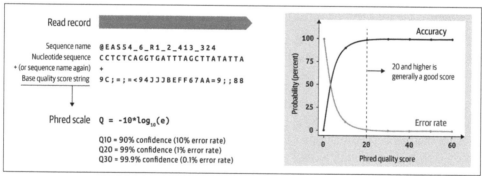

Figure 2-13. FASTQ and Phred scale.

The *Phred scale* is a log-based scale for expressing very small quantities such as probabilities and likelihoods in a form that is more intuitive than the original decimal number; most Phred-scaled values we'll encounter are between 0 and 1,000 and rounded to the nearest integer.

When the read data is mapped to a reference genome, it is transformed into Sequence Alignment Map (*https://oreil.ly/1fqR5*) (SAM), the format of choice for *mapped* read data because it can hold mapping and alignment information, as illustrated in Figure 2-14. It can additionally hold a variety of other important metadata that analysis programs might need to add on top of the sequence itself. Like FASTQ, SAM is a structured text file format (*https://oreil.ly/QPcIK*), and because it holds more information, those files can become very large indeed. You'll generally encounter both formats in their compressed forms; FASTQ is typically simply compressed using the gzip utility, whereas SAM can be compressed using specialized utilities into either Binary Alignment Map (BAM) or CRAM, which offers additional degrees of lossy compression.

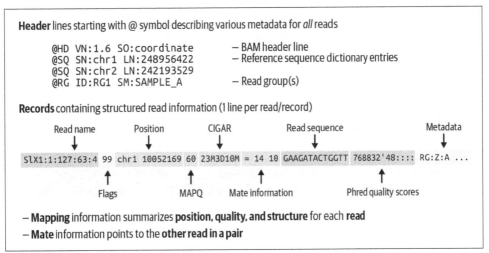

Figure 2-14. Key elements of the SAM format: file header and read record structure.

 Unlike SAM and BAM, CRAM is not a real acronym; it was so named to follow the SAM/BAM naming pattern and as a play on the verb *to cram,* but the *CR* does not spell out to anything meaningful.

In the SAM/BAM/CRAM format, the read's local alignment is encoded in the Concise Idiosyncratic Gapped Alignment Report (CIGAR) string. Yes, someone really wanted it to spell out *CIGAR.* Briefly, this system uses single-letter operators to represent the alignment relationship of one or more bases to the reference with a numerical prefix that specifies how many positions the operator describes in such a way that the detailed alignment can be reconstructed based solely on the starting position information, the sequence, and the CIGAR string.

For example, Figure 2-15 shows a read that starts with one soft-clipped (ignored) base (1S), then four matching positions (4M), one deletion gap (1D), another two matches (2M), one insertion gap (1I), and, finally, one last match (1M). Note that this is a purely structural description that is not intended to describe sequence content: in the first stretch of matches (4M), the third base does not match the reference, so all it means is that "there is a base in position 4."

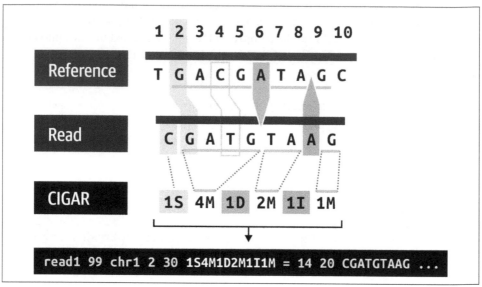

Figure 2-15. The CIGAR string describes the structure of the read alignment.

Let's come back briefly to that soft clip shown in Figure 2-15, which we qualified as "ignored." Soft clips happen at the beginning and/or at the end of a read when the mapper isn't able to align one or more bases to its satisfaction. The mapper basically gives up and uses the soft clip symbol to indicate that those bases do not align to the reference and should be ignored in downstream analysis. As you'll see in Chapter 5, some of the data that ends up in soft clips is recoverable and can be useful.

Technically, it is possible to store read data in BAM format without including mapping information, as we describe in Chapter 6.

Note that the long-term storage of large amounts of sequence data is a problem that preoccupies large sequencing centers and many funding agencies, and has prompted various groups to propose alternative formats to FASTQ and SAM/BAM/CRAM. However, none has so far managed to gain traction with the wider community.

The amount of read data produced for a single experiment varies according to the type of library as well as the target coverage of the experiment. *Coverage* is a metric that summarizes the number of reads overlapping a given position of the reference genome. Roughly, this rolls up to the number of copies of the original DNA represented in the DNA library. If this confuses you, remember that the library preparation process involves shredding DNA isolated from many cells in a biological sample, and the shredding is done randomly (for most library preparation techniques), so for any given position in the genome, we expect to obtain read pairs from multiple overlapping fragments.

Speaking of libraries, let's take a closer look at our options on that front.

Types of DNA Libraries: Choosing the Right Experimental Design

There are about as many ways to prepare DNA libraries as there are molecular biology reagents companies. Generally, though, we can distinguish three main approaches that determine the genomic territory made available to sequencing and the major technical properties of the resulting data: *amplicon generation*, *whole genome preparation*, and *target enrichment*, which is used to produce gene panels and exomes, as depicted in Figure 2-16.

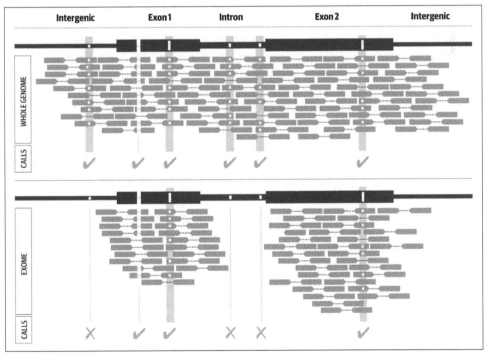

Figure 2-16. Experimental design comparison between whole genome (top) and exome (bottom).

The whole genome library shown in Figure 2-16 will lead to sequence coverage across all regions (exon, intron and intergenic), while the exome library will only lead to coverage across the targeted regions, typically corresponding to exons.

Amplicon preparation

This family of DNA preparation techniques relies on classic polymerase chain reaction (PCR) amplification (either using specifically designed primers or using restriction enzyme sites) to produce *amplicons*, fragments of DNA present in large numbers of copies that start and stop at the same positions. This means that the territory made available for sequencing is typically very small and dependent on the starting and stopping points we predetermine with our primers or restriction enzymes.

This approach is fairly economical for small-scale experiments and does not require a reference genome, making it a popular option for the study of nonmodel organisms. However, it produces data with enormous PCR-related technical biases. As a result, this approach lends itself very poorly to variant discovery analysis when done with the GATK and related tools; thus, we don't cover it further in this book.

Whole genome preparation

As the name implies, the goal of whole genome preparation is to sequence the entire genome. It involves putting the DNA through a mechanical shearing process that rips it into a multitude of small fragments and then "filtering" the fragments based on their size to keep only those within a certain range, typically 200 to 800 base pairs. Because a DNA sample contains many copies of the genome and the shearing process is random, we expect that the resulting fragments will be tiled redundantly across the entire genome. If we sequence each fragment once, we should see every base in the genome multiple times.

In practice, some parts of the genome are still very difficult to make available to the sequencing process. Within our cells, our DNA isn't simply floating around, loosely unfurled; much of it is tightly rolled up and packaged with protein complexes. We can relax it to a point with biochemical persuasion, but some areas remain quite difficult to "liberate." In addition, some parts of the chromosomes such as the centromere (the middle bit) and telomeres (the bits at the ends) contain vast stretches of highly repetitive DNA that is difficult to sequence and assemble faithfully. The human genome reference is actually incomplete in those areas, and specialized techniques are necessary to probe them. All this to say, when we talk about whole genome sequencing, we mean all of the parts that we can readily access. Nevertheless, even with those limitations, the amount of territory covered by whole genome sequencing is enormous.

Target enrichment: gene panels and exomes

Target enrichment approaches are an attempt at a best-of-both-worlds solution: they allow us to focus on the subset of the genome that we care about, but in a way that is more efficient than amplicon techniques and, to some extent, less prone to technical issues. Briefly, the first step is the same as for whole genome preparation: the genome is sheared into fragments, but then the fragments are filtered not based on size but on whether they match a location we care about. This involves using *bait* sequences, which are designed very much like PCR primers but immobilized on a surface to *capture* targets of interest, typically a subset of genes (for gene panels) or all annotated exons (for exomes). The result is that we produce patches of data that can inform us about only those regions that we managed to capture.

One often overlooked downside of this very popular approach is that the manufacturers of commercial kits for generating panel and exome data use different sets of baits, sometimes with important differences in the targeted regions, as illustrated in Figure 2-17. This causes batch effects and makes it more difficult to compare the results of separate experiments that used different target regions.

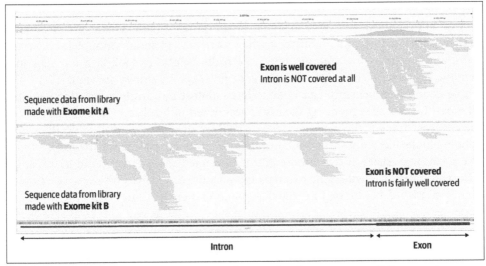

Figure 2-17. Different exome preparation kits can lead to important differences in coverage location and quantity.

Exome sequencing is sometimes referred to as *Whole Exome Sequencing* (WES) out of a misguided attempt to contrast it with *Whole Genome Sequencing* (WGS). It's misguided, because the contrast is already in the name: the point of the exome is that it's a *subset* of the whole genome. Calling it the whole exome is simply unnecessary and arguably inaccurate, to boot.

Whole genome versus exome

So which one is better? The most balanced thing we can say is that these approaches yield qualitatively and quantitatively different information, and each comes with specific limitations. Exome sequencing has proved extremely popular because for a long time it was much cheaper and faster than genomes, and for many common applications such as clinical diagnostics, it was usually sufficient to provide the necessary information. However, this approach suffers from substantial technical issues, including reference bias and unevenness of coverage distribution, which introduce many confounding factors into downstream analyses.

From a purely technical perspective, the data produced by WGS is superior to targeted datasets. It also empowers us to discover biologically important sequence conformations that lurk in the vast expanses of genomic territory that do not encode genes. Thanks to falling costs, whole genome sequencing is now becoming popular, though the downside is that the much larger amount of data produced is more challenging to manage. For example, a standard human exome sequenced at 50X coverage takes about 5 GB to store in BAM format, whereas the same person's whole genome sequenced at 30X coverage takes about 120 GB.

If you have not worked with exomes or whole genomes before, the difference between them can seem a bit abstract when we're talking about amounts of data, file sizes, and coverage distributions. It can be really helpful to look at a few files in a genome browser to see what this means in practice. As shown in Figure 2-18, the whole genome sequencing data is distributed almost uniformly across the genome, whereas the exome sequencing data is concentrated in localized piles. If you look at the coverage track in the genome viewer, you'll see the coverage profile for the whole genome looks like a distant mountain range, whereas the profile for the exome looks like a volcanic island in the middle of a flat ocean. This is an easy and surprisingly effective way to distinguish the two data types by the naked eye. We show you how to view sequencing data in the Integrated Genome Viewer (IGV) in Chapter 4.

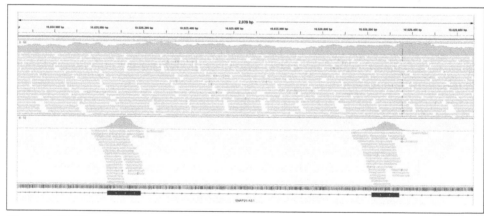

Figure 2-18. Visual appearance of whole genome sequence (WGS, top) and exome sequence (bottom) in a genome browser.

Most of the exercises in this book use whole genome sequencing data, but we also provide pointers to resources for analyzing exome data, and in Chapter 14 we go through a case study involving exomes.

Data Processing and Analysis

Finally, we get to the computational part! Now, you might think that as soon as we have the sequence data in digital form, the real fun begins. However, there is a twist: the data produced by the sequencers is not directly usable for variant discovery analysis: it's a huge pile of short reads, and we don't know where they came from in the original genome. In addition, there are sources of bias and errors that are likely to cause problems in downstream analysis if left untreated. So we're going to need to apply a few preprocessing steps before we can get to the genomic analysis proper. First, we map the read data to a reference genome and then we apply a couple of data cleanup operations to correct for technical biases and make the data suitable for analysis. Next, we'll finally be able to apply our variant discovery analysis workflows. In this section, we go over the core concepts involved in genome mapping (aka alignment) and variant calling, but we leave the rest for Chapter 6. We also touch briefly on data quality issues that can interfere in variant discovery analysis.

Mapping Reads to the Reference Genome

Ideally, we'd like to reconstruct the entire genome of the individual we're working on to compare it to the reference, but we don't have the sample in one piece—it's more like a million pieces, as illustrated in Figure 2-19!

Well, actually 360 million for a standard 30X WGS, and 30 million for a standard 50X Exome, according to the Lander-Waterman equation (*https://oreil.ly/NCMd4*). It's quite the jigsaw puzzle. And, by the way, the pieces overlap 30 to 50 times (depending on target depth).

So the goal of the mapping step is to match each read to a specific stretch of the reference genome. The output is a BAM file that has mapping and alignment information encoded in each read record, as described in "High-Throughput Sequencing Data Generation" on page 32.

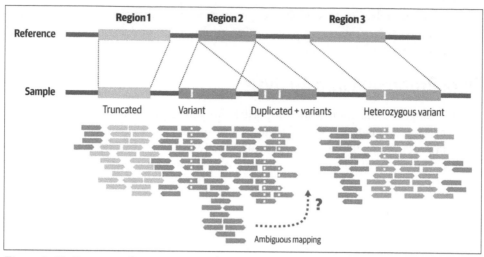

Figure 2-19. Sequence divergence introduces mapping challenges and ambiguity.

This is often called *alignment* because it outputs detailed alignment information for each read, but it is arguably more appropriate to call it *mapping*. Genome mapping algorithms are primarily designed to solve that fundamental problem of identifying the best matching location for a very small string of letters (typically ~150 to ~250) against an enormously huge one (about three billion for humans, not counting alternate contigs). Among other neat tricks, this involves using a scoring system that penalizes gaps. However, this has drawbacks when it comes to producing the fine-grained letter-by-letter alignment. The mapper works on one read at a time, so it is not well equipped to deal with naturally occurring sequence features like large insertions and deletions, and it will often prefer to clip sequences off the end of a read rather than introduce a large indel. We look at an example of this in Chapter 5 when we introduce GATK and its flagship germline short variant caller, HaplotypeCaller. What is important to understand at this point is that we trust the mapper to place the reads in the appropriate places most of the time, but we can't always trust the exact sequence alignments it produces.

As noted in Figure 2-19, an additional complication is that some genomic regions are copies of ancestral regions that were duplicated. The sequence of copies may be more or less divergent depending on how long ago the duplication happened and how many other mutation events produced variants in the copies. This makes the mapper's job more difficult because the existence of competing copies introduces ambiguity as to where a particular read might belong. Thankfully, most short read sequencing is now done using paired-end sequencing, which helps resolve ambiguity by providing additional information to the mapper. Based on the protocol used for library preparation, we know that we can expect DNA fragments to be within a certain size range, so the mapper can evaluate the likelihood of possible mapping locations based on the distance between the two reads in the pair, as shown in Figure 2-20. This does not apply for long-read sequencing technologies such as PacBio and Oxford Nanopore, for which the longer length of the reads typically leaves less opportunity for ambiguity by reading all the way through repeat regions.

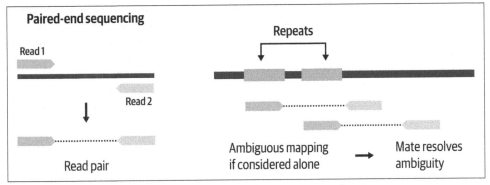

Figure 2-20. Paired-end sequencing helps resolve mapping ambiguity.

 Fragments of DNA generated during library preparation are sometimes called *inserts* because older preparation techniques involved inserting each fragment into a small circular piece of DNA called a *plasmid*. As a result, the relevant quality-control metric is usually called *insert size* even though nowadays the library preparation no longer involves insertion into plasmid vectors.

As noted a moment ago, a few other processing steps are needed to clean up the aligned data before we would normally move on to variant calling. We cover them later; for now, it's time to finally tackle the beast that is going to follow us through just about every chapter in this book.

Variant Calling

The term *variant calling* is thrown around a lot and is sometimes used to refer to the entire variant discovery process. Here we're defining it narrowly as the specific step in which we identify potential variants by looking for evidence of a particular type of variation in the available sequence data. The output of this step is a raw callset, or set of calls, in which each call is an individual record for a potential variant, describing its position and various properties. The standard format for this is the VCF, which we describe in the next sidebar.

Variant Data Format

The standard format for variant data is *Variant Call Format* (VCF), as shown in Figure 2-21. It's a tab-separated columnar text format in which each variant is represented on one line, indexed by its genomic location. The first eight columns describe variant-level information, including its positional information as well as a list of reference and alternate alleles observed at that position. The last variant-level column aggregates arbitrary annotations (defined in the header) as a single string of elements separated by semicolons. The most commonly encountered annotations are variant context statistics that describe the quality of the evidence supporting the variant call and functional predictions of variant effect. Sample-specific information is stored in additional columns; one column encoding a sequence of keys for retrieving field values and then one subsequent column per sample, with no upper limit on the number of samples that can theoretically be included.

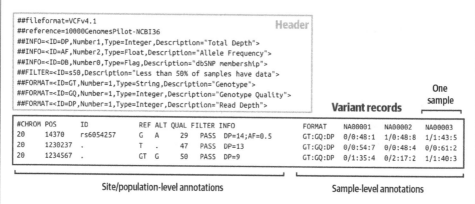

Figure 2-21. Basic structure of a VCF file.

You will have ample opportunity throughout the book to work with VCF files and get to know the format, so we won't go into it further here. For more immediate reading, see the VCF documentation (*https://oreil.ly/iKicj*) on the GATK website.

The nature of the evidence we use to identify variants is different depending on the type of variant we're looking for.

For short sequence variants (SNVs and indels), we're comparing the specific bases in short stretches of sequence. In principle, it's quite straightforward: we produce a pileup of reads, compare them to the reference sequence, and identify where there are mismatches. Then we simply count the numbers of reads supporting each allele to determine genotype: if it's half and half, the sample is heterozygous; if it's all one or the other, the sample is homozygous reference or homozygous variant. For example, the IGV screenshot in Figure 2-22 shows a probable 10-base deletion (left), a homozygous variant SNP (middle right), and a heterozygous variant SNP (rightmost) in whole genome sequence data.

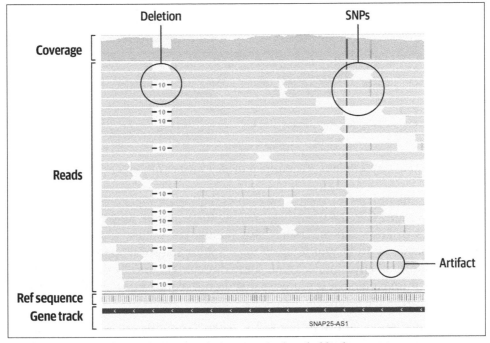

Figure 2-22. Pileup of reads in IGV showing several probable short variants.

For copy-number variants, we're looking at the relative number of sequence reads produced across multiple regions. Again, it's a fairly simple reasoning on the surface: if a region of the genome is duplicated in a sample, when we go to sequence that sample, we're going to generate twice as many reads that align to that region, as demonstrated in Figure 2-23.

And for structural variants, we're looking at a combination of both base-pair resolution sequence comparisons as well as relative amounts of sequence.

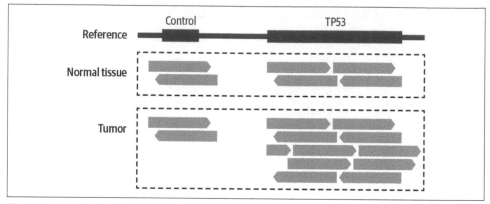

Figure 2-23. Relative amounts of coverage provide evidence for copy-number modeling.

In practice, the process for identifying each of the variant types confidently is rather more complicated than the simple logic that we just outlined, largely because of various sources of error and uncertainty that muddy the waters, which we touch on shortly. That's why we'll want to apply more robust statistical approaches that take various sources of error into account; some through empirically derived heuristics, and others through modeling processes that do not require prior knowledge of what kinds of errors might be present in the data.

We can rarely say definitively, "this call is correct" or "this call is an error," so we express results in probabilistic terms. For each variant call output by the program, the caller will assign scores that represent how confident we can be in the result. In the earlier indel example, the variant quality (QUAL) score of 50 means "there is a 0.001% chance that this variant is not real" and the genotype quality (GT) score of 17 for the second sample means "there is a 1.99% chance that this genotype assignment is incorrect." (We cover the distinction between variant quality and genotype quality in more detail in Chapter 5.)

Variant calling tools typically use some internal cutoffs to determine when to emit a variant call or not emit one depending on these scores. This avoids cluttering the output with calls that are grossly unlikely to be real. However, it's important to understand that those cutoffs are usually designed to be very permissive in order to achieve a high degree of sensitivity. This is good because it minimizes the chance of missing real variants, but it does mean that we should expect the raw output to still contain a number of false positives—and that number can be quite large. For that reason, we need to filter the raw callset before proceeding to further downstream analysis.

Filtering variants is ultimately a classification problem: we're trying to distinguish variants that are more likely to be real from those that are more likely to be artifacts. The more clearly we can make this distinction, the better chance we have of eliminating most of the artifacts without losing too many real variants. Therefore, we'll evaluate every filtering method based on how favorable a trade-off it enables us to make, expressed in terms of sensitivity, or recall (percentage of the real variants that we're able to identify), and specificity, or precision (the percentage of the variants we called that are actually real), as depicted in Figure 2-24.

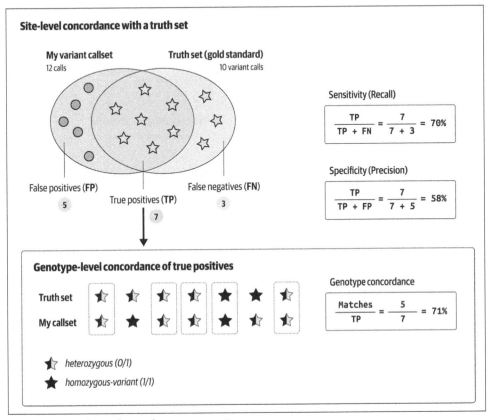

Figure 2-24. Cheat sheet of variant metrics.

There are many approaches to filtering variants, and the choice of what is appropriate depends not only on the variant class we're looking at, but also on the type and quality of the prior knowledge we have about genomic variation in the organism we study. For example, in humans we have extensive catalogs of variants that have been previously identified and validated, which we can use as a basis for filtering variant calls in new samples, so we can use modeling approaches that rely on the availability of training data from that same organism. In contrast, if we're looking at a nonmodel organism that is being studied for the very first time, we do not have that luxury and will need to fall back on other methods.

Data Quality and Sources of Error

A lot of things can go wrong (Figure 2-25) during the entire process, from initial sample collection all the way to the application of computational algorithms, with multiple sources of error and confounding factors creeping in at each stage. Let's review the most common, which are important for understanding many of the procedures that we'll apply in the course of genomic analysis.

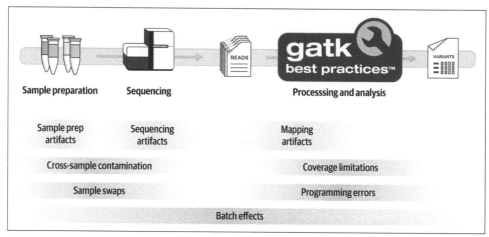

Figure 2-25. Common sources of error in variant discovery.

Contamination and sample swaps

At the earliest stages of sample collection and preparation, we are vulnerable to several types of contamination. The most difficult to prevent is contamination by foreign material, especially in the case of cheek swabs and saliva samples. Think about what's going on in your own mouth right now. When is the last time you brushed your teeth? What did you have for lunch? Any chance you have a bit of a cold at the moment, maybe a runny nose? If we were to try to sequence your genome based on a saliva sample or cheek swab, whatever is in your mouth is going to be sequenced along with your own DNA. At the very least, this means some of the resulting sequence data will be useless noise; in worst-case scenarios, it can cause errors in the downstream analysis results.

Sample Contamination Stories

A few years ago, the Broad Genomics facility processed a group of samples that failed quality control checks with a very particular failure mode. A substantial proportion of reads were mapped in just a few places, accumulating to huge depths, with only a subset of the read sequence aligning to the genome reference. The QC team investigated and found that the reads came from bacterial DNA, with short stretches of sequence that are highly conserved throughout the tree of life. These sequences match some human sequence well enough to align to the human reference, albeit with some read clipping on either side. Going back to the original samples, the team realized that the cheek swabs were from sick children who had very high bacterial load in their mouths due to heavy nasal congestion. The GATK methods development team addressed the issue by adding a data processing filter that weeds out sequence reads that are clipped at both ends.

The following year, the GATK methods development team received another fun request from the Broad Genomics facility: could it please come up with a way to identify cow, pig, chicken, and fish DNA and remove them from human samples? To make a long story short, people were sometimes eating lunch immediately before providing saliva samples, and leftover food fragments were being sequenced along with the diner's DNA. Vegetarians could get away with this because plant DNA is different enough that it simply doesn't map to the human genome, but the animals we eat are close enough to us that some of their DNA does map and aligns well enough to the human reference, which can then introduce errors in the variant-calling process. So members of the team got to work creating human-animal hybrids *in silico*. They added specific amounts of animal sequence data to human data so that they could determine how much should be filtered out. Then, the team adapted an existing method used for metagenomic analysis to identify the contaminating data and exclude it from the dataset, which successfully addressed the original problem. The resulting tool, called PathSeq, can be used to preprocess samples in this way, but it can also be used for identifying foreign DNA from other sources, such as viral DNA in blood or tissue samples.

Then there's cross-sample contamination, which people don't like to talk about because it implies something went wrong either with the quality of equipment or the skill of the person doing the work. But the reality is that cross-contamination between samples happens often enough that it should be addressed in the analysis. It can happen during the initial sample collection, especially when sampling is done in the field in less than ideal conditions, or during library preparation steps. The result in either case is that some genetic material from person A ends up in the tube for sample B. When the level of contamination is low, it's typically not a problem for germline variant analysis, but as you'll see in Chapter 7, it can have major implications for somatic analysis. Also in Chapter 7, we discuss how we can detect and handle cross-sample contamination in the somatic short variant pipeline.

Finally, sample swaps generally come down to mislabeling—either the label ending up on the wrong tube, or handwritten labels with ambiguous lettering—or metadata handling errors during processing. To detect any sample swaps that occur during processing, the best practice recommendation is to put inbound samples through a fingerprinting procedure and then check processed data against the original fingerprints to detect any swaps.

Biochemical, physical, and software artifacts

Various physical and biochemical effects that occur during sample preparation (such as shearing and oxidation) can cause DNA damage that ultimately translates into having the wrong sequence in the output data. Library preparation protocols that involve a PCR amplification step introduces similar errors because our good friend, the polymerase, is wonderfully productive but prone to incorporating the wrong base. And, of course, sometimes the sequencing machine fails to read the light signals bouncing off the DNA accurately, producing either low-confidence BCLs or outright errors. So the data we have might not always be of the highest quality.

Some biochemical properties of sequence context affect the efficiency of library preparation, flowcell loading, and sequencing processes. This results in producing artificially inflated coverage in some areas and inadequate coverage in others. For example, and as Figure 2-26 shows, we know that stretches of DNA with high GC content (areas with lots of G and C bases) will generally show less coverage compared to the rest of the genome. Highly repetitive regions are also associated with coverage abnormalities.

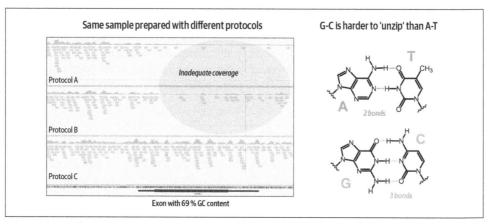

Figure 2-26. Some biochemical properties of the DNA itself cause biases in certain regions.

Then, after the data has been generated and we go to process it, we run into other sources of error. For example, mapping artifacts caused by the biological complexity of the genome (spoiler: it's a mess) can lead to reads ending up in the wrong place, supporting apparent variants that are not real. Shortcuts in algorithms made to improve performance can cause small approximations to balloon into errors. And although we're sad to say this, software bugs are inevitable.

The cherry on the cake, finally, consists of batch effects—because we can deal with a certain amount of error as long as it's consistent across all the samples we're comparing, but the minute we begin comparing results from samples that were processed differently, we're in the danger zone. If we're not careful, we could end up ascribing biological impact to an artifact that just happens to be present in a subset of our samples because of how they were processed.

Functional Equivalence Pipeline Specification

The data preprocessing phase that takes place before variant calling is probably the part of genomic analysis that has the most heterogeneity in terms of valid alternative implementations. There are many ways you can "do it right" depending on how you want to handle the logistics side, whether you care more about cost or runtime, and so on. For each step, several alternative software tools (some open source, some proprietary) are available for you to use—especially for mapping. It's almost a rite of passage for newly minted bioinformatics developers to write a new mapping algorithm.

However, there is a dark side to this abundance of choice. Differences between implementations can cause subtle batch effects in downstream analyses that compare datasets produced by those variant pipelines, sometimes with important consequences for the scientific results. That is why historically the Broad Institute teams have always chosen to redo the preprocessing on genomic data coming in from external sources. They would systematically revert any aligned read data back to an unmapped state and scrub it clean of any tagging or any reversible modifications. Then, they would put the reverted data through their internal preprocessing pipeline.

That strategy is unfortunately no longer sustainable given the huge increase in the amount of data being produced and used in large aggregation projects such as gnomAD (*https://gnomad.broadinstitute.org*). To address this issue, the Broad Institute teamed up with several of the other largest genome sequencing and analysis centers in North America (New York Genome Center, Washington University, University of Michigan, Baylor College of Medicine) to define a standard for standardizing pipeline implementations. The idea is that any data processed through a pipeline that conforms with the specification will be compatible, so that any datasets aggregated from multiple conformant sources can be analyzed together without fear of batch effects.

The consortium published the Functional Equivalence specification (*https://oreil.ly/_yLwF*) and encourages all genome centers as well as anyone else who is producing data, even at a small scale, to adopt the specification in order to promote interoperability and eliminate batch effects from joint studies.

Wrap-Up and Next Steps

This concludes our genomics primer; you now know everything you need on the scientific side to get started with GATK and variant discovery. Next up, we do the equivalent exercise on the computational side with a technology primer that takes you through the key concepts and terminology you need to know to run the tools involved.

## CHAPTER 3

Computing Technology Basics for Life Scientists

In an ideal world, you wouldn't need to worry too much about computing infrastructure when you're pursuing your research. In fact, in later chapters we introduce you to systems that are specifically designed to abstract away the nitty-gritty of computing infrastructure in order to help you focus on your science. However, you will find that a certain amount of terminology and concepts are unavoidable in the real world. Investing some effort into learning them will help you to plan and execute your work more efficiently, address performance challenges, and achieve larger scale with less effort. In this chapter, we review the essential components that form the most common types of computing infrastructure, and we discuss how their strengths and limitations inform our strategies for getting work done efficiently at scale. We also go over key concepts such as parallel computing and pipelining, which are essential in genomics because of the need for automation and reproducibility. Finally, we introduce virtualization and lay out the case for cloud infrastructure.

The first few sections in this chapter are aimed at readers who have not had much training, if any, in informatics, programming, or systems administration. If you are a computational scientist or an IT professional, feel free to skip ahead until you encounter something that you don't already know. The last two sections, which together cover pipelining, virtualization, and the cloud, are more specifically focused on the problems that we tackle in this book and should be informative for all readers regardless of background.

Basic Infrastructure Components and Performance Bottlenecks

Don't worry; we're not going to make you sit through an exhaustive inventory of computer parts. Rather, we've put together a short list of the components, terminology, and concepts that you're most likely to encounter in the course of your work. In relation to each of these, we've summarized the main performance challenges and the strategies that you might need to consider to use them effectively.

Let's begin with a brief overview of the types of processors that you might come across in scientific computing today.

Types of Processor Hardware: CPU, GPU, TPU, FPGA, OMG

At its simplest, a *processor* is a component in your computer that performs computations. There are various types of processors, with the most common being the *central processing unit* (CPU) that serves as the main processor in general-use computers, including personal computers such as laptops. The CPU in your laptop may have multiple cores, subunits that can process operations more or less independently.

In addition to a CPU, your personal computer also has a *graphical processing unit* (GPU) that processes the graphical information for display on your screen. GPUs came into the limelight with the development of modern video games, which require extremely fast processing to ensure smooth visual rendering of game action. In essence, the GPU solution outsources the rather specific type of processing involved in mathematical calculations like matrix and vector operations from the CPU to a secondary processing unit that specializes in handling certain types of calculations that are applied to graphical data very efficiently. As a result, GPUs are also becoming a popular option for certain types of scientific computing applications that involve a lot of matrix or vector operations.

The third type of processor you should know about is called a *field-programmable gate array* (FPGA), which, despite breaking with the *PU naming convention, is also a type of processing unit; however, it's unlikely that you'll find one in your laptop. What's interesting about FPGAs is that unlike GPUs, FPGAs were not developed for a specific type of application; quite the contrary, they were developed to be adaptable for custom types of computations. Hence "field-programmable" as part of their name.

On GCP, you might also come across something called a *tensor processing unit* (TPU), which is a kind of processor developed and branded by Google for machine learning applications that involve tensor data. A *tensor* is a mathematical concept used to represent and manipulate multiple layers of data related to vectors and matrices. Consider that a vector is a tensor with one dimension, and a matrix is a tensor with two dimensions; more generally, tensors can have arbitrary numbers of dimensions beyond that, so they are very popular in machine learning applications. TPUs belong to a category of processors called application-specific integrated circuit (*https:// oreil.ly/bz4mv*) (ASIC), which are custom designed for specialized uses rather than general use.

Now that you have the basic types of processors down, let's talk about how they are organized in typical high-performance computing setups.

Levels of Compute Organization: Core, Node, Cluster, and Cloud

When you move beyond personal computers and into the world of high-performance computing, you'll hear people talk about cores, nodes, and either clusters or clouds, as illustrated in Figure 3-1. Let's review what these mean and how they relate to one another.

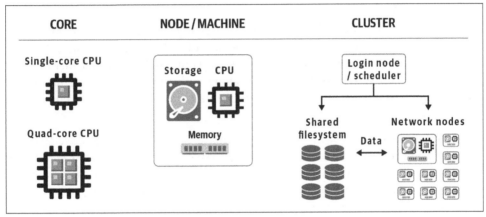

Figure 3-1. Levels of compute organization.

Low level: core

A *core* is the smallest indivisible processing unit within a machine's, or node's, processor unit, which can comprise one or more cores. If your laptop or desktop is relatively recent, its CPU probably has at least two cores, and is therefore called *dual-core*. If it has four, it's a *quad-core*, and so on. High-end consumer machines can have more than that; for example, the latest Mac Pro has a twelve-core CPU (which should be called dodeca-core if we follow the Latin terminology) but the CPUs on professional-grade machines can have tens or hundreds of cores, and GPUs typically have an order of magnitude more, into the thousands. Meanwhile, TPUs have core counts in the single digits like consumer CPUs, and FPGAs break the mold entirely: their cores are defined by how they are programmed, not by how they are built.

Mid level: node/machine

A *node* is really just a computer that is part of a cluster or cloud. It is analogous to the laptop or desktop computer that most of us interact with primarily in our day-to-day work, except without the dedicated monitor and peripherals we are used to seeing associated with personal computers. A node is also sometimes simply called a *machine*.

Top level: cluster and cloud

A cluster and a cloud are both a collection of machines/nodes.

A *cluster* is an HPC structure composed of nodes networked together to some extent. If you have access to a cluster, the chances are that either it belongs to your institution, or your company is renting time on it. A cluster can also be called a *server farm* or a *load-sharing facility*.

A *cloud* is different from a cluster in that in its resting state, its nodes are not explicitly networked together. Rather, it is a collection of independent machines that are available to be networked (or not) depending on your needs. We cover that in more detail in the final section of this chapter, along with the concept of virtualization, which gives us virtual machines (VMs), and containerization, which gives us Docker containers.

For now, however, we move on to the very common concern of how to use a given set of computing infrastructure effectively, which typically revolves around identifying and solving key computational bottlenecks. As with the rest of this chapter, an in-depth exploration of this topic would be beyond the scope of this book, so we're aiming simply to familiarize you with the key concepts and terminology.

Addressing Performance Bottlenecks

You'll occasionally find that some computing operations seem slow and you'll need to figure out how to make them go faster (if possible). The solutions available to you will depend on the nature of the bottleneck you're facing.

At a very high level, following are the main operations that the computer typically has to perform (not necessarily in a linear order):

1. Read some data into memory from the permanent storage where it resides at rest

2. Have the processor execute instructions, transforming data and producing results

3. Write results back to the permanent storage

Data storage and I/O operations: hard drive versus solid state

Steps 1 and 3 are called *I/O operations* (I/O stands for input/output). You might hear people describe some software programs as being "I/O-bound," which means the part of the program that takes the longest is reading and writing data to and from relatively slow storage. This is typically the case for simple programs that do things like file format conversions, in which you're just reading in some data and writing it out in a different shape, and you're not doing any real computing (i.e., there's little to no math involved). In those cases, you can speed up operation by using faster storage drives; for example, solid-state drives (SSDs) rather than hard-disk drives (HDDs). The key difference between them is that HDDs have physical disks called platters that spin and an armature that reads data from and writes it to the platter via magnetics—like a tiny high-tech turntable—whereas SSDs have no moving parts. That makes SSDs less prone to physical malfunctions and also quite a bit faster at accessing data.

If you're working with a networked infrastructure in which the storage drives are not directly connected to the computing nodes, you will also be limited by the speed at which data can be transferred over the network connections. That can be determined by hardware factors as pedestrian as the kind of cables used to connect the network parts. Although you might not notice the difference when computing on small files, you definitely will notice it when running on whole genomes; and even on a network with very fast transfer speeds, transferring whole genomes will consume some noticeable time.

Memory: cache or crash

Step 2 is where your program is taking data and applying some kind of transformation or calculation, aka the interesting part. For a lot of applications, the calculation requires holding a lot of information in memory. In those cases, if your machine doesn't have enough memory, it might resort to *caching*, which is a way of using local storage space as a substitute for real memory. That allows you to keep working, but now your processes become I/O bound because they need to copy data back and forth to slow storage, which takes you back to the first bottleneck. In extreme cases, the program can stall indefinitely, fail to complete, or crash. Sometimes, it's possible for a developer to rewrite the program to be smarter about the information it needs to see concurrently, but when it's not, the solution is to simply add more memory. Fortunately, unlike memory in humans, computer memory is just hardware, and it comes relatively cheap.

Specialized hardware and code optimizations: navigating the trade-offs

At times, the nature of the program requires the processor itself to do a lot of heavy lifting. For example, in the widely used GATK tool `HaplotypeCaller`, an operation can calculate genotype likelihoods; we need to compute the likelihood of every single sequence read given each candidate allele using a hidden Markov model (HMM) called *PairHMM* (don't worry if this sounds like gibberish at the moment—it's just a bunch of genomics-specific math). In some areas of the genome, that leads us to do millions of computations per site across a very large number of sites. We know from performance profiling tests, which record how much time is spent in processing for each operation in the program, that PairHMM is by far the biggest bottleneck for this tool. We can reduce this bottleneck in some surface-level ways; for example, by making the program skip some of the computations for cases in which we can predict they will be unnecessary on uninformative. After all, the fastest way to calculate something is to not calculate it at all.

Being lazy gets us only so far, however, so to get to the next level, we need to think about the kind of processor we can (or should) use for the work we need to do. Not just because some processors run faster than others, but also because it's possible to write program instructions in a way that is very specific to a particular type and architecture of processor. If done well, the program will be extra efficient in that context and therefore run faster. That is what we call *code optimization*, and more specifically *native* code optimization because it must be written in a low-level language that the processor understands "natively" without going through additional translation layers.

Within a type of processor like CPUs, different manufacturers (e.g., Intel and AMD) develop different *architectures* for different generations of their products (e.g., Intel Skylake and Haswell), and these different architectures provide opportunities for optimizing the software. For example, the GATK package includes several code modules corresponding to alternative implementations of the PairHMM algorithm that are optimized for specific Intel processor architectures. The program automatically activates the most appropriate version when it finds itself running on Intel processors, which provides some useful speed gains.

However, the benefits of hardware optimizations are most obvious across processor types; for example, if you compare how certain algorithms perform when implemented to run on FPGAs instead of CPUs. The Illumina DRAGEN toolkit (originally developed by Edico Genome) includes implementations of tools like `HaplotypeCaller` that are optimized to run on FPGAs and as a result are much faster than the original Java software version.

The downside of hardware-optimized implementations is that by definition, they require specialized hardware. This can be a big problem for the many research labs that rely on shared institutional computing systems and don't have access to other hardware. In contrast, applications written in Java, like GATK, can run on a wide range of hardware architectures because the Java Virtual Machine (JVM) translates the application code (called *bytecode* in the Java world) into instructions appropriate for the machine. This *separation of concerns* (SoC) between the bytecode of Java and what actually is executed on the machine is called an *abstraction layer* and it's incredibly convenient for everyone involved. Developers don't need to worry about exactly what kind of processor we have in our laptops, and we don't need to worry about what kind of processor they had in mind when they wrote the code. It also guarantees that the software can be readily deployed on standard off-the-shelf hardware, which makes it usable by anyone in the world.

Sometimes, you'll need to choose between different implementations of the same algorithms depending on what is most important to you, including how much you prize speed over portability and interoperability. Other times, you'll be able to enjoy the best of both worlds. For example, the GATK team at the Broad Institute has entered into a collaboration with the DRAGEN team at Illumina, and the two teams are now working together to produce unified DRAGEN-GATK pipelines that will be available both as a free open source version (via Broad) and as a licensed hardware-accelerated version (via Illumina). A key goal of the collaboration is to make the two implementations functionally equivalent—meaning that you could run either version and get the same results within a margin of error considered to be insignificant. This will benefit the research community immensely in that it will be possible to combine samples analyzed by either pipeline into downstream analyses without having to worry about batch effects, which we discussed briefly in the previous chapter.

Why GATK Is Written in Java

For GATK, the analysis toolkit that we work with throughout most of this book, the choice of Java was deliberate: it makes development much easier for people who are not software engineers. Considering that the majority of the GATK methods development team are primarily data scientists, it's more important to enable them to focus on developing algorithms to answer complex scientific questions rather than code optimizations for specific hardware.

But the good news is that we can still take advantage of native code optimizations where it matters! Third-party teams that specialize in the kind of hardcore software engineering involved have contributed libraries of GATK code that are optimized for specific hardware. The GATK engine has a plug-in system that allows us to swap out these code libraries at runtime (i.e., when we run a tool command), so we can, for example, run HaplotypeCaller with a version of the PairHMM algorithm optimized for a specific processor architecture.

For convenience, the library selection is automatically done by default. So as an end user, you don't need to know what kind of hardware you're working with. The engine will try to recognize what architecture it is running on, and if it finds a match in the available native code libraries, it switches to the correct one. If it doesn't find a match, it falls back to the default Java implementation.

Between 2013 and 2014, a subset of the GATK team experimented with rewriting the GATK engine and key tools in C++, a more "low-level" language that allows the programmer to specify processor-level instructions. Codenamed Foghorn, the project was motivated on one hand by the emerging need to rewrite GATK with a more streamlined framework, and on the other, by claims that getting rid of the Java layer and using low-level capabilities like cache manipulation would deliver substantial speed improvements. In practice, the developers found that, although they were able to wring some speed improvements out of the C++ implementation, these were not as large as expected. More important, the pace of development was massively slower because the tool developers had to spend a lot more time solving software engineering problems instead of focusing on the data science.

The project was abandoned in late 2014, and the team recommitted to developing GATK in Java going forward. However, the effort spent on Foghorn was not all in vain: many lessons learned in the course of that project went on to serve as foundations for the development of GATK4.

Parallel Computing

When you can't go faster, go parallel. In the context of computing, *parallel computing*, or *parallelism*, is a way to make a program finish sooner by performing several operations in parallel rather than sequentially (i.e., waiting for each operation to finish before starting the next one). Imagine that you need to cook rice for 64 people, but your rice cooker can make enough rice for only 4 people at a time. If you need to cook all of the batches of rice sequentially, it's going to take all night. But if you have eight rice cookers that you can use in parallel, you can finish up to eight times sooner.

This is a simple idea but it has a key requirement: you must be able to break the job into smaller tasks that can be performed independently. It's easy enough to divide portions of rice because rice itself is a collection of discrete units. But you can't always make that kind of division: for example, it takes one pregnant woman nine months to grow a baby, but you can't do it in one month by having nine women share the work.

The good news is that most genomic analyses are more like rice than like babies—they essentially consist of a series of many small independent operations that can be parallelized. So how do we get from cooking rice to executing programs?

Parallelizing a Simple Analysis

Consider that when you run an analysis program, you're just telling the computer to execute a set of instructions. Suppose that we have a text file and we want to count the number of lines in it. The set of instructions to do this can be as simple as this:

Open the file; count the number of lines in it; tell us the number; close the file.

Note that "tell us the number" can mean writing it to the console or storing it somewhere for use later on—let's not worry about that right now.

Now suppose that we want to know the number of words on each line. The set of instructions would be as follows:

> Open the file; read the first line; count the number of words; tell us the number; read the second line; count the number of words; tell us the number; read the third line; count the number of words; tell us the number.

And so on until we've read all the lines, and then finally we can close the file. It's pretty straightforward, but if our file has a lot of lines, it will take a long time, and it will probably not use all the computing power we have available. So, to parallelize this program and save time, we just cut up this set of instructions into separate subsets, like this:

- Open the file; index the lines.
- Read the first line; count the number of words; tell us the number.
- Read the second line; count the number of words; tell us the number.
- Read the third line; count the number of words; tell us the number.
- [Repeat for all lines.]
- Collect final results and close the file.

Here, the "read the Nth line" steps can be performed in parallel because they are all independent operations.

You'll notice that we added a step, "index the lines." That's a little bit of preliminary work that allows us to perform the "read the Nth line" steps in parallel (or in any order we want) because it tells us how many lines there are and, importantly, where to find each one within the file. It makes the entire process much more efficient. As you will see in the following chapters, tools like GATK require index files for the main data files (reference genome, read data and variant calls). The reason is to have that indexing step already done so that we can have the program look up specific chunks of data by their position in the genome.

Anyway, that's the general principle: you transform your linear set of instructions into several subsets of instructions. There's usually one subset that has to be run first and one that has to be run last, but all the subsets in the middle can be run at the same time (in parallel) or in whatever order you want.

From Cores to Clusters and Clouds: Many Levels of Parallelism

So how do we go from rice cookers to parallelizing the execution of a genomic analysis program? Overall, the action of parallelizing computing operations consists of sending subsets of the work we want done to multiple cores for processing. We can do that by splitting up the work across the cores of a single multicore machine, or we can dispatch work to other machines if we have access to a cluster or cloud. In fact, we can combine the two ideas and dispatch work to multicore machines, in which the work is further split up among each machine's cores. Going back to the rice-cooking example, it's as if instead of cooking the rice yourself, you hired a catering company to do it for you. The company assigns the work to several people, who each have their own cooking station with multiple rice cookers. Now, you can feed a lot more people in the same amount of time! And you don't even need to clean the dishes.

Whether we want to distribute the work across multiple cores on a single machine or across multiple machines, we're going to need a system that splits up the work, dispatches jobs for execution, monitors them for completion, and then compiles the results. Several kinds of systems can do that, falling broadly into two categories: internal or external to the analysis program itself. In the first case, the parallelization happens "inside" the program that we're running: we run that program's command line, and the parallelization happens without any additional "wrapping" on our part. We call that *multithreading*. In the second case, we need to use a separate program to run multiple instances of the program's command line. An example of an external parallelization is writing a script that runs a given tool separately on the data from each chromosome in a genome and then combines the result with an additional merge step. We call that approach *scatter-gather*. We cover that in more detail in the next section when we introduce workflow management systems. In Figure 3-2, you can see how we can use multithreading and scatter-gather parallelism in the course of an analysis.

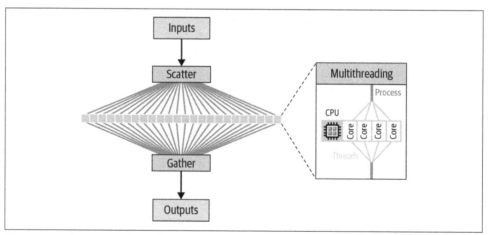

Figure 3-2. Scatter-gather allows parallel execution of tasks on different CPU cores (on a single machine or multiple machines, depending on how it's implemented).

Spark: Multithreading on Steroids

Spark (*https://spark.apache.org*) is a powerful open source framework for data analysis that was originally developed at the University of California, Berkeley, and is now maintained by the Apache Software Foundation. It is a huge topic in and of itself, worthy of (and already covered by) multiple books; here we just want to call out a couple of specific applications for which Spark makes a big difference in terms of speed and scalability of genomic analysis.

In GATK, Spark serves as the internal parallelization system that distributes execution across multiple cores. In the simplest case, it will distribute work across local cores of a multicore machine; but it is also capable of distributing work to multiple machines of a Spark-compatible cluster. This functionality is also used in another analysis toolkit called Hail (*https://hail.is*), in the context of interactive data analysis that happens mostly downstream of GATK.

Trade-Offs of Parallelism: Speed, Efficiency, and Cost

Parallelism is a great way to speed up processing on large amounts of data, but it has overhead costs. Without getting too technical at this point, let's just say that parallelized jobs need to be managed, you need to set aside memory for them, regulate file access, collect results, and so on. So it's important to balance the costs against the benefits, and avoid dividing the overall work into too many small jobs. Going back to our earlier example, you wouldn't want to use a thousand tiny rice cookers that each boil a single grain of rice. They would take far too much space on your countertop, and the time required to distribute each grain and then collect it when it's cooked would more than negate any benefits from parallelizing in the first place.

More generally, although it's tempting to think of parallelism as a way to make things go faster, it's important to remember that the impression of speed is entirely subjective to the observer. In reality, the computation being run on each piece of data is not going any faster. We're just running more computations at the same time, and we're limited by the parallel capacity of our computing setup (typically measured in number of nodes or cores) as well as hardware limitations like I/O and network speeds. It's more realistic to think of parallelism as a way to optimize available resources in order to finish tasks *sooner*, rather than making individual tasks run *faster*.

This distinction might seem pedantic given that, from the point of view of the human at the keyboard, the elapsed time (often called *wall-clock time*; that is, "the time shown by the clock on the wall") is shorter. And isn't that what we all understand as going faster? However, from the point of view of the resources we utilize, if we add up the time spent doing computation by the processor across all the cores we use, we might find that the overall job takes *more* time to complete compared to purely sequential execution because of the overhead costs.

That brings us to another important question: what is the monetary cost of utilizing those resources? If we're working with a dedicated machine that is just sitting there with multiple cores and nothing else to do, the parallelization is still absolutely worth it, even with the overhead. We'll want to parallelize the heck out of everything we do on that machine in order to maximize efficiency. However, when we start working in the cloud environment, as we do in Chapter 4, and we need to start paying for itemized resources as we go, we'll want to look more carefully at the trade-offs between minimizing wall-clock time and the size of the bill.

Pipelining for Parallelization and Automation

Many genomic analyses involve running a lot of routine data-processing operations that need to be parallelized for efficiency and automated to reduce human error. We do this by describing the workflow in a machine-readable language, which we then can feed into a workflow management system for execution. We go over how this works in practice in Chapter 8, but first let's set the stage by introducing basic concepts, definitions, and key software components. As we go, please keep in mind that this field currently has no such thing as a one-size-fits-all solution, and it's ultimately up to you to review your needs and available options before picking a particular option. However, we can identify general principles to guide your selection, and we demonstrate these principles in action using the open source pipelining solution that is recommended by the GATK development team and used in production at the Broad Institute. As with most of this book, the goal here is not to prescribe the use of specific software, but to show through working examples how all of this fits together in practice.

Scripts, Workflows, and Pipelines

The terms *script*, workflow, and *pipeline* are often used interchangeably but are not entirely equivalent. Very generically, a script is a set of instructions in written form, whereas a workflow is a kind of script comprising multiple steps that are connected. In this book, we mostly talk about *writing and running workflows*, though we might occasionally use *script* or *workflow script* to refer specifically to the written form.

The term *pipeline* is often used either as a substitute for workflow or as shorthand for the real-world operation of a particular workflow script (or set of scripts that are run in sequence), as in "the Broad's production pipeline," which refers to a combination of infrastructure, process, and artifacts involved in executing workflows in a routine, highly automated manner. We limit our usage of the word pipeline to the latter, along with *pipelining* as a convenient shorthand for the action of producing and running analysis workflows.

One tricky aspect is that we have dozens of scripting languages and workflow management systems to choose from in the bioinformatics world—likely hundreds if you look at a wider range of fields. It can be difficult to compare them directly because they tend to be developed with a particular type of audience in mind, leading to very different modalities of user experience. They are often tuned for particular use cases and are sometimes optimized to operate on certain classes of infrastructure. We often see one solution that is preferred by one group prove to be particularly difficult or frustrating to use for another. These various solutions are also generally not interoperable, meaning that you can't take a workflow script written for one workflow management system and run it unmodified on the next one over. This lack of standardization is a topic of both humor and desperation in just about every field of research known to humankind, as is illustrated in Figure 3-3.

Figure 3-3. XKCD comic on the proliferation of standards (source: https://xkcd.com/927).

In recent years, we have seen some high-profile initiatives such as the Global Alliance GA4GH emerge with the explicit mission of developing common standards and consolidating efforts around a subset of solutions that have interoperability as a core value. For example, the GA4GH Cloud Work Stream has converged on a small set of workflow languages for its driver projects, including CWL, Nextflow, and WDL, which we use in this book. At the same time, given the recognition that no single language is likely to satisfy all needs and preferences, several groups are working to increase interoperability by building support for multiple workflow languages into their workflow management systems. The workflow management system we use in this book, Cromwell, supports both WDL and CWL, and it could be extended to support additional languages in the future.

Workflow Languages

In principle, we could write our workflows in almost any programming language we like; but in practice, some are more amenable than others for describing workflows. Just like natural languages, programming languages also exhibit a fascinating diversity and can be classified in various ways including grammar, mode of execution, and the programming paradigms that they support.

From a practical standpoint, we begin by making a distinction between all-purpose programming languages, which are intended to be usable for a wide range of applications, and domain-specific languages (DSLs) that are, as the name indicates, specifically designed for a particular domain or activity. The latter are typically preloaded with things like specially formulated data structures (i.e., ways to represent and manipulate data that "understand" the nature of the underlying information) and convenience functions that act as shortcuts; for example, handling domain-specific file formats, applying common processing actions, and so on. As a result, a DSL can be an attractive option if your needs fit well within the intended scope of the language, especially if your computational background is limited, given that the DSL typically enables you to get your work done without having to learn a lot of programming concepts and syntax.

On the other hand, if your needs are more varied or you are used to having the more expansive toolbox of a general-purpose language at your disposal, you might find yourself uncomfortably constrained by the DSL. In that case, you might prefer to use a general-purpose language, especially one enriched with domain-specific libraries that provide relevant data structures and convenience functions (e.g., Python with Biopython and related libraries). In fact, using a general-purpose language is more likely to enable you to use the same language for writing the data-processing tasks themselves and for managing the flow of operations, which is how many have traditionally done this kind of work. What we're seeing now in the field, however, is a move toward separation of description and content, which manifests as increased adoption of DSLs specifically designed to describe workflows as well as of specialized workflow management systems. This evolution is strongly associated with the push for interoperability and portability.

Popular Pipelining Languages for Genomics

When we look at the cross-section of people who find themselves at the intersection of bioinformatics and genomics, we see a wide range of backgrounds, computational experience, and needs. Some come from a software engineering background and prefer languages that are full featured and highly structured, offering great power at the cost of accessibility. Some come from systems administration and believe every problem can be solved with judicious application of Bash, sed, and awk, the duct tape of the Unix-verse. On the "bio" side of the fence, the more computationally trained tend to feel most at home with analyst favorites like Python and R, which have been gaining ground over old-time classics Perl and MATLAB; some also tend to gravitate toward DSLs. Meanwhile wetlab-trained researchers might find themselves baffled by all of this, on initial contact at least. (Author's note and disclaimer: Geraldine identifies as one of the initially baffled, having trained as a traditional microbiologist and eventually learned the rudiments of Perl and Python in a desperate bid to escape the wetlab workbench. Spoiler: it worked!)

Based on recent polling, some of the languages that are most popular with workflow authors in the genomics space are SnakeMake and Nextflow. Both are noted for their high degree of flexibility and ease of use. Likewise, CWL and WDL are picking up steam because of their focus on portability and computational reproducibility. Of the two, CWL is more frequently preferred by people who have a technical background and enjoy its high level of abstraction and expressiveness. In contrast, WDL is generally considered to be more accessible to a wide audience.

At the end of the day, when it comes to picking a workflow language, we look at four main criteria: what kind of data structures the language supports (i.e., how we can represent and pass around information), how it enables us to control the flow of operations, how accessible it is to read and write for the intended audience, and how it affects our ability to collaborate with others. Whatever we choose, it's unlikely that we can satisfy everyone's requirements. However, if we were to boil all this down to just one recommendation, it would be this: if you want your workflow scripts to be widely used and understood in your area of research, pick a language that is open and accessible enough to newcomers yet scales well enough to the ambitions of the more advanced. And, of course, try to pick a language that you can run across different workflow management systems and computing platforms, because you never know what environment you or your collaborators might find yourselves in next.

Workflow Management Systems

Many workflow management systems exist, but in general they follow the same basic pattern. First, the workflow engine reads and interprets the instructions laid out in the workflow script, translating the instruction calls into executable jobs that are associated with a list of inputs (including data and parameters). It then sends out each job with its list of inputs to another program, generally called a *job scheduler*, that is responsible for orchestrating the actual execution of the work on the designated computing environment. Finally, it retrieves any outputs produced when the job is done. Most workflow management systems have some built-in logic for controlling the flow of execution; that is, the order in which they dispatch jobs for execution and for determining how they deal with errors and communicate with the compute infrastructure.

Examples of Job Schedulers

If you have used on-premises HPC before, you might have used a job scheduler such as Sun GridEngine (SGE), Slurm, or HTCondor. On GCP, we'll use an alternative called Pipelines API (PAPI) to marshal cloud resources with minimal effort. On AWS, there is a similar system called AWS Batch.

Another important advance for increasing portability and interoperability of analyses is the adoption of container technology, which we cover in detail in the last section of this chapter. For now, assume that a container is a mechanism that allows you to encapsulate all software requirements for a particular task, from the deepest levels of the operating system (OS) all the way to library imports, environment variables, and accessory configuration files.

Virtualization and the Cloud

Up to this point, we have been assuming that whether you're working with a single computer or a cluster, you're dealing with "real" physical machines that are each set up with a given OS and software stack, as represented in Figure 3-4 A. Unfortunately, interacting with that kind of system has several disadvantages, especially in a shared environment like an institutional cluster. As an end user, you typically don't have a choice regarding the OS, environment, and installed software packages. If you need to use something that isn't available, you can ask an administrator to install it, but they might decline your request or the package you want might not be compatible with existing software. For the system administrators on the other side of the help-desk, it can be a headache to keep track of what users want, manage versions, and deal with compatibility issues. Such systems take effort to update and scale.

That is why most modern systems use various degrees of *virtualization*, which is basically a clever bit of abstraction that makes it possible to run multiple different software configurations on top of the same hardware through virtual machines (VMs) and containers as represented in Figure 3-4 B and C respectively. These constructs can be utilized in many contexts, including optionally on local systems (you can even use containers on your laptop!), but they are absolutely essential for cloud infrastructure.

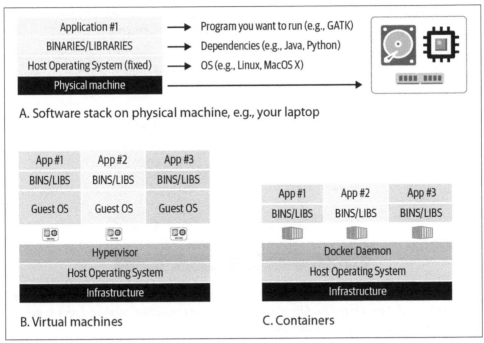

A. Software stack on physical machine, e.g., your laptop

B. Virtual machines

C. Containers

Figure 3-4. A) The software stack installed on a physical machine; B) a system hosting multiple VMs; C) a system hosting multiple containers.

VMs and Containers

A VM is an infrastructure-level construct that includes its own OS. The VM sits on top of a virtualization layer that runs on the actual OS of the underlying physical machine(s). In the simplest case, VMs can be run on a single physical machine, with the effect of turning that physical machine into multiple servers that share the underlying resources. However, the most robust systems utilize multiple physical machines to support the layer of VMs, with a complex layer between them that manages the allocation of physical resources. The good news is that for end users, this should not make any difference—all you need to know is that you can interact with a particular VM in isolation without worrying about what it's sitting on.

A container is similar in principle to a VM, but it is an application-level construct that is much lighter and more mobile, meaning that it can be deployed easily to different sites, whereas VMs are typically tied to a particular location's infrastructure. Containers are intended to bundle all the software required to run a particular program or set of programs. This makes it a lot easier to reproduce the same analysis on any infrastructure that supports running the container, from your laptop to a cloud platform, without having to go through the pain of identifying and installing all the software dependencies involved. You can even have multiple containers running on the same machine, so you can easily switch between different environments if you need to run programs that have incompatible system requirements.

If you're thinking, "These both sound great; which one should I use?" here's some good news: you can use both in combination, as illustrated in Figure 3-5.

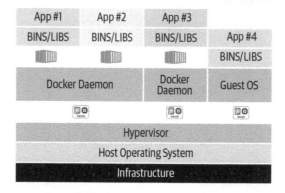

Figure 3-5. A system with three VMs: the one on the left is running two containers, serving App #1 and App #2; the middle is running a single container, serving App #3; the right is serving App #4 directly (no container).

There are several registries for sharing and obtaining containers, including Docker Hub (*https://hub.docker.com*), Quay.io (*https://quay.io*), and GCR (*https:// cloud.google.com/container-registry*), Google's general-purpose container registry in GCP. In the registry, the container is packaged as an *image*. Note that this has nothing to do with pictures; here the word *image* is used in the same software-specific way that refers to a special type of file. You know how sometimes when you need to install new software on your computer, the download file is called a *disk image*? That's because the file you download is in a format that your OS is going to treat as if it were a physical disk on your machine. This is basically the same thing. To use a container, you first tell the Docker program to download, or *pull*, a container image file from a registry—for example, Docker Hub (more on Docker shortly)—and then you tell it to initialize the container, which is conceptually equivalent to booting up a VM. And after the container is running, you can run any software within it that is installed on its system. You can also install additional packages or perform additional configurations as needed. Figure 3-6 illustrates the relationship between container, image, and registry.

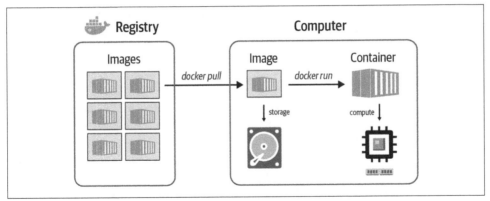

Figure 3-6. The relationship between registry, image, and container.

The most widely used brand of container systems is Docker, produced by the company of the same name. As a result of Docker's ubiquitousness, people will often say "a docker" instead of "a container," much like when "xerox" became a replacement for "copy machines" (in the US at least) because of the dominance of the Xerox company. However, docker with a lowercase *d* is also the command-line program that you install on your machine to run Docker containers. Similarly, although the action of bundling a software tool, package, or analysis into a Docker container should rightly be called "containerizing," people often call it "dockerizing," as in, "I dockerized my Python script." Dockerizing a tool involves writing a script called a Dockerfile that describes all installations and environment configurations necessary to *build* the Docker image, as demonstrated in Figure 3-7.

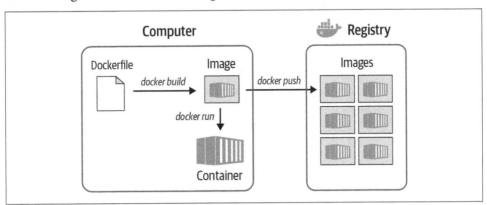

Figure 3-7. The process for creating a Docker image.

As noted earlier, it is possible to use containers in various contexts, including local machines, HPC, and the cloud. One important restriction is that Docker specifically is usually not allowed in shared environments like most institutions' HPCs, because it requires a very high level of access permissions called *root*. In that kind of setting, system administrators will prefer *Singularity*, an alternative system that achieves the same results. Fortunately, it is possible to run Docker containers within a Singularity system.

Introducing the Cloud

Finally, we get to the topic many of you have been waiting for: what is this cloud thing anyway? The surprisingly easy answer is that the cloud is a bunch of computers that you can rent. In practice, that means that as a user, you can easily launch a VM and select how much RAM, storage, and what CPUs you want. You want a VM with 1 TB of RAM and 32 CPUs for genome assembly? No problem! Most of the VMs in the cloud are running some form of Linux as the OS, which you get to choose when you launch it, and are typically accessed using a remote shell via Secure Shell (SSH).

Although some VMs include free storage, this is typically ephemeral and will go away when you stop and start your VM. Instead, use block storage (a persistent device) to store data, scripts, and so on on your VM. You can think of these very much like a USB thumb drive that you can have "plugged" into your VM whenever you like. Even when you terminate your VM, files on block storage will be OK and safely saved. Files can also be stored in object store—think of this more like Google Drive or Dropbox, where files can be read and written by multiple VMs at the same time, but you don't typically use these as a normal filesystem. Instead, they are more akin to an SSH File Transfer Protocol (SFTP) server for sharing files between VMs, where you transfer files through a utility to and from the object storage system. The final basic component of a cloud is networking. With a virtual networking service, you can control who has access to your VMs, locking it down tightly to ensure that only you (or others you trust) have access.

Clouds are not fluffy

When you think about clouds, they are fluffy, distant, and elusive, not at all concrete, real things that you can touch, feel, and capture. Unlike their namesake, the cloud infrastructure that most of us use directly (or indirectly) today ultimately is composed of real, physical computers racked up and blinking away in huge datacenters. What makes it different, though, from previous models for compute (and rings true to their name) is its ephemeral nature. Just like clouds coming and going—popping up, dumping their rain, and then blowing away—cloud computing is transient for the end user. The cloud allows you as a researcher, developer, or analyst to request computational infrastructure when you need it, to use it for computing as long as you need it, and then you can release all the resources when you're done.

This approach is great because it saves time and money insomuch as you can spin up a lot of resources at once, get your work done, and spin these back down, saving on the costs or running hardware continuously. You don't need to think too much about where the servers are racked, how they are configured, the health of the hardware, power consumption, or myriad other infrastructure concerns. These are all *abstracted away* from you and are taken care of without you having to think about it too much. What you focus on, instead, is the computational work that you need to perform, the resources you need to do it, and how to most effectively use these resources both from a time and money perspective.

Evolution of cloud infrastructure and services

Amazon launched the first widely successful commercial public cloud service in 2006, but the basic idea has been around for a long time. Mainframes in the 1960s were often rented for use, which made a ton of sense, given the massive costs of buying and operating them. Regardless of the invention of the personal computer, the idea of renting computing infrastructure has cropped up again over and over in the intervening decades. In academic groups and industry, the concept of shared grid computing in the 1990s and 2000s was the more modern equivalent of rented mainframe time. Groups banded together to build cheap but powerful Linux-based HPC clusters that were often centrally managed and allocated out to multiple groups based on some sort of financial split.

Today's public clouds are different, though, in the level of abstraction. Hence, the adoption of the fluffy, amorphous name to reflect the fact that an understanding of the underlying details is not required in order to run large-scale analysis on clouds. When working with a given cloud, you might know the general region of the world that hosts your infrastructure (e.g., North Virginia for AWS us-east-1), but many of the details are hidden from you: how many people are using the underlying hardware of your VM, where the datacenter is really located, how the network is set up, and so on. What you do know are key details that affect service cost and job execution time, like how many CPUs are available, how much RAM the VM has, the uptime guarantees of the file storage system, and the regulations the system conforms to.

There are now many public cloud providers—clouds available to anyone who can pay for the service. The most dominant currently in the Western hemisphere are AWS, Microsoft Azure, and GCP. Each provides a similar mix of services that range from simple VMs rentable by the hour (or minute), file storage services, and networking to more specialized services such as Google's Cloud TPU service, which allows you to perform accelerated machine learning operations. The important feature, though, is that these resources are provided as services: you use what you need per hour, minute, second, or API call, and are charged accordingly.

Pros and cons of the cloud

One of the major advantages that many people point to when discussing the cloud is cost. When building a datacenter, the fixed costs are enormous. You must hire people to rack and maintain physical servers, monitor the network for intrusion, deal with power fluctuations, backups, air conditioning, and so on. Honestly, it is a lot of work! For a datacenter that supports hundreds of users, the costs associated with maintaining the infrastructure can be worth it. But many researchers, developers, analysts, and others are realizing that they don't need to have hundreds of computers always available and running, just waiting for a task. Instead, it makes a lot more sense to use a cloud environment in which you can do local work on your laptop, without extensive resources, and then, when your analysis is ready, you can scale up to hundreds or thousands of machines. Commercial public clouds allow you to easily *burst* your capacity and do a huge analysis when you need to, as opposed to waiting weeks, months, or even years for a dedicated local cluster to finish your tasks. Likewise, you don't need to pay for the maintenance of local infrastructure for all the time you spend developing your algorithms and perfecting your analysis locally.

Finally, as a public cloud user, you have full control of your environment. Need a specific version of Python? Do you have a funky library that compiles only if very specific tool chains are installed? No problem! The cloud lets you have full control over your VMs, something that a shared, local infrastructure would never allow. Even with this control, when they are set up following cloud vendor best practices, public cloud solutions are invariably more secure than on-premises infrastructure because of the vast amount of resources dedicated to security services in these environments and the isolation between users afforded by virtualization.

Although the public cloud platforms are amazing, powerful, flexible and, in many cases, can be used effectively to save a ton of money in the long run, there are some disadvantages to look out for. If you are looking to always process a fixed number of genomes produced by your sequencing group per month, the public cloud might be less attractive and it would make more sense to build a small local compute environment for this very predictable workload of data produced locally. This is assuming, of course, that you have IT professionals who can act as administrators. Another consideration is expertise. Using the cloud demands a certain level of expertise, and an unsuspecting novice user might accidentally use VMs with weak passwords, set up data storage buckets with weak security, share credentials in an insecure way, or just be totally lost in the process of managing a fleet of Linux VMs. Even these potential downfalls, though, are generally outweighed by the benefits of working flexibly on commercial cloud environments for many people.

Categories of Research Use Cases for Cloud Services

The basic components of the cloud described in the previous section are really just the tip of the iceberg. Many more services are available on the main commercial cloud platforms. In fact, there are far too many services, some universal and some unique to a particular cloud, than we can describe here. But let's take a look at how researchers might use the services or the cloud most commonly. Table 3-1 provides an overall summary.

Table 3-1. An overview of the types of usage of cloud infrastructure

Usage type	Cloud environment	Description	Positives	Negatives
Lightweight development	Google Cloud Shell	Using a simple-to-launch free VM for editing code and scripts	• Free • Extremely easy to launch and log in to	• Extremely limited VM
Intermediate analysis and development	Single VM	Launching a single VM, logging in, performing development and analysis work	• Can control the resources on the VM launched • VMs can be powerful enough to perform realistic analysis	• Launching a VM requires more configuration • Larger VMs have increased costs
Batch analysis	Multiple VMs via batch system	Using a system like AWS Batch or Google Cloud Pipelines API to launch many VMs and analyze data in parallel	• Allows for parallel, scaled-up analysis • Workflow management systems like Cromwell support these with little effort	• Increased costs and complexity
Framework analysis	Multiple VMs via a framework	Using Spark, Hadoop, or other framework for data analysis	• These frameworks allow for specialty analysis	• Increased costs and complexity

Lightweight development: Google Cloud Shell

The cloud is a fantastic place for software development. Even though many research-ers will want to use their own laptops or workstation for development, there can be some really compelling reasons for using the cloud as a primary development envi-ronment, especially for testing. On GCP, for example, you can use the Google Cloud Shell from the Google Cloud Console for light development and testing. This is a free (yes, *free!*) VM with one virtual CPU core and 5 GB of storage that you can use just by clicking the terminal icon in the web console. This is a fantastic environment for some light coding and testing; just remember to copy code off of your free instance (using Git, for example) because there are quotas for total runtime per week, and, if you don't use the service for a while, your 5 GB volume might get cleaned out. Still, this is a great option for quickly getting started with the cloud and performing light-weight tasks. You just need a web browser, and the GCP tools are all preinstalled and configured for you. Many other tools that you might want to work with are already installed as well, including Git and Docker, along with languages like Java and Python. You'll have a chance to try it out early on in the next chapter.

Intermediate development and analysis: single VM

Although the Google Cloud Shell is great for many purposes, easy to use, and free, sometimes you might need a bit more power, especially if you want to test your code or analysis at the next scale up, so you spin up your own dedicated VM. This is perhaps the most commonly used option because of the mix of flexibility and simplicity it offers: you can customize your VM, ensuring you have enough CPU cores, RAM, and local storage to accomplish your goal. Unlike the Google Cloud Shell, you must pay for each hour or minute you run this VM; however, you have full control over the nature of the VM. You might use this for software or algorithm development, testing your analysis approach, or spinning up a small fleet of these VMs to perform analysis on multiple VMs simultaneously. Keep in mind, however, that if you are manually launching these VMs, fewer tools will be preinstalled on them and ready to go for you. That makes using utilities such as Git and Docker very helpful for moving your analysis tasks from VM to VM. You'll have a chance to use this extensively in Chapter 4 through Chapter 7.

Batch analysis: multiple VMs via batch services

This approach is really the sweet spot for most users who are aware of it. Although you might use your laptop or Google Cloud Shell for software and script development, and one or more VMs for testing them on appropriately sized hardware, you ultimately don't want to manually manage VMs if your goal is to scale up your analysis. Imaging running 10,000 genome alignments at the same time; you need systems that can batch up the work, provision VMs automatically for you, and turn the VMs off when your work is done. Batch systems are designed just for this task; Google Cloud, for example, offers the Google Cloud Pipelines API, which you can use to submit a large batch of multiple jobs simultaneously. The service will take care of spinning up numerous VMs to perform your analysis and then automatically clean them up after collecting the output files. This is extremely convenient if you need to perform noninteractive analysis on a ton of samples. You'll see in Chapter 8 through Chapter 11 that workflow engines like Cromwell are designed to take advantage of these batch services, which take care of all the details of launching batch jobs. That makes it much easier for you to focus on the details of the analysis you're performing rather than on the infrastructure involved.

Framework analysis: multiple VMs via framework services

The final approach that many researchers will use involves interactive, iterative analysis. In genomics, you can use a batch system to perform large-scale alignment and variant calling but, after you have VCF files for your variants, you might choose to move to a Spark cluster, RStudio, Jupyter Notebook, or any of a large number of analytical environments for subsequent analysis. In Chapter 12, we explore how this works in Terra, which you can use to easily create a custom environment for data processing with a Jupyter interface for interactive analysis, generating plots for your publications, and sharing results with others.

Wrap-Up and Next Steps

In this chapter, we completed the primer topics, which gave you a background on genomics (Chapter 2) and computing technologies (this chapter). We delved into the nitty-gritty details of computer hardware, parallel computing, and virtualization and gave you a glimpse of the power of using workflow execution systems to scale out your analysis on the cloud. In Chapter 4, we take our first baby steps to the cloud environment and show you how to get started with your own VMs running in GCP.

First Steps in the Cloud

In the previous two chapters, we took you through the essentials of genomics and computing technology. Our goal was to make sure you have enough of a grounding in both domains regardless of whether you're coming to this more from one side or the other—or perhaps even from another domain altogether; if so, welcome! And hang in there.

We realize that those first two chapters might have felt very passive since there were no hands-on exercises involved. So here's the good news: you're finally going to get to do some hands-on work. This chapter is all about getting you oriented and comfortable with the GCP services that we use throughout this book. First, we walk you through creating a GCP account and running simple commands in Google Cloud Shell. After that, we show you how to set up your own VM in the cloud, get Docker running on it, and set up the environment that you'll use in Chapter 5 to run GATK analyses. Finally, we show you how to configure the IGV to access data in Google Cloud Storage. After you have all that set up, you'll be ready to do some actual genomics.

Setting Up Your Google Cloud Account and First Project

You can sign up for an account on GCP by navigating to *https://cloud.google.com* and following the prompts. We are purposely light on the details here because the interface for account setup has been known to change. At a high level, though, your goals are to establish a new Google Cloud account, set up a billing account, accept the free trial credits (if you're eligible), and create a new project that links to your billing account.

If you don't already have a Google identity of some kind, you can create one with your regular email account; you don't need to use a Gmail account. Keep in mind also that if your institution uses G Suite, your work email might already be associated with a Google identity even if the domain name is not *gmail.com*.

After you've signed up, make your way to the GCP console (*https://oreil.ly/T4nVl*), which provides a web-based graphical interface for managing cloud resources. You can access most of the functionality offered in the console through a pure command-line interface. In the course of the book, we show you how to do some things through the web interface and some through the command line, depending on what we believe is most convenient and/or typical.

Understanding the GCP Free Tier Program

GCP currently offers a free trial with $300 credit (*https://oreil.ly/REJoI*) spendable over the first 12 months of your account, plus additional services that are advertised as "Always Free." We take advantage of some of these free services, such as the Google Cloud Shell, in the course of this chapter, but we also use some services that cost money, so we recommend that you take advantage of the free trial program if you can. Based on our testing, the free trial should be sufficient to work through all of the book examples without having to pay anything.

You'll notice that even the free trial account requires that you provide billing information, which can be either a credit card or a bank account (which can take a few days to clear). The GCP website (*https://oreil.ly/A9jCr*) states that this is necessary to limit abuse by robots and make sure only real people are creating free trial accounts. It also states that Google will not charge anything to your account automatically when the free trial ends. According to the program documentation, you will have the opportunity to upgrade your account by expressly allowing GCP to charge your account, but if you do not do so, your account will automatically be downgraded to access only the free services.

Creating a Project

Let's begin by creating your first project, which is necessary to organize your work, set up billing, and gain access to GCP services. In the console, go to the "Manage resources" (*https://oreil.ly/2oA64*) page and then, at the top of the page, select Create Project. As shown in Figure 4-1, you need to give your project a name, which must be unique within the entire GCP. You can also select an organization if your Google identity is associated with one (which is usually the case if you have an institutional/work G Suite account), but if you just created your account, this might not be applicable to you at the moment. Having an organization selected means new projects will be associated with that organization by default, which allows for central management of projects. For the purposes of these instructions, we assume that you're setting up your account for the first time and there isn't a preexisting organization linked to it.

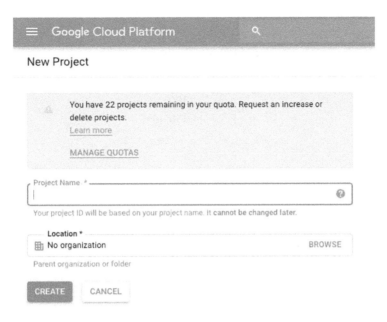

Figure 4-1. Creating a new project.

Checking Your Billing Account and Activating Free Credits

If you followed the sign-up process outlined in the previous section and activated your free trial, the system will have set up billing information for you as part of the overall account creation process. You can check your billing information in the Billing section of the console (*https://oreil.ly/X8G6K*), which you can also access at any time from the sidebar menu.

If you're eligible for the free credits program, one of the panels on the billing overview page will summarize the number of credits and days you have left to spend them. Note that if yours is displaying a blue Upgrade button, as shown in Figure 4-2, your trial has not yet started and you need to activate it in order to take advantage of the program. You might also see a "Free trial status" banner at the top of your browser window with a blue Activate button. Someone at GCP is working really hard to not let you walk away from free money, so click either of those buttons to start the process and receive your free credits.

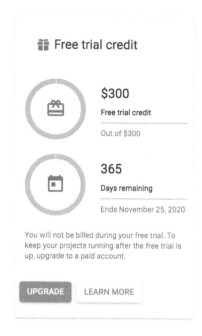

Figure 4-2. The panel in the Billing console summarizing free trial credits availability.

More generally, the billing overview page provides summaries of how much money (or credits) you have spent so far as well as some basic forecasting. That being said, it's important to understand that the system does not show you costs in real time: there is some lag time between the moments when you use chargeable resources and when the costs are updated on your billing page.

Many people who make the move to the cloud report that keeping track of their spending is one of the most difficult parts of the process. It's also the one that causes them the most anxiety because it can be very easy to spend large sums of money pretty quickly in the cloud if you're not careful. One feature offered by GCP that we find particularly useful in this respect is the "Budgets & alerts" settings, as depicted in Figure 4-3. This allows you to set email alerts that will notify you (or whoever is the billing administrator on your account) when you exceed certain spending thresholds. To be clear, this won't stop anything from running or prevent you from starting any new work that would push you over the threshold, but at least it will let you know where you stand.

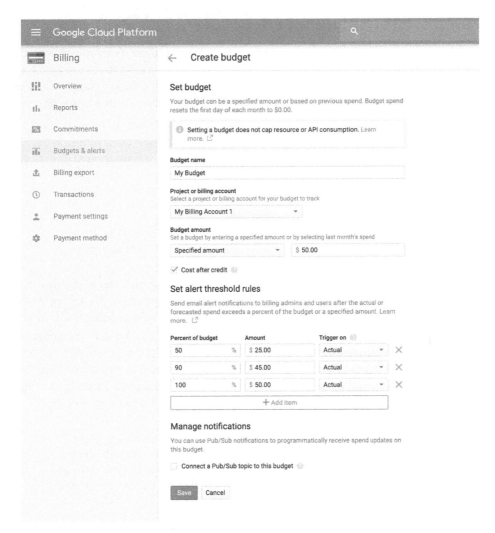

Figure 4-3. Budget and alert threshold administration.

To access the billing notifications feature, on the main menu on the GCP console, choose Billing, select the billing account you just created, and then look for the Budgets and alerts option. After you select it, you will be able to set up a new budget using the Create budget form shown in Figure 4-3. You can create multiple budgets and set multiple triggers for different percentages of the budget if you want warnings as you approach your budget amount. But as we just mentioned, keep in mind that it is still only a notification service and will not prevent you from incurring additional charges.

Running Basic Commands in Google Cloud Shell

Now that you've established your account, set up billing, and created your project, the next step is to log in to your first VM. For our exercises here, we use Google Cloud Shell, which does not require any configuration to get started and is completely free, although it comes with a few important limitations that we discuss in a moment.

Logging in to the Cloud Shell VM

To create a secure connection to a Cloud Shell VM using the SSH protocol, in the upper-right corner of the console, click the terminal icon:

This launches a new panel in the bottom on the console; if you want, you can also pop the terminal out to its own window. This gives you shell access to your own Debian-based Linux VM provisioned with modest resources, including 5 GB of free storage (mounted at *$HOME*) on a persistent disk. Some basic packages are preinstalled and ready to go, including the Google Cloud SDK (aka gcloud), which provides a rich set of command-line-based tools for interacting with GCP services. We'll use it in a few minutes to try out some basic data management commands. In the meantime, feel free to explore this Debian VM, look around, and see what tools are installed.

> Be aware that weekly usage quotas limit how much time you can spend running the Cloud Shell; as of this writing, it's 50 hours per week. In addition, if you don't use it regularly (within 120 days, as of this writing), the contents of the disk that provides you with free storage might end up being deleted.

When you log in to Cloud Shell for the first time, it prompts you to specify a Project ID using the aforementioned gcloud utility:

```
Welcome to Cloud Shell! Type "help" to get started.
To set your Cloud Platform project in this session use "gcloud config set project
[PROJECT_ID]"
```

You can find your Project ID on the Home page of the console, as shown in Figure 4-4.

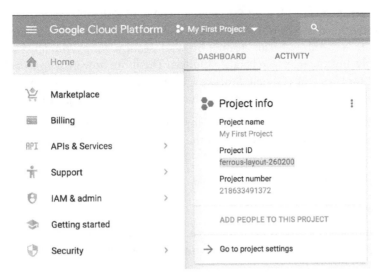

Figure 4-4. Location of the Project ID in the GCP console.

When you have your Project ID, run the following command in the Cloud Shell, substituting your own Project ID for the one shown here:

```
genomics_book@cloudshell:~$ gcloud config set project ferrous-layout-260200
Updated property [core/project].
genomics_book@cloudshell:~ (ferrous-layout-260200)$
```

Notice that your command prompt now includes your Project ID. It is quite long, so going forward, we'll show only the last character in the prompt—in this case, the dollar sign ($)—when we demonstrate running commands. For example, if we list the contents of the working directory using the `ls` command, it will look like this:

```
$ ls
README-cloudshell.txt
```

And, hey, there's already something here: a *README* file, which, as the name indicates, really wants you to read it. You can do so by running the `cat` command:

```
$ cat README-cloudshell.txt
```

This displays a welcome message that summarizes some usage instructions and recommendations for getting help. And with that, you're ready to use Cloud Shell to begin interacting with basic GCP services. Let's get cracking!

Using gsutil to Access and Manage Files

Now that we have access to this extremely simple-to-launch and free (if fairly limited) VM, let's use it to see whether we can access the bundle of example data provided with this book. The data bundle resides in Google Cloud Storage (GCS), which is a form of *object store* (i.e., it's used for storing files) with units of storage called *buckets*. You can view the contents of GCS buckets and perform basic management tasks on them via the web through the storage browser section (*https://oreil.ly/sqrkr*) of the GCP console, but the interface is fairly limited. The more powerful approach is to use the gcloud tool, gsutil (Google Storage Utilities), from the command line. You can access buckets through their GCS path, which is just their name prefixed with *gs://*.

As an example, the path for the public storage bucket for this book is *gs://genomics-in-the-cloud*. You can list the contents of the bucket by typing the following command in your cloud shell:

```
$ gsutil ls gs://genomics-in-the-cloud
gs://genomics-in-the-cloud/hello.txt
gs://genomics-in-the-cloud/v1/
```

There should be a file called *hello.txt*. Let's use the gsutil version of the Unix command cat, which allows us to read the content of text files to see what this *hello.txt* file contains:

```
$ gsutil cat gs://genomics-in-the-cloud/hello.txt
HELLO, DEAR READER!
```

You can also try copying the file to your storage disk:

```
$ gsutil cp gs://genomics-in-the-cloud/hello.txt .
Copying gs://genomics-in-the-cloud/hello.txt...
/ [1 files][   20.0 B/   20.0 B]
Operation completed over 1 objects/20.0 B.
```

If you list the contents of your working directory by using ls again, you should now have a local copy of the *hello.txt* file:

```
$ ls
hello.txt README-cloudshell.txt
```

While we're playing with gsutil, how about we do something that will be useful later: create a storage bucket of your own, so that you can store outputs in GCS. You'll need to substitute my-bucket in the command shown here because bucket names must be unique across all of GCS:

```
$ gsutil mb gs://my-bucket
```

If you didn't change the bucket name or you tried a name that was already taken by someone else, you might get the following error message:

```
Creating gs://my-bucket/...
ServiceException: 409 Bucket my-bucket already exists.
```

If this is the case, just try something else that's more likely to be unique. You'll know it worked when you see the Creating *name*... in the output and then get back to the prompt without any further complaint from gsutil. When that's done, you're going to create an environment variable that will serve as an alias for your bucket name. That way, you'll save yourself some typing and you'll be able to copy and paste subsequent commands without having to substitute the bucket name every time:

```
$ export BUCKET="gs://my-bucket"
```

You can run the echo command on your new variable to verify that your bucket name has been stored properly:

```
$ echo $BUCKET
gs://my-bucket
```

Now, let's get you comfortable with using gsutil. First, copy the *hello.txt* file to your new bucket. You can do either directly from the original bucket:

```
$ gsutil cp gs://genomics-in-the-cloud/hello.txt $BUCKET/
```

Or, you can do it from your local copy; for example, if you made modifications that you want to save:

```
$ gsutil cp hello.txt $BUCKET/
```

Finally, as one more example of basic file management, you can decide that the file should reside in its own directory in your bucket:

```
$ gsutil cp $BUCKET/hello.txt $BUCKET/my-directory/
```

As you can see, the gsutil commands are set up to be as similar as possible to their original Unix counterparts. So, for example, you'll also be able to use -r to make the cp and mv commands recursive to apply to directories. For large file transfers, you can use a few cloud-specification optimizations to speed up the process, like the gsutil -m option, which parallelizes file transfers. Conveniently, the system will usually inform you in the terminal output when you could take advantage of such optimizations, so you don't need to go and memorize the documentation before getting underway.

Managing Buckets Through the GCP Console

If you'd rather manage your storage buckets through a web interface, go to *Storage* (*https://oreil.ly/NGmdr*), which should list all buckets that belong to your default project, as shown in Figure 4-5. To create a new bucket, at the top center of the page, click Create Bucket.

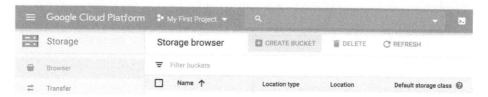

Figure 4-5. GCP console storage browser.

This opens a reasonably simple configuration form. The most important thing to do here is to choose a good name because the name you choose must be unique across all of Google Cloud—so be creative! If you choose a name that is already in use, the system will let you know when you click Continue in the configuration form, as demonstrated in Figure 4-6.

Figure 4-6. Naming your bucket.

When you have a unique name, the system will let you proceed to the next step by expanding the menu options. These allow you to customize the storage location and access controls for your bucket, but for the time being, feel free to just accept the defaults and click Create. Doing so will take you back to the list of buckets, which should at this point include your newly created one. You can click its name to view its contents—but of course it's still empty, so the view won't be particularly exciting, as illustrated in Figure 4-7.

The interface offers a few basic management options like deleting buckets and files as well as uploading files and folders. Note that you can even drag and drop files and folders from your local machine into the bucket contents window, which is stunningly easy (go ahead, try it), but it's not something you can expect to do very often in the course of your genomics work. In the real world, you're more likely to use the gsutil command-line utility. One of the advantages of using the command-line path is that you can save those commands as a script, for provenance and so that your steps can be reproduced if needed.

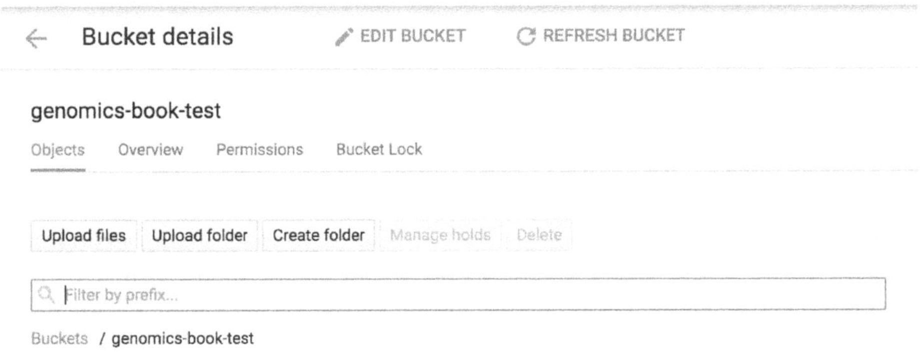

genomics-book-test

Objects   Overview   Permissions   Bucket Lock

Upload files   Upload folder   Create folder   Manage holds   Delete

🔍 Filter by prefix...

Buckets / genomics-book-test

Figure 4-7. Viewing the contents of your bucket.

Pulling a Docker Image and Spinning Up the Container

Cloud Shell is the gift that keeps on giving: the Docker application (which we introduced in Chapter 3) comes preinstalled, so you can go ahead and get started with that, too! We're going to use a simple Ubuntu container to illustrate basic Docker functionality. Although a Docker image is available for GATK—and that's what we're going to use for a good chunk of the next few chapters—we're not going to use it here because it's rather large, so it takes a little while to get going. We wouldn't actually be able to run any realistic analyses with it in the free Cloud Shell because of the small amount of CPU and memory resources allocated for this free VM.

The first thing to do to learn how to use Docker containers in this context is to...well, avoid the online Docker documentation! Seriously. Not because it's bad, but because the majority of those documents are written mainly for people who want to run web applications in the cloud. If that's what *you* want to do, more power to you, but you're reading the wrong book. What we're providing here are tailored instructions that will teach you how to use Docker to run research software in containers.

As just noted, we're going to use a very generic example: an image containing the Ubuntu Linux OS. It's an official image that is provided as part of the core library in a public container image repository, Docker Hub, so we just need to state its name. You'll see later that images contributed by the community are prefixed by the contributor's username or organization name. While still in your Cloud Shell terminal (it doesn't matter where your working directory is), run the following command to retrieve the Ubuntu image from the Docker Hub library of official (certified) images:

```
$ docker pull ubuntu
Using default tag: latest
```

```
latest: Pulling from library/ubuntu
7413c47ba209: Pull complete
0fe7e7cbb2e8: Pull complete
1d425c982345: Pull complete
344da5c95cec: Pull complete
Digest: sha256:d91842ef309155b85a9e5c59566719308fab816b40d376809c39cf1cf4de3c6a
Status: Downloaded newer image for ubuntu:latest
docker.io/library/ubuntu:latest
```

The pull command fetches the image and saves it to your VM. The version of the container image is indicated by its tag (which can be anything the image creator wants to assign) and by its sha256 hash (which is based on the image contents). By default, the system gives us the latest version that is available because we did not specify a particular tag; in a later exercise, you'll see how to request a specific version by its tag. Note that container images are typically composed of several modular *slices*, which are pulled separately. They're organized so that the next time you pull a version of the image, the system will skip downloading any slices that are unchanged compared to the version you already have.

Now let's start up the container. There are three main options for running it, but the tricky thing is that there is usually only one correct way to do it *as its author intended*, and it's difficult to know what that is if the documentation doesn't specify it (which is soooo often the case). Confused? Let's walk through the cases to make this a bit more concrete, and you'll see why we're putting you through this momentary frustration and mystery—it's to save you potential misery down the road.

First option
  Just run it!

  ```
  $ docker run ubuntu
  ```

Result
  A short pause, then your command prompt comes back. No output. What happened? Docker did in fact spin up the container, but the container wasn't configured to *do* anything under those conditions, so it basically shrugged and shut down again.

Second option
  Run it with a command appended:

  ```
  $ docker run ubuntu echo "Hello World!"
  Hello World!
  ```

Result
  It echoed Hello World!, as requested, and then shut down again. OK, so now we know that we can pass commands to the container, and if it's a command that is recognized by something in there, it will be executed. Then, when any and all

commands have been completed, the container will shut down. A bit lazy, but reasonable.

Third option

Run it interactively by using the -it option:

```
$ docker run -it ubuntu /bin/bash
root@d84c079d0623:/#
```

Result

Aha! A new command prompt (Bash in this case)! But with a different shell symbol: # instead of $. This means that the container is running and you are in it. You can now run any command that you would normally use on an Ubuntu system, including installing new packages if you like. Try running a few Unix commands such as ls or ls -la to poke around and see what the container can do. Later in the book, particularly in Chapter 12, we go into some of the implications of this, including practical instructions for how to package and redistribute an image you've customized in order to share your own analysis in a reproducible way.

When you're done poking around, type **exit** at the command prompt (or press Ctrl +D) to terminate the shell. Because this is the main process the container was running, terminating it will cause the container to shut down and return to the Cloud Shell itself. To be clear, this will shut down the container *and any commands that are currently running*.

If you're curious: yes, it is possible to step outside of the container without shutting it down; this is called *detaching*. To do so, press Ctrl+P+Q instead of using the exit command. You'll then be able to jump back into the container at any time—provided that you can identify it. By default, Docker assigns your container a universally unique identifier (UUID) as well as a random human-readable name (which tend to sound a bit silly). You can run docker ps to list currently running containers or docker ps -a to list containers that have been created. This displays a list of containers indexed by their container IDs that should look something like this:

```
$ docker ps -a
CONTAINER ID      IMAGE    COMMAND               CREATED        STATUS
PORTS             NAMES
c2b4f8a0c7a6      ubuntu   "/bin/bash"           5 minutes ago  Up 5 minutes
vigorous_rosalind
9336068da866      ubuntu   "echo 'Hello World!'" 8 minutes ago  Exited (0) 8 minutes ago
objective_curie
```

We're showing that two entries correspond to the last two invocations of Docker, each with a unique identifier, the CONTAINER ID. We see the container with ID c2b4f8a0c7a6 that is currently running was named vigorous_rosalind and has a status of Up 5 minutes. You can tell that the other container, objective_curie, is not running because its status is Exited (0) 8 minutes ago. The names we see here were randomly assigned (We swear! What are the odds?), so they're admittedly not terribly meaningful. If you have multiple containers running at the same time, this can become a bit confusing, so you'll want a better way to identify them. The good news is that you can give them a meaningful name by adding `--name=meaningful_name` immediately after `docker run` in your initial command, substituting `meaningful_name` with the name that you want to give the container.

To enter the container, simply run `docker attach c2b4f8a0c7a6` (substituting your container ID), press Enter, and you will find yourself back at the helm (your keyboard might be labeled Return instead of Enter). You can open a second command tab in Cloud Shell if you'd like to be able to run commands outside the container alongside the work you're doing inside the container. Note that you can have multiple containers running at the same time on a single VM—that's one of the great advantages of the container system—but they will be competing for the CPU and memory resources of the VM, which in Cloud Shell are rather minimal. Later in this chapter, we show you how to spin up VMs with beefier capabilities.

Mounting a Volume to Access the Filesystem from Within the Container

Having completed the previous exercise, you are now able to retrieve and run an instance of any container image shared in a public repository. Many commonly used bioinformatics tools, including GATK, are available preinstalled in Docker containers. The idea is that knowing how to use them out of a Docker container means you won't need to worry about having the correct OS or software environment. However, there's still one trick that we need to show you in order to make that really work for you: how to access your machine's filesystem from within the container by *mounting a volume*.

What does that last bit mean? By default, when you're inside the container, you can't access any data that resides on the filesystem outside of the container. The container is a closed box. There are ways to copy things back and forth between the container and your filesystem, but that becomes tedious really fast. So we're going to follow the easier path, which is to establish a link between a directory outside the container in a way that makes it appear as if it were within the container. In other words, we're going to poke a hole in the container wall, as shown in Figure 4-8.

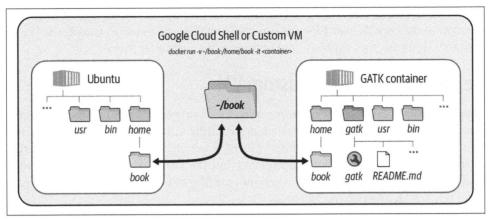

Figure 4-8. Mounting a directory from your Google Cloud Shell VM into a Docker container: Ubuntu container used in this chapter (left); GATK container introduced in Chapter 5 (right).

As an example, let's create a new directory called *book* in our Cloud Shell VM's home directory, and put the *hello.txt* file from earlier inside it:

```
$ mkdir book
$ mv hello.txt book/
$ ls book
hello.txt
```

So this time, let's run the command to spin up our Ubuntu container by using the -v argument (where v is for volume), which allows us to specify a filesystem location and a mount point within the container:

```
$ docker run -v ~/book:/home/book -it ubuntu /bin/bash
```

The -v ~/book_data:/home/book part of the command links the location you specified to the path */home/book* directory within the Docker container. The /home part of the path is a directory that already exists in the container, whereas the book part can be any name you choose to give it. Now, everything in the book directory on your filesystem can be accessed from within the Docker container's */home/book* directory:

```
# ls home/book
hello.txt
```

Here, we're using the same name for the mount point as for the actual location we're mounting because it's more intuitive that way, but you could use a different name if you wanted. Note that if you give your mount point the name of a directory or file that already exists with that path in the container, it will "squash" the existing path, meaning that path will not be accessible for as long as the volume is mounted.

A few other Docker tricks are good to know, but for now, this is enough of a demonstration of the core Docker functionality that you're going to use in Chapter 5. We go into the details of more sophisticated options as we encounter them.

Setting Up Your Own Custom VM

Now that you've successfully run some basic file-management commands and got the hang of interacting with Docker containers, it's time to move on to bigger and better things. The Google Cloud Shell environment is excellent for quickly getting started with some light coding and execution tasks, but the VM allocated for Cloud Shell is really underpowered and will definitely not cut the mustard when it comes to running real GATK analyses in Chapter 5.

In this section, we show you how to set up your own VM in the cloud (sometimes called an *instance*) using Google's Compute Engine service, which allows you to select, configure, and run VMs of whatever size you need.

Creating and Configuring Your VM Instance

First, go to the *Compute Engine* (*https://oreil.ly/sGeug*) or access the page through the sidebar menu on the left, as shown in Figure 4-9.

Figure 4-9. Compute Engine menu showing the VM instances menu item.

Click the VM Instances link in this menu to go to an overview of running images. If this is a new account, you won't have any running. Notice at the top that there's an option for Create Instance. Click that, and let's walk through the process of creating a new VM with just the resources you need.

Next, in the top menu bar, click Create Instance, as shown in Figure 4-10. This brings up a configuration form, as shown in Figure 4-11.

Figure 4-10. Create a VM instance.

Figure 4-11. The VM instance configuration panel.

Follow the step-by-step instructions in the subsections that follow to configure the VM. There are tons of options and the process can be quite confusing if you don't have experience with the terminology, so we mapped out the simplest path through the configuration form that will allow you to run all of the command exercises in the first few chapters of this book. Please make sure that you use exactly the same settings as shown here unless you really know what you're doing.

Name your VM

Give your VM a name; for example, `genomics-book`, as shown in Figure 4-12. This must be unique within your project, but unlike bucket names, it does not need to be unique across GCP. Some people like to use their username so that others with access to the project can instantly identify who created the resource.

Figure 4-12. Name your VM instance.

Choose a region (important!) and zone (not so important)

There are different physical locations for the cloud. Like most commercial cloud providers, GCP maintains datacenters in many parts of the world and provides you with the option to choose which one you want to use. Regions are the top-level geographical distinction, with names that are reasonably descriptive (like `us-west2`, which refers to a facility in Los Angeles). Each region is further divided into two or more zones designated by single letters (`a`, `b`, `c`, etc.), which correspond to separate datacenters with their own physical infrastructure (power, network, etc.), though in some cases they might share the same building.

This system of regions and zones plays an important role in limiting the impact of localized problems like power outages, and all major cloud providers use some version of this strategy. For more on this topic, see this entertaining blog post (*https://oreil.ly/pZUl6*) by Kyle Galbraith about how cloud regions and zones (in his case, on AWS) could play an important role in the event of a zombie apocalypse.

> The ability to choose specific regions and zones for your projects is increasingly helpful for dealing with regulatory restrictions on where human-subjects data can be stored because it allows you to specify a compliant location for all storage and compute resources. However, some parts of the world are not yet well covered by cloud services or are covered differently by the various cloud providers, so you might need to factor in available datacenter locations when choosing a provider.

To choose a region for your project, you can consult the full list of available Google Cloud regions and zones (*https://oreil.ly/D4Iqa*) and make a decision based on geographic proximity. Alternatively, you can use an online utility that measures how close you *effectively* are to each datacenter in terms of network response time, like *http://www.gcping.com*. For example, if we run this test from the small town of Sunderland in western Massachusetts (results in Table 4-1), we find that it takes 38 milliseconds to get a response from the us-east4 region located in Northern Virginia (698 km away), versus 41 milliseconds from the northamerica-northeast1 region located in Montreal (441 km away). This shows us that geographical proximity does not correlate directly with network region proximity. As an even more striking example, we find that we are quite a bit "closer" to the europe-west2 region in London (5,353 km away), with a response time of 102 milliseconds, than to the us-west2 region in Los Angeles (4,697 km away) which gives us a response time of 180 milliseconds.

Table 4-1. Geographical distance and response time from Sunderland, MA

Region	Location	Distance (km)	Response (ms)
us-east4	Northern Virginia, US	698	38
northamerica-northeast1	Montreal	441	41
europe-west2	London	5,353	102
us-west2	Los Angeles	4,697	180

This brings us back to our VM configuration. For the Region, we're going to be using us-east4 (Northern Virginia) because it's closest to the one of us who travels least (Geraldine), and for the Zone we just randomly choose us-east4-a. You need to make sure that you choose *your* region based on the preceding discussion, both for your own benefit (it will be faster) and to avoid clobbering that one datacenter in Virginia in the unlikely event that all 60,000 registered users of the GATK software begin working through these exercises at the same time—though that's one way to test the vaunted "elasticity" of the cloud.

Select a machine type

This is where you can configure the resources of the VM you're about to launch. You can control RAM as well as CPUs. For some instance types (available under Customize) you can even select VMs with GPUs, which are used to accelerate certain programs. The hitch is that what you select here will determine how much you'll be billed per second of the VM's uptime; the bigger and beefier the machine, the more it will cost you. The right side of the page should show how the hourly and monthly cost changes when you change the machine type. Note also that you're billed for how long the VM is online, not for how much time you spend actually using it. We cover strategies for limiting costs later, but keep that in mind!

Here, select `n1-standard-2`; this is a fairly basic machine that's not going to cost much at all, as shown in Figure 4-13.

Figure 4-13. Selecting a machine type.

Specify a container? (nope)

We're not going to fill this out. This is useful if you want to use a very specific setup using a custom container image that you've preselected or generated yourself. In fact, we could have preconfigured a container for you and skipped a bunch of setup that's coming next. But then you wouldn't have the opportunity to learn how to do those things for yourself, would you? So, for now, let's just skip this option.

Customize the boot disk

Like Machine Type, this is another really useful setting. You can define two things here: the OS that you want to use and the amount of disk space you want. The former is especially important if you need to use a particular type and version of OS. And, of course, the latter is important if you don't want to run out of disk space halfway through your analysis.

By default, the system proposes a particular flavor of Linux OS, accompanied by a paltry 10 GB of disk space, as shown in Figure 4-14. We're going to need a bigger boat.

Figure 4-14. Choosing a boot disk size and image.

To access the settings menu for this, click Change. This opens a new screen with a menu of predefined options. You can also make your own custom images, or even find more images in Google Cloud Marketplace (*https://oreil.ly/sjiIf*).

For our immediate purposes, we prefer Ubuntu 18.04 LTS, which is the most recent version of Ubuntu's long-term release, as of this writing. It might not be as bleeding edge as Ubuntu 19.04, but the LTS, which stands for *long-term support*, guarantees that it's being maintained for security vulnerabilities and package updates for five years from release. This Ubuntu image has a ton of what we already need, ready to go and installed, including various standard Linux tools and the GCP SDK command-line tools, which we will rely on quite heavily.

Select Ubuntu in the Operating System menu, then select Ubuntu 18.04 LTS in the version menu, as shown in Figure 4-15.

Figure 4-15. Selecting a base image.

At the bottom of the form, you can change the Boot disk Size to give yourself more space. As shown in Figure 4-16, go ahead and select 100 GB instead of the default 10 GB (the data we're going to be working with can easily take up a lot of space). You can bump this up quite a bit more, depending on your dataset size and needs. Although you can't easily adjust it after the VM launches, you do have the option of adding block storage volumes to the running instance after launch—think of it as the cloud equivalent of plugging in a USB drive. So if you run out of disk space, you won't be totally stuck.

Figure 4-16. Setting the boot disk size.

After you've done all this, click Select; this closes the screen and returns you to the instance creation form where the "Boot disk" section should match the screenshot in Figure 4-17.

Figure 4-17. The updated boot disk selection.

At the bottom of the form, click Create. This returns you to the page that lists Compute Engine VM instances, including your newly created VM instance. You might see a spinning icon in front of its name while the instance is being created and booted up, and then a green circle with a checkmark will appear when it is running and ready for use, as shown in Figure 4-18.

Figure 4-18. Viewing the VM status.

And voilà, your VM is ready for action.

Logging into Your VM by Using SSH

There are several ways that you can access the VM after it's running, which you can learn about in the GCP documentation. We're going to show you the simplest way to do it, using the Google Cloud console and the built-in SSH terminal. It's hard to beat: as soon as you see a green checkmark in the Google Cloud console, you can simply click the SSH option to open a drop-down menu, as illustrated in Figure 4-19. Select the option "Open in a browser window," and a few seconds later you should see an SSH terminal open to this VM.

Figure 4-19. Options for SSHing into your VM.

This opens a new window with a terminal that allows you to run commands from within the VM instance, as shown in Figure 4-20. It might take a minute to establish the connection.

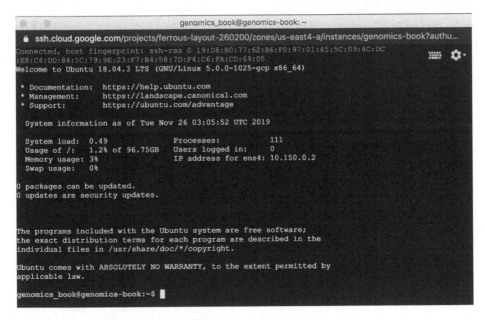

Figure 4-20. VM instance terminal.

Feel free to look around and get to know your brand-new VM; you're going to spend a lot of time with it in the course of the next few chapters (but, like, in a good way).

Checking Your Authentication

You're probably itching to run something interesting, but let's begin by making sure your account credentials are set up properly so you can use the GCP command-line tools, which come preinstalled on the image we chose. In the SSH terminal, run the following command:

```
$ gcloud init
Welcome! This command will take you through the configuration of gcloud.
Your current configuration has been set to: [default]
You can skip diagnostics next time by using the following flag:
  gcloud init --skip-diagnostics
Network diagnostic detects and fixes local network connection issues.
Checking network connection...done.
Reachability Check passed.
Network diagnostic passed (1/1 checks passed).
Choose the account you would like to use to perform operations for
this configuration:
[1] XXXXXXXXXXX-compute@developer.gserviceaccount.com
[2] Log in with a new account
Please enter your numeric choice:
```

The line that starts with [1] shows you that by default, GCP has you logged in under a service account: the domain is *@developer.gserviceaccount.com*. This is fine for running tools within your VM, but if you want to be able to manage resources, including copying files out to GCS buckets, you need to do so under an account with the relevant permissions. It is possible to grant this service account all the various permissions that you'll need for these exercises, but that would lead us a bit further into the guts of GCP account administration than we'd like to go at this juncture—we want to get you doing genomics work ASAP! So instead, let's just use the original account that you used to create the project at the beginning of this chapter, given that it already has those permissions as a project owner.

To log in with that account, press 2 at the prompt. This triggers some interaction with the program; GCP will warn you that using your personal credentials on a VM is a security risk because if you give someone else access to the VM, they will be able to use your credentials:

```
You are running on a Google Compute Engine virtual machine.
It is recommended that you use service accounts for authentication.
You can run:
 $ gcloud config set account `ACCOUNT`
to switch accounts if necessary.
Your credentials may be visible to others with access to this
virtual machine. Are you sure you want to authenticate with
your personal account?
Do you want to continue (Y/n)?
```

The solution: don't share access to your personal VM.[1]

If you type Y for yes, the program will give you a link:

```
Go to the following link in your browser:
    https://accounts.google.com/o/oauth2/auth?redirect_uri=<...>
Enter verification code:
```

When you click the link or copy and paste it into your browser, you are presented with a Google login page. Log in with the same account you used for the GCP to get your authentication code and then copy and paste that back into your terminal window. The gcloud utility will confirm your login identity and ask you to select the project ID you want to use from the list of projects you have access to. It will also offer the option to set your preferred compute and storage zone, which should match what you set earlier when you created the VM. If you're not seeing what you expect in the project ID list, you can always double-check the resource management page in the GCP console (*https://oreil.ly/T50ev*).

Copying the Book Materials to Your VM

Throughout the next few chapters, you're going to run real GATK commands and workflows on your VM, so you need to retrieve the example data, source code, and a couple of software packages. We've bundled most of that in a single place: a Cloud Storage bucket called genomics-in-the-cloud. The only piece that is separate is the source code, which we provide in GitHub.

First, you're going to copy the data bundle from the bucket to your VM using gsutil, the GCP storage utility that we already used earlier in the Cloud Shell portion of this

---

1 Keep in mind that if you create accounts for other users in your GCP project, they will be able to SSH to your VMs as well. It is possible to further restrict access to your VMs in a shared project, but that is beyond the simple introduction we're presenting here.

chapter. In your VM's terminal window, make a new directory called **book**, and then run the `gsutil` command to copy the book data bundle to the storage space associated with your VM:

```
$ mkdir ~/book
$ gsutil -m cp -r gs://genomics-in-the-cloud/v1/* ~/book/
```

This will copy about 10 GB of data to your VM's storage, so it might take a few minutes even with the `-m` flag enabling parallel downloads. As you'll see later, it is possible to run some analysis commands directly on files in Cloud Storage without copying them first, but we want to keep things as simple as possible in the beginning.

Now, go ahead and retrieve the source code from the public repository on GitHub (*https://oreil.ly/genomics-repo*). We're making the code available there because it's a highly popular platform for sharing code under *version control*, and we're committed to providing long-term maintenance for the code we use in the book. To get a copy on your VM, first use `cd` to move into the newly created *book* directory and then use the `git clone` command to copy the contents of the repository:

```
$ cd ~/book
$ git clone https://github.com/broadinstitute/genomics-in-the-cloud.git code
```

This creates a directory (*~book/code*) that includes all the sample code we use throughout the book. Not only that, but it will be set up as an active Git repository, so you can get the latest changes by running the `git pull` command in the code directory, as follows:

```
$ cd ~/book/code
$ git pull
```

With that, you should now have the latest and greatest version of the book code. To find out what has changed since the original publication, check out the *README* text file in the code directory.

Using Git and GitHub for Version Control

You don't need to be familiar with Git in order to work through the examples in this book, but we do encourage you to consider taking some time at your convenience to learn how to use it at a basic level. It's becoming increasingly common to use Git as a version control system for scientific code, so you will run into it again when exploring other people's work, and you might benefit from using it to make your own work computationally reproducible. An excellent place to start is with Software Carpentry's lesson on using Git for version control (*https://oreil.ly/yMLtP*).

Installing Docker on Your VM

You're going to be working with Docker on your VM, so let's make sure that you can run it. If you simply run the command docker in the terminal, you'll get an error message because Docker does not come preinstalled on the VM:

```
$ docker
Command 'docker' not found, but can be installed with:
snap install docker     # version 18.09.9, or
apt  install docker.io
See 'snap info docker' for additional versions.
```

The error message helpfully points out how to remedy the situation using a preinstalled package called snap, but we're actually going to use a slightly different way of installing Docker: we're going to download and run a script from the Docker website that will largely automate the installation process. This way, you'll know what to do if you find yourself needing to install Docker somewhere that doesn't have a built-in package manager option.

Run the following command to install Docker on the VM:

```
$ curl -sSL https://get.docker.com/ | sh
# Executing docker install script, commit: f45d7c11389849ff46a6b4d94e0dd1ffebca
32c1 + sudo -E sh -c apt-get update -qq >/dev/null
...
Client: Docker Engine - Community
Version:          19.03.5
...
If you would like to use Docker as a non-root user, you should now consider
adding your user to the "docker" group with something like:
 sudo usermod -aG docker genomics_book
Remember that you will have to log out and back in for this to take effect!
WARNING: Adding a user to the "docker" group will grant the ability to run
         containers which can be used to obtain root privileges on the
         docker host.
         Refer to https://docs.docker.com/engine/security/security/#docker-daemon-
         attack-surface for more information.
```

This might take a little while to complete, so let's take that time to examine the command in a bit more detail. First, we're using a convenient little utility called curl (short for *Client URL*) to download the installation script from the Docker website URL we provided, with a few command parameters (-sSL) that instruct the program to follow any redirection links and save the output as a file. Then, we use the pipe character (|) to hand that output file over to a second command, sh, which means "run that script that we just gave you." The first line of output lets you know what it's doing: Executing docker install script (we omitted parts of the preceding output for brevity).

When it finishes, the script prompts you to run the usermod command in the example that follows in order to grant yourself the ability to run Docker commands without using sudo each time. Invoking sudo docker can result in output files being owned by root, making it difficult to manage or access them later, so it's really important to do this step:

```
$ sudo usermod -aG docker $USER
```

This does not produce any output; we'll test in a minute whether it worked properly. First, however, you need to log out of your VM and then back in again. Doing so will make the system reevaluate your Unix group membership, which is necessary for the change you just made to take effect. Simply type **exit** (or press Ctrl+D) at the command prompt:

```
$ exit
```

This closes the terminal window to your VM. Go back to the GCP console, find your VM in the list of Compute Engine instances, and then click SSH to log back in again. This probably feels like a lot of hoops to jump through, but hang in there; we're getting to the good part.

Setting Up the GATK Container Image

When you're back in your VM, test your Docker installation by pulling the GATK container, which we use in the very next chapter:

```
$ docker pull us.gcr.io/broad-gatk/gatk:4.1.3.0
4.1.3.0: Pulling from us.gcr.io/broad-gatk/gatk
ae79f2514705: Pull complete
5ad56d5fc149: Pull complete
170e558760e8: Pull complete
395460e233f5: Pull complete
6f01dc62e444: Pull complete
b48fdadebab0: Pull complete
16fb14f5f7c9: Pull complete
Digest: sha256:e37193b61536cf21a2e1bcbdb71eac3d50dcb4917f4d7362b09f8d07e7c2ae50
Status: Downloaded newer image for us.gcr.io/broad-gatk/gatk:4.1.3.0
us.gcr.io/broad-gatk/gatk:4.1.3.0
```

As a reminder, the last bit after the container name is the version tag, which you can change to get a different version than what we've specified here. Note that if you change the version, some commands might no longer work. We can't guarantee that all code examples are going to be future-compatible, especially for the newer tools, some of which are still under active development. As noted earlier, for updated materials, see this book's GitHub repository (*https://oreil.ly/genomics-repo*).

The GATK container image is quite large, so the download might take a little while. The good news is that next time you need to pull a GATK image (e.g., to get another release), Docker will pull only the components that have been updated, so it will go faster.

 Here we're pulling the GATK image from the Google Container Repository (GCR) because GCR is on the same network as the VM we're running on, so it will be faster than pulling it from Docker Hub. However, if you're working on a different platform, you might find it faster to pull the image from the GATK repository on Docker Hub. To do so, change the us.gcr.io/broad-gatk part of the image path to just **broadinstitute**.

Now, remember the instructions you followed earlier in this chapter to spin up a container with a mounted folder? You're going to use that again to make the book directory accessible to the GATK container:

```
$ docker run -v ~/book:/home/book -it us.gcr.io/broad-gatk/gatk:4.1.3.0 /bin/bash
```

You should now be able to browse the book directory that you set up in your VM from within the container. It will be located under */home/book*. Finally, to double-check that GATK itself is working as expected, try running the command gatk at the command line from within your running container. If everything is working properly, you should see some text output that outlines basic GATK command-line syntax and a few configuration options:

```
# gatk
Usage template for all tools (uses --spark-runner LOCAL when used with a Spark tool)
    gatk AnyTool toolArgs
Usage template for Spark tools (will NOT work on non-Spark tools)
    gatk SparkTool toolArgs  [ -- --spark-runner <LOCAL | SPARK | GCS> sparkArgs ]
Getting help
    gatk --list      Print the list of available tools
    gatk Tool --help  Print help on a particular tool
Configuration File Specification
     --gatk-config-file                PATH/TO/GATK/PROPERTIES/FILE
gatk forwards commands to GATK and adds some sugar for submitting spark jobs
  --spark-runner <target>     controls how spark tools are run
    valid targets are:
    LOCAL:      run using the in-memory spark runner
    SPARK:      run using spark-submit on an existing cluster
                --spark-master must be specified
                --spark-submit-command may be specified to control the Spark submit command
                arguments to spark-submit may optionally be specified after --
    GCS:        run using Google cloud dataproc
                commands after the -- will be passed to dataproc
                --cluster <your-cluster> must be specified after the --
                spark properties and some common spark-submit parameters will be translated
                to dataproc equivalents
  --dry-run     may be specified to output the generated command line without running it
  --java-options 'OPTION1[ OPTION2=Y ... ]''   optional - pass the given string of options to
                the java JVM at runtime.
                Java options MUST be passed inside a single string with space-separated values
```

We discuss what that all means in loving detail in Chapter 5; for now, you're done setting up the environment that you'll be using to run GATK tools over the course of the next three chapters.

Stopping Your VM…to Stop It from Costing You Money

The VM you just finished setting up is going to come in handy throughout the book; you'll come back to this VM for many of the exercises in the next few chapters. However, as long as it's up and running, it's costing you either credits or actual money. The simplest way to deal with that is to stop it: put it on pause whenever you're not actively using it.

You can restart it on demand; it just takes a minute or two to get it back up and running, and it will retain all environment settings, the history of what you ran previously, and whatever data you have in local storage. Note that you will be charged a small fee for that storage even while the VM is not running and you're not getting charged for the VM itself. In our opinion, this is well worth it for the convenience of being able to come back to your VM after an arbitrary amount of time and just pick up your work where you left off.

To stop your VM, in the GCP console, go to the VM instances management page, as shown previously. Find your instance and click the vertical three-dot symbol on the right to open the menu of controls, and then select Stop, as shown in Figure 4-21. The process might take a couple of minutes to complete, but you can safely navigate away from that page. To restart your instance later on, just follow the same steps but click Start in the control menu.

Figure 4-21. Stopping, starting, or deleting your VM instance.

Alternatively, you can delete your VM entirely, but keep in mind that deleting the VM will delete all locally stored data too, so make sure you save anything you care about to a storage bucket first.

Configuring IGV to Read Data from GCS Buckets

Just one more small step remains before you move on to the next chapter: we're going to install and configure a genome browser called Integrated Genome Viewer (IGV) that can work directly with files in GCP. That will allow you to examine sequence data and variant calls without needing to copy the files to your local machine.

First, if you don't have it installed yet on your local machine, get the IGV program from the website (*https://oreil.ly/bEPS_*) and follow the installation instructions. If you already have a copy, consider updating it to the latest version; we are using 2.7.2 (macOS version). Once you have the application open, choose View > Preferences from the top menu bar, as shown in Figure 4-22.

Figure 4-22. Selecting the Preferences menu item.

This opens the Preferences pane, shown in Figure 4-23.

In the Preferences pane, Select the "Enable Google access" checkbox, click Save, and then quit IGV and reopen it to force a refresh of the top menu bar. You should now see a Google menu item that was not there previously; click it and select Login, as shown in Figure 4-24, to set up IGV with your Google account credentials.

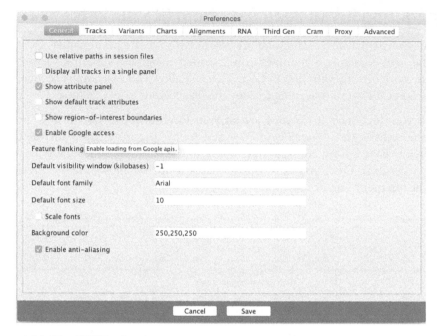

Figure 4-23. The IGV Preferences pane.

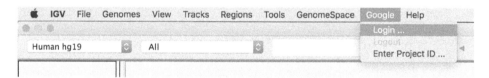

Figure 4-24. Selecting the Google Login menu item.

This will take you to a Google login page in your web browser; follow the prompts to allow IGV to access relevant permissions on your Google account. When this is complete, you should see a web page that simply says OK. Let's switch back to IGV and test that it works. From the top-level menu, click Files > Load from URL, as shown in Figure 4-25, making sure not to select one of the other options by mistake (they look similar, so it's easy to get tripped up). Make sure also that the reference drop-down menu in the upper-left corner of the IGV window is set to "Human hg19."

If you're confused about what is different between the human references, see the notes in "The Reference Genome as Common Framework" on page 21 about hg19 and GRCh38.

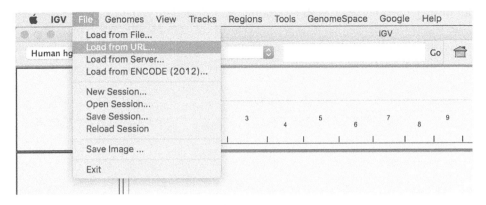

Figure 4-25. The Load from URL menu item.

Finally, enter the GCS file path for one of the sample BAM files we provide in the book data bundle in the dialog window that pops up (e.g., *mother.bam*, as shown in Figure 4-26), and then click OK. Remember, you can get a list of files in the bucket by using `gsutil` from your VM or from Cloud Shell, or you can browse the contents of the bucket by using the Google Cloud console storage browser (*https://oreil.ly/ 1iQmv*). If you use the browser interface to get the path to the file, you'll need to compose the GCS file path by stripping off the first part of the URL before the bucket name; for instance, remove *https://console.cloud.google.com/storage/browser* and replace that with `gs://`. Do the same for the BAM's accompanying index file, which should have the same filename and path but ends in *.bai.*[2]

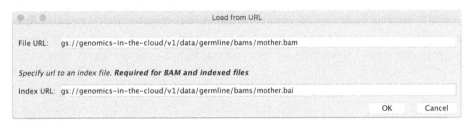

Figure 4-26. The Load from URL dialog box.

This will make the data available to you in IGV as a new data track, but by default nothing will be loaded in the main viewer. To check that you can view data, in the search window, enter the genomic coordinates **20:9,999,830-10,000,170** and then click Go. These coordinates will take you to the 10 millionth DNA base ±170 on the

---

2 For example, *https://console.cloud.google.com/storage/browser/genomics-in-the-cloud/v1/data/germline/bams/ mother.bam* becomes *gs://genomics-in-the-cloud/v1/data/germline/bams/mother.bam*.

20th human chromosome, as shown in Figure 4-27, where you'll see the left-side edge of the slice of sequence data that we provide in this sample file. We explain in detail how to interpret the visual output of IGV in Chapter 5, when we use it to investigate the result of a real (small) analysis.

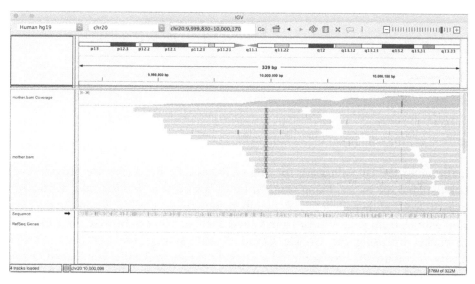

Figure 4-27. IGV view of a BAM file located in a GCS bucket.

IGV retrieves only small slices of data at a time, so the transfer should be very fast unless you have a particularly slow internet connection. Do keep in mind, however, that GCP, like all commercial cloud providers, will charge an egress fee (*https:// oreil.ly/rktm2*) for transferring data out of the cloud. On the bright side, it's a small fee, proportional to the amount of data you transfer. So the cost of viewing slices of data in IGV is trivial—on the order of fractions of pennies—and it is definitely preferable to what it would cost to transfer the entire file for offline browsing!

Who Pays the Egress Fee?

By default, Google buckets are set up so that their owner pays any egress fees. However, it is possible to modify the bucket configuration to a setting called `requester-pays`, which means that the person requesting the data will be charged instead. To access data that resides in a bucket set to `requester-pays`, you would need to give IGV a Project ID (using the Google menu shown in Figure 4-24) for billing purposes. You don't need to do that in the context of this book, because the bucket we are using is hosted by the Broad Institute Data Sciences Platform, which is generously footing the bill for hosting the example data and covering any egress charges.

You can view the contents of other data files, like VCF files, using the same set of operations, as long as the files are stored in a GCP bucket. Unfortunately, it means that this won't work for files that are on the local storage of your VM, so anytime you want to examine one of those, you'll need to copy it to a bucket first. You're going to get really friendly with gsutil in no time.

Oh, one last thing while you have IGV open: click the little yellow callout bubble in the IGV window toolbar, which controls the behavior of the detail viewer, as shown in Figure 4-28. Do yourself a favor and switch the setting from Show Details on Hover to Show Details on Click. Whichever action you choose will trigger the appearance of little dialog that gives you detailed information about any part of the data that you either click or hover over; for example, for a sequence read, it will give you all the mapping information as well as the full sequence and base qualities. You can try it out now with the data you just loaded. As you'll see, the detail display functionality in and of itself is very convenient, but the "on Hover" version of this behavior can be a bit overwhelming when you're new to the interface; hence, our recommendation to switch to "on Click."

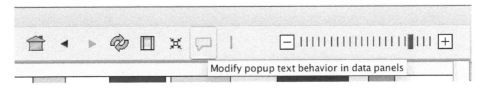

Figure 4-28. Changing the behavior of the detail viewer from "on Hover" to "on Click."

Wrap-Up and Next Steps

In this chapter, we showed you how to get started with GCP resources, from creating an account, using the super-basic Cloud Shell, and then graduating to your own custom VM. You learned how to manage files in GCS, run Docker containers, and administer your VM. Finally, you retrieved the book data and source code, finished setting up your custom VM to work with the GATK container, and set up IGV to view data stored in buckets. In Chapter 5, we get you started with GATK itself, and before you know it, you'll be running real genomics tools on example data in the cloud.

First Steps with GATK

In Chapter 4, we set you up to do work in the cloud. This chapter is all about getting you oriented and comfortable with the GATK and related tools. We begin by covering the basics, including computational requirements, command-line syntax, and common options. We show you how to spin up the GATK Docker container on the GCP VM you set up in Chapter 4 so that you can run real commands at scale with minimal effort. Then we work through a simple example of variant calling with the most widely used GATK tool, `HaplotypeCaller`. We explore some basic filtering mechanisms to give you a feel for working with variant calls and their context annotations, which play an important role in filtering results. Finally, we introduce the real-world GATK Best Practices workflows, which are guidelines for getting the best results possible out of your variant discovery analyses.

Getting Started with GATK

GATK is an open source software package developed at the Broad Institute. As its full name suggests, GATK is a *toolkit*, not a single tool—where we define *tool* as the individual functional unit that you will invoke by name to perform a particular data transformation or analysis task. GATK contains a fairly large collection of these individual tools, some designed to convert data from one format to another, to collect metrics about the data or, most notably, to run actual computational analyses on the data. All of these tools are provided within a single packaged executable. In fact, starting with version 4.0, which is sometimes referred to as *GATK4*, the GATK executable also bundles all tools from the Picard toolkit, another popular open source genomics package that has historically focused on sequence metrics collection and file format operations.

The names of GATK tools are almost all boring (except maybe Funcotator), and most of them are self-explanatory, which is as it should be. With more than two hundred tools in the toolkit, no one wants to deal with cutesy names that give no indication as to what each tool does.

Note that because the toolkit is fully open source, the public release includes tools that are under active development or experimentation; those tools are marked as such in the documentation and typically include a disclaimer in their output logs.

GATK tools are primarily designed to be run from the command line. In various scenarios, you can use them without ever running the commands directly; for example, either through a graphical user interface (GUI) that wraps the tools or in scripted workflows. In fact, most large-scale and routine uses of GATK are typically done through workflows, as we discuss in more detail in Chapter 8. Yet, to get started with the toolkit, we like to peel away all these layers and just run individual commands in the terminal, to demonstrate basic usage and provide visibility into how you can customize the tools' operation.

Operating Requirements

Because GATK is written in Java, you can run it almost anywhere as long as you have the proper version of Java installed on your machine (as of this writing, JRE or JDK 1.8), except on Microsoft Windows, which is not supported. No installation is necessary in the traditional sense; in its simplest form, all you need to do is download the GATK package, place its contents in a convenient directory on your hard drive (or server filesystem) and, optionally, add the location of the GATK wrapper script to your path.

However, a subset of tools (especially the newer ones that use fancy machine learning libraries) have additional R and/or Python dependencies, which can be quite painful to install. Also, this exposes you to annoying conflicts with the requirements of other tools that you might need to use. For example, suppose that you are using a version of GATK that requires NumPy 1.13.3, but you also need to use a different tool that requires NumPy 1.12.0. This could quickly become a nightmare, where you update a package to enable one analysis only to find later that another of your scripts no longer runs successfully.

You can use a package manager like Conda (*https://docs.conda.io*) to mitigate this problem. There is an official GATK Conda script that you can use to replicate the environment required for GATK, but frankly, it's not a perfect solution.

The best way to deal with this is to use a container, which provides a fully self-contained environment that encapsulates all requirements and dependencies, as described in Chapter 3 (theory) and Chapter 4 (practice). Every GATK release comes with a container image that matches that version's requirements exactly, and it's just a `docker pull` away. That means no more time and effort wasted setting up and managing your work environment—a huge win for portability and reproducibility of complex analyses.

Meet Mo, the mascot of the GATK nightly builds, the process that generates a new build of the GATK every night with the day's latest developments.

Mo looks sleepy because he was up all night running tests and working to make sure the bleeding-edge types could try out the very latest code with minimal hassle. Mo's blue plumage, fez, and bowtie are an homage to the British TV show character Doctor Who as played by actor Matt Smith. Because, why not? Bow ties are cool (though GATK does prefer BWA for short read alignments). He is named after the first person ever to ask whether the owl had a name (he didn't).

Command-Line Syntax

Syntax is admittedly not the most riveting topic, but it's essential to learn as soon as possible how to properly phrase instructions to the program. So let's begin by reviewing the basics. You're not going to run any commands just yet, but we'll get there soon.

As mentioned earlier, GATK is a Java-based command-line toolkit. Normally, the syntax for any Java-based command-line program is

```
$ java -jar program.jar [program arguments]
```

where `java` invokes the Java system on your machine to create a Java Virtual Machine (JVM). The JVM is a virtual environment in which your program will run. Then, the command specifies the JAR file; JAR stands for *Java ARchive*, because it's a collection of Java code ZIPped into a single file. Finally, the command specifies the arguments that are intended for the program itself.

However, since version 4.0, the GATK team provides a Python wrapper script that makes it easier to compose GATK command lines without needing to think about the Java syntax. The wrapper script takes care of calling the appropriate JAR and sets some parameters that you would otherwise need to specify manually.

So, instead of what we just showed you, the basic syntax for calling GATK is now:

```
$ gatk ToolName [tool arguments]
```

Here, *ToolName* is the name of the tool that you want to run; for example, `Haploty peCaller`, the germline short variant caller. The tool name argument is *positional*, meaning that it must be the first argument provided to the GATK command. After that, everything else is usually nonpositional, except for a few exceptions: some tools have pairs of arguments that go together, and Spark-related arguments (which we get to shortly) always come last. Tool names and arguments are all case sensitive.

In some situations, you might need to specify parameters for the JVM, like the maximum amount of memory it can use on your system. In the generic Java command, you would place them between `java` and `-jar`; for example:

```
$ java -Xmx4G -XX:+PrintGCDetails -jar program.jar [program arguments]
```

In your GATK commands, you pass them in through a new argument, `--java-options`, like this:

```
$ gatk --java-options "-Xmx4G -XX:+PrintGCDetails" ToolName [tool arguments]
```

Finally, you can expect the majority of GATK argument names to follow *POSIX convention*, in which the name is prefixed by two dashes (`--`) and where applicable, words are separated by single dashes (`-`). A minority of very commonly used arguments accept a short name, prefixed by a single dash (`-`), that is often a single capital letter. Arguments that do not follow these conventions include Picard tool arguments and a handful of legacy arguments.

Multithreading with Spark

Earlier versions of the GATK used some homegrown, hand-rolled code to provide multithreading functionality. If you've been using old versions of GATK, you might be familiar with the engine arguments `-nt` and `-nct`. Starting with GATK4, all that is gone; the GATK team chose to replace it with Spark, which has emerged as a much more robust, reliable, and performant option.

The relevant Spark library is bundled inside GATK itself, so there is no additional software to install. It works right out of the box, on just about any system—contrary to surprisingly widespread belief, you do not need a special Spark cluster to run Spark-enabled tools. However, some tools have computational requirements that exceed the capabilities of single-core machines. As you'll see at several points in this book, one of the advantages of cloud infrastructure is that we're able to pick and choose suitable machines on demand, depending on the needs of a given analysis.

Which GATK Tools Are Spark Enabled?

Not all GATK tools are Spark enabled. Some are not Spark enabled at all, some exist in a non-Spark version and a Spark-enabled version, and some exist only in a Spark-enabled version. This can be a bit confusing, so here is some guidance for navigating your options. Generally, the name of Spark-enabled tools end with the suffix `*Spark`, though this convention is not strictly enforced. When in doubt, check the individual tool documentation pages under the Tool Index (*https://oreil.ly/qjW9v*), which specify whether the tool supports Spark as well as any limitations such as experimental or beta status.

GATK tools that existed in version 3.8 and earlier
> Most of the "classic" GATK tools that predate GATK4 are *not* Spark enabled. A subset of these also exist in Spark-enabled form, which are distinguished by the suffix `*Spark` appended to the name of the original tool. Some of these are still in an experimental state and should not be used in high-stakes work such as in clinical analyses. See the log output and the tool documentation for any applicable warnings and notes on development status.
>
> *Example:* `HaplotypeCaller` versus `HaplotypeCallerSpark`

Original Picard tools versus Spark-enabled reimplementations
> No Spark-enabled tools are in the standalone Picard package. However, a subset of GATK tools that were originally in Picard were reimplemented in GATK4 and were Spark enabled in the process. These are also identified with the suffix `*Spark` appended to their original Picard name. Several of these are now considered sufficiently mature for production work and are significantly faster (in wall-clock time) compared to the original.
>
> *Example:* `MarkDuplicates` versus `MarkDuplicatesSpark`, and `SortSam` versus `SortSamSpark`.

Natively Spark-enabled tools
> Many of the tools that were written for GATK4 were given native Spark support. Those that we are aware of have the `*Spark` suffix in their name, but a few tools might not follow this naming convention. Again, when in doubt, check the tool documentation page.

Example: `PathSeqScoreSpark`, `FindBreakpointEvidenceSpark`

Spark pipelines

One of the best ways to take full advantage of Spark's strengths is to wire up multiple tools into a complete pipeline that runs on Spark from start to finish. GATK developers have been experimenting with this approach and have produced several such pipelines, which are listed as individual tools in the Tool Index (*https:// oreil.ly/M39ID*) and are recognizable by the `*PipelineSpark` suffix appended to their name. Most of these are still considered experimental; the only exception currently is `PathSeqPipelineSpark`.

Example: `StructuralVariationDiscoveryPipelineSpark`, `ReadsPipelineSpark`

Computational requirements

All GATK tools, whether Spark enabled or not, can in principle be run on a "normal" single-core machine, but some tools might not perform well under those conditions because of the scale of the computations they need to perform and the corresponding resource requirements. See the tool documentation for the specific requirements of each tool.

To use Spark multithreading, the basic syntax is fairly straightforward; we just need to add a few arguments to the regular GATK command line. Note that when we add Spark arguments, we first add a `--` separator (two dashes), which signals to the parser system that what follows is intended for the Spark subsystem.

In the simplest case, called *local execution*, the program executes all the work on the same machine where we're launching the command. To do that, simply add the argument `--spark-master 'local[*]'` to the base command line. The `--spark-master local` part instructs the Spark subsystem to send all work to the machine's own local cores, and the value in brackets specifies the number of cores to use. To use all available cores, replace the number by an asterisk. Here is an example command showing a Spark-enabled tool being run on four cores of a normal local machine:

```
gatk MySparkTool \
    -R data/reference.fasta \
    -I data/sample1.bam \
    -O data/variants.vcf \
    -- \
    --spark-master 'local[4]'
```

If you have access to a Spark cluster, the Spark-enabled tools are going to be extra happy, though you might need to provide additional parameters to use them effectively. We don't cover that in detail, but here are examples of the syntax you would use to send the work to your cluster depending on your cluster type:

• Run on the cluster at 23.195.26.187, port 7077:

```
--spark-runner SPARK --spark-master spark://23.195.26.187:7077
```

- Run on the cluster called my_cluster in Google Dataproc:

```
--spark-runner GCS --cluster my_cluster
```

That covers just about all you need to know about Spark for now, because using the cloud allows us to sidestep some of the issues that made Spark so valuable in other settings.

Running GATK in Practice

By this point, you're probably itching to run some actual GATK commands, so let's get started. We're going to assume that you are using the VM that you set up in Chapter 4 and that you have also already pulled and tested the GATK container image as instructed in that same chapter.

If your VM is still up and running, go ahead back into it using SSH, unless you never even left. If it's stopped, head on over to the GCP console VM instance management page (*https://oreil.ly/sGeug*), start your VM, and then log back into it.

Docker setup and test invocation

As a reminder, this was the command that you ran in Chapter 4:

```
$ docker run -v ~/book:/home/book -it us.gcr.io/broad-gatk/gatk:4.1.3.0 /bin/bash
```

If your container is still running, you can hop back in and resume your session. As we showed in Chapter 4, you just need to determine the container ID by running docker ps and then run docker attach ID. If you had shut down your container last time, simply run the full command again.

When the container is ready to go, your terminal prompt will change to something like this:

```
(gatk) root@ce442edab970:/gatk#
```

This indicates that you are logged into the container. Going forward, we use the hash sign (#) as the terminal prompt character to indicate when you are running commands within the container.

You can use the usual shell commands to navigate within the container and check out what's in there:

```
# ls
GATKConfig.EXAMPLE.properties  gatk-package-4.1.3.0-spark.jar  gatkdoc
README.md                      gatk-spark.jar                  gatkenv.rc
gatk                           gatk.jar                        install_R_packages.R
gatk-completion.sh             gatkPythonPackageArchive.zip    run_unit_tests.sh
gatk-package-4.1.3.0-local.jar gatkcondaenv.yml                scripts
```

As you can see, the container includes several *.jar* files as well as the `gatk` wrapper script, which we use in our command lines to avoid having to deal with the *.jar files* directly. The `gatk` executable is included in the preset user path, so we can invoke it from anywhere inside the container. For example, you can invoke GATK's help/usage output by simply typing the `gatk` command in your terminal:

```
# gatk
Usage template for all tools (uses --spark-runner LOCAL when used with a Spark tool)
    gatk AnyTool toolArgs
Usage template for Spark tools (will NOT work on non-Spark tools)
    gatk SparkTool toolArgs  [ -- --spark-runner <LOCAL | SPARK | GCS> sparkArgs ]
Getting help
    gatk --list      Print the list of available tools
    gatk Tool --help  Print help on a particular tool
Configuration File Specification
    --gatk-config-file              PATH/TO/GATK/PROPERTIES/FILE

gatk forwards commands to GATK and adds some sugar for submitting spark jobs

    --spark-runner <target>     controls how spark tools are run
     valid targets are:
    LOCAL:      run using the in-memory spark runner
    SPARK:      run using spark-submit on an existing cluster
                --spark-master must be specified
                --spark-submit-command may be specified to control the Spark submit command
                arguments to spark-submit may optionally be specified after --
    GCS:        run using Google cloud dataproc
                commands after the -- will be passed to dataproc
                --cluster <your-cluster> must be specified after the --
                spark properties and some common spark-submit parameters will be translated
                to dataproc equivalents
    --dry-run       may be specified to output the generated command line without running it
    --java-options 'OPTION1[ OPTION2=Y ... ]''   optional - pass the given string of options to
                    the java JVM at runtime.
                    Java options MUST be passed inside a single string with space-separated values
```

You might recall that you already ran this when you tested your Docker setup in Chapter 4. Hopefully the outputs make more sense now that you are now an expert in GATK syntax.

To get help or usage information for a specific tool—for example, the germline short variant caller `HaplotypeCaller`–try this:

```
# gatk HaplotypeCaller --help
```

This is convenient for looking up an argument while you're working, but most GATK tools spit out a lot of information that can be difficult to read in a terminal. You can access the same information in a more readable format in the Tool Index (*https://oreil.ly/9OiEJ*) on the GATK website.

Running a real GATK command

Now that we know we have the software ready to go, let's run a real GATK command on some actual input data. The inputs here are files from the book data bundle that you copied to your VM in Chapter 4.

Let's begin by moving into the *home/book/data/germline* directory and creating a new directory named *sandbox* to hold the outputs of the commands that we're going to run:

```
# cd /home/book/data/germline
# mkdir sandbox
```

Now you can run your first real GATK command! You can either type it as shown in the example that follows, on multiple lines, or you can put everything on a single line if you remove the backslash (\) characters:

```
# gatk HaplotypeCaller \
    -R ref/ref.fasta \
    -I bams/mother.bam \
    -O sandbox/mother_variants.vcf
```

This uses `HaplotypeCaller`, GATK's current flagship caller for germline SNPs and indels, to do a basic run of germline short variant calling on a very small input dataset, so it should take less than two minutes to run. It will spit out a lot of console output; let's have a look at the most important bits.

The header, which lets you know what you're running:

```
Using GATK jar /gatk/gatk-package-4.1.3.0-local.jar
Running:
    java -Dsamjdk.use_async_io_read_samtools=false -Dsamjdk.use_async_io_wri
te_samtools=true -Dsamjdk.use_async_io_wri
te_tribble=false -Dsamjdk.compression_level=2 -jar /gatk/gatk-package-4.1.3.0-
local.jar HaplotypeCaller -R ref/ref.fas
ta -I bams/mother.bam -O sandbox/mother_variants.vcf
09:47:17.371 INFO  NativeLibraryLoader - Loading libgkl_compression.so from
jar:file:/gatk/gatk-package-4.1.3.0-local.
jar!/com/intel/gkl/native/libgkl_compression.so
09:47:17.719 INFO  HaplotypeCaller - -------------------------------------------
-------------------
09:47:17.721 INFO  HaplotypeCaller - The Genome Analysis Toolkit (GATK) v4.1.3.0
09:47:17.721 INFO  HaplotypeCaller - For support and documentation go to
https://software.broadinstitute.org/gatk/
09:47:17.722 INFO  HaplotypeCaller - Executing as root@3f30387dc651 on Linux
v5.0.0-1011-gcp amd64
09:47:17.723 INFO  HaplotypeCaller - Java runtime: OpenJDK 64-Bit Server VM
v1.8.0_191-8u191-b12-0ubuntu0.16.04.1-b12
09:47:17.724 INFO  HaplotypeCaller - Start Date/Time: August 20, 2019 9:47:17 AM
UTC
```

The progress meter, which informs you how far along you are:

```
09:47:18.347 INFO  ProgressMeter - Starting traversal
09:47:18.348 INFO  ProgressMeter -        Current Locus  Elapsed Minutes
Regions Processed   Regions/Minute
09:47:22.483 WARN  InbreedingCoeff - InbreedingCoeff will not be calculated; at
least 10 samples must have called geno
types
```

```
09:47:28.371 INFO  ProgressMeter -          20:10028825          0.2
33520         200658.5
09:47:38.417 INFO  ProgressMeter -          20:10124905          0.3
34020         101709.1
09:47:48.556 INFO  ProgressMeter -          20:15857445          0.5
53290         105846.1
09:47:58.718 INFO  ProgressMeter -          20:16035369          0.7
54230          80599.5
09:48:08.718 INFO  ProgressMeter -          20:21474713          0.8
72480          86337.1
09:48:18.718 INFO  ProgressMeter -          20:55416713          1.0
185620        184482.4
```

And the footer, which signals when the run is complete and how much time it took:

```
09:48:20.714 INFO  ProgressMeter - Traversal complete. Processed 210982 total
regions in 1.0 minutes.
09:48:20.738 INFO  VectorLoglessPairHMM - Time spent in setup for JNI call:
0.045453468000000004
09:48:20.739 INFO  PairHMM - Total compute time in PairHMM
computeLogLikelihoods(): 6.333675601
09:48:20.739 INFO  SmithWatermanAligner - Total compute time in java
Smith-Waterman: 6.18 sec
09:48:20.739 INFO  HaplotypeCaller - Shutting down engine
[August 20, 2019 9:48:20 AM UTC]
org.broadinstitute.hellbender.tools.walkers.haplotypecaller.HaplotypeCaller done.
Elapsed time: 1.06 minutes.
Runtime.totalMemory()=717225984
```

If you see all of that, hurray! You have successfully run a proper GATK command that called variants on a real person's genomic sequence data. We dig into the output results and how to interpret them in more detail later in this chapter, but for now you can rejoice at having achieved this first milestone of running GATK in the cloud.

If, however, anything went wrong and you're not seeing the expected output, start by checking your spelling and all file paths. Our experience suggests that typos, case errors, and incorrect file paths are the leading cause of command-line errors worldwide. Human brains are wired to make approximations and often show us what we expect to see, not what is actually there, so it's really easy to miss small errors. And remember that case is important! Make sure to use a terminal font that differentiates between lowercase l and uppercase I.

For additional help troubleshooting GATK issues, you can reach out to the GATK support forum staff (*https://oreil.ly/YfsSM*). Make sure to mention that you are following the instructions in this book to provide the necessary context.

Running a Picard command within GATK4

As mentioned earlier, the Picard toolkit is included as a library within the GATK4 executable. This is convenient because many genomics pipelines involve a mix of

---

both Picard and GATK tools. Having both toolkits in a single executable means that you have fewer separate packages to keep up-to-date, and it largely guarantees that the tools will produce compatible outputs because they are tested together before release.

Running a Picard tool from within GATK is simple. In a nutshell, you use the same syntax as for GATK tools, calling on the name of the tool that you want to run and providing the relevant arguments and parameters in the rest of the command after that. For example, to run the classic quality control tool ValidateSamFile, you would run this command:

```
# gatk ValidateSamFile \
    -R ref/ref.fasta \
    -I bams/mother.bam \
    -O sandbox/mother_validation.txt
```

If you previously used Picard tools from the standalone executable produced by the Picard project maintainers, you'll notice that the syntax we use here (and with GATK4 in general) has been adapted from the original Picard style (e.g., I=sample1.bam) to match the GATK style (e.g., -I sample1.bam). This was done to make command syntax consistent across tools, so that you don't need to think about whether any given tool that you want to use is a Picard tool or a "native" GATK tool when composing your commands. This does mean, however, that if you have existing scripts that use the standalone Picard executable, you will need to convert the command syntax to use them with the GATK package.

Getting Started with Variant Discovery

Enough with the syntax examples; let's call some variants!

What follows is not considered part of the GATK Best Practices, which we introduce in the last section of this chapter. At this stage, we are just presenting a simplified approach to variant discovery in order to demonstrate in an accessible way the key principles of how the tools work. In Chapters 6 and 7, we go over the GATK Best Practices workflows for three major use cases, and we highlight specific steps that are particularly important with additional details.

Calling Germline SNPs and Indels with HaplotypeCaller

We're going to look for germline short sequence variants because that's the most common type of variant analysis in both research and clinical applications today, using the HaplotypeCaller tool. We briefly go over how the HaplotypeCaller works, and then we do a hands-on exercise to examine how this plays out in practice.

In the hands-on exercises in this chapter and the next, we use a version of the human genome reference (b37) that is not the most recent (hg38). This is because a lot of

GATK tutorial examples were developed when b37 (with its almost-twin hg19) was still the reigning reference, and the relevant materials have not yet been regenerated with the new reference. Many researchers in the field have not yet moved to the newer reference because doing so requires a lot of work and validation effort. For this book, we chose to accept the reality of this hybrid environment rather than trying to sweep it out of sight. Learning to be aware of the existence of different references and navigating different sets of resource data is an important part of the discipline of computational genomics.

HaplotypeCaller in a nutshell

HaplotypeCaller is designed to identify germline SNPs and indels with a high level of sensitivity and accuracy. One of its distinguishing features is that it uses local *de novo* assembly of haplotypes to model possible variation. If that sounds like word salad to you, all you really need to know is that whenever the program encounters a region showing signs of variation, it completely realigns the reads in that region before proceeding to call variants. This allows the HaplotypeCaller to gain accuracy in regions that are traditionally difficult to call; for example, when they contain different types of variants close to one another. It also makes the HaplotypeCaller much better at calling indels than traditional position-based callers like the old UnifiedGenotyper and Samtools mpileup.

HaplotypeCaller operates in four distinct stages, illustrated in Figure 5-1 and outlined in the list that follows. Note that this description refers to established computational biology terms and algorithms that we chose not to explain in detail here; if you would like to learn more about these concepts, we recommend consulting the GATK documentation and forum for pointers to appropriate study materials.

1. *Define active regions*

    The tool determines which regions of the genome it needs to operate on based on the presence of significant evidence for variation, such as mismatches or gaps in read alignments.

2. *Determine haplotypes by reassembly of the active region*

    For each active region, the program builds a De Bruijn–like assembly graph to realign the reads within the bounds of the active region. This allows it to make a list of all the possible haplotypes supported by the data. The program then realigns each haplotype against the reference sequence using the Smith-Waterman algorithm in order to identify potentially variant sites.

3. *Determine likelihoods of the haplotypes given the read data*

    For each active region, the program then performs a pairwise alignment of each read against each haplotype using the PairHMM algorithm. This produces a matrix of likelihoods of haplotypes given the read data. These likelihoods are then marginalized to obtain the likelihoods of alleles per read for each potentially variant site.

4. Assign sample genotypes

For each potentially variant site, the program applies Bayes' rule, using the likelihoods of alleles given the read data, to calculate the posterior probabilities of each genotype per sample given the read data observed for that sample. The most likely genotype is then assigned to the sample.

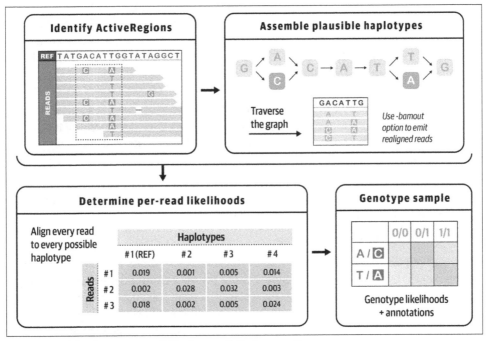

Figure 5-1. The four stages of HaplotypeCaller's operation.

The final output of HaplotypeCaller is a VCF file containing variant call records, which include detailed genotype information and a variety of statistics that describe the context of the variant and reflect the quality of the data used to make the call.

Running HaplotypeCaller and examining the output

Let's run HaplotypeCaller in its simplest form on a single sample to become familiar with its operation. This command requires a reference genome file in FASTA format (-R) as well as one or more files with the sequence data to analyze in BAM format (-I). It will output a file of variant calls in VCF format (-O). Optionally, the tool can take an interval or list of intervals to process (-L), which we use here to restrict analysis to a small slice of data in the interest of time (this can also be used for the purpose of parallelizing execution, as you'll see a little later in this book):

```
# gatk HaplotypeCaller \
    -R ref/ref.fasta \
    -I bams/mother.bam \
    -O sandbox/mother_variants.200k.vcf \
    -L 20:10,000,000-10,200,000
```

After the command has run to completion (after about a minute), we need to copy the output file to a bucket so that we can view it in IGV, as we demonstrated at the end of Chapter 4. The simplest way to do this is to open a second SSH terminal to the VM from the GCP console so that we can run `gsutil` commands, which is more convenient to do from outside the container. If at any point you lose track of which window is which or where you should be running a given command, remember that the VM prompt ends in a dollar sign ($) and the container prompt ends in a hash (#). Just match them up, and you'll be OK.

 Alternatively, you can copy the files directly from within the container, but first you need to run `gcloud init`, as described in Chapter 4, from within the container, which has its own authentication scope.

In the second SSH terminal window that is now connected to your VM, move from the home directory into the germline *sandbox* directory:

```
$ cd ~/book/data/germline/sandbox
```

Next, you're going to want to copy the contents of the *sandbox* directory to the storage bucket that you created in Chapter 4, using the `gsutil` tool. However, it's going to be annoying if you need to constantly replace the name we use for our bucket with your own, so let's take a minute to set up an environment variable. It's going to serve as an alias so that you can just copy and paste our commands directly from now on.

Create the BUCKET variable by running the following command, replacing the *my-bucket* part with the name of your own bucket, as you did in Chapter 4 (last time!):

```
$ export BUCKET="gs://my-bucket"
```

Now, check that the variable was set properly by running `echo`, making sure to add the dollar sign in front of BUCKET:

```
$ echo $BUCKET
gs://my-bucket
```

Finally, you can copy the file that you want to view to your bucket:

```
$ gsutil cp mother_variants.200k.vcf* $BUCKET/germline-sandbox/
Copying file://mother_variants.200k.vcf [Content-Type=text/vcard]...
Copying file://mother_variants.200k.vcf.idx [Content-Type=application/octet-
stream]...
```

```
- [2 files][101.1 KiB/101.1 KiB]
  Operation completed over 2 objects/101.1 KiB.
```

When the copy operation is complete, you're going to go to your desktop IGV application and load the reference genome, the input BAM file, and the output VCF file. We covered that procedure in Chapter 4, so you might feel comfortable doing that on your own. But just in case it's been a while, let's go through the main steps together briefly. Feel free to refer to the screenshots in Chapter 4 if you need a refresher.

All of the actions in the following numbered list are to be performed in the IGV application unless otherwise specified:

1. In IGV, in the top menu, check that your Google login is active.

2. On the drop-down menu on the left, verify that you have the "Human hg19" reference selected. Technically, our data is aligned to the b37 reference genome, but for pure viewing purposes (not analysis), the two are interchangeable.

3. From the top menu bar, select File > Load from URL and then paste the following file path for the BAM file into the top field:

   gs://genomics-in-the-cloud/v1/data/germline/bams/mother.bam

   You can leave the second field blank if you're using IGV version 2.7.2 or later; IGV will automatically retrieve the index file. Click OK.

4. Get the file path for the output VCF file that you just copied to your bucket. You can get it by running **echo $BUCKET/germline-sandbox/mother_variants. 200k.vcf** in your VM terminal.

5. From the top menu bar, select File > Load from URL and then, in the first field, paste the VCF file path that you got from the previous step. You can leave the second field blank again. Click OK.

At this point, you should have all the data loaded and ready to go, but you don't see anything yet because by default IGV shows you the full span of the genome, and at that scale, the small slice of data we're working with is not visible. To zoom in on the data, in the location box, enter the coordinates **20:10,002,294-10,002,623** and then click Go. You should see something very similar to Figure 5-2.

The variant track showing the contents of the VCF file is displayed on top, and below is the read data track showing the contents of the BAM file. Variants are indicated as small vertical bars (colored in red in Figure 5-2), and sequence reads are represented as gray horizontal bars with a pointy bit indicating the read orientation. In the read data, we see various mismatches (small colored bars) and gaps (purple I-shaped symbols for insertions; deletions would show up as black horizontal bars).

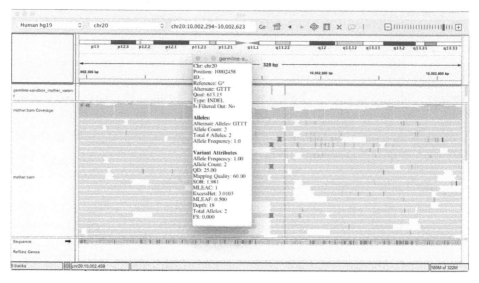

Figure 5-2. The original BAM file and output VCF file loaded in IGV.

We see that `HaplotypeCaller` called two variants within this region. You can click each little red bar in the variant track to get more details on each one. On the right, we have a homozygous SNP with excellent support in the read data. That seems likely to be correct. On the left, we have a homozygous insertion of three T bases, yet we see only a small proportion of reads with the insertion point symbol that support this call. How is this possible, when so few reads seem to support an insertion at this position?

Your first reflex when you encounter something like this (a call that doesn't make sense, especially if it's an indel) should be to turn on the display of *soft-clipped* sequences in the genome viewer you're using. *Soft clips* are segments of reads that are marked by the mapper (in this case BWA) as being not useful, so they are hidden by default by IGV. The mapper introduces soft clips in the CIGAR string of a read when it encounters bases, typically toward the end of the read, that it is not able to align to the reference to its satisfaction based on the scoring model it employs. At some point, the mapper gives up trying to align the bases, and just puts in the S operator to signify that downstream tools should disregard the following bases, even though they are still present in the read record. Some mappers even remove those bases, which is called *hard clipping*, and uses the H operator instead of S.

Soft clips are interesting to us because they often occur in the vicinity of insertions, especially large insertions. So let's turn on soft clips in IGV in the Preferences. In the top menu bar, select View > Preferences and click the Alignments tab to bring up the relevant settings. Select the checkbox labeled "Show soft-clipped bases," as shown in Figure 5-3, and then click Save to close the Preferences pane.

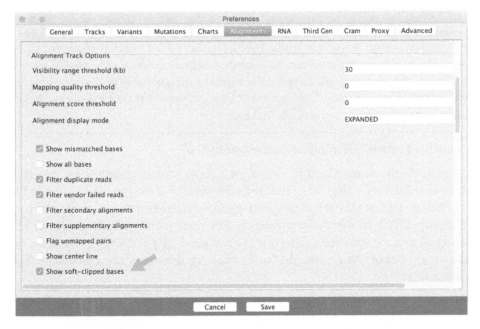

Figure 5-3. IGV alignment settings.

As we see in Figure 5-4, with soft-clip display turned on, the region we've been looking at lights up with mismatching bases! That's a fair amount of sequence that we were just ignoring because BWA didn't know what to do with it. `HaplotypeCaller`, on the other hand, will have taken it into account when it called variants because the first thing it does within an active region is to throw out all the read alignment information and build a graph to realign them, as we discussed earlier.

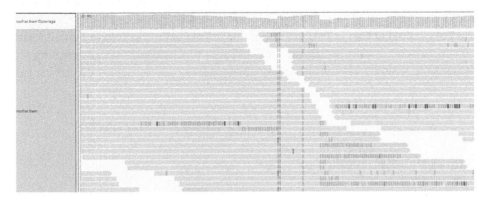

Figure 5-4. Turning on the display of soft clips shows a lot of information that was hidden.

So now we know that HaplotypeCaller had access to this additional information and somehow decided that it supported the presence of an insertion at this position. But the extra sequence we see here is unaligned, so how do we check HaplotypeCaller's work? The tool has a parameter called -bamout, which makes HaplotypeCaller output a BAM file, or a *bamout*, showing the result of the assembly graph-based realignment it performs internally. That's great because it means that we can see exactly what HaplotypeCaller saw when it made that call.

Generating an output BAM to troubleshoot a surprising call

Let's generate the bamout so that we can get a better understanding of what HaplotypeCaller thinks it's seeing here. Run the command that follows, which should look very similar: it's basically the same as the previous one except we changed the name of the output VCF (to avoid overwriting the first one), we narrowed the interval we're looking at, and we specified a name for the realigned output BAM file using the new -bamout argument. As a reminder, we're running this from the */home/book/data/ germline* directory:

```
# gatk HaplotypeCaller \
    -R ref/ref.fasta \
    -I bams/mother.bam \
    -O sandbox/mother_variants.snippet.debug.vcf \
    -bamout sandbox/mother_variants.snippet.debug.bam \
    -L 20:10,002,000-10,003,000
```

Because we're interested in looking at only that particular region, we give the tool a narrowed interval with -L 20:10,002,000-10,003,000, to make the runtime shorter.

After that has run successfully, copy the bamout (*sandbox/mother_variants.snippet.debug.bam*) to your bucket and then load it in IGV, as we've previously described. It should load in a new track at the bottom of the viewer, as shown in Figure 5-5. You should still be zoomed in on the same coordinates (20:10,002,294-10,002,623), and have the original mother.bam track loaded for comparison. If you have a small screen, you can switch the view setting for the BAM files to Collapsed by right-clicking anywhere in the window where you see sequence reads.

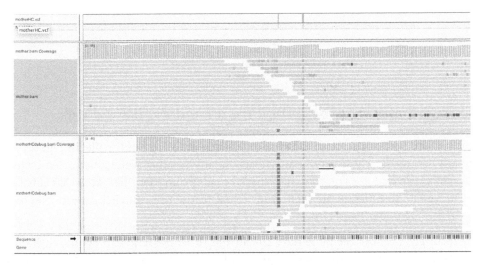

Figure 5-5. Realigned reads in the bamout file (bottom track).

The bottom track shows us that after realignment by HaplotypeCaller, almost all the reads support the insertion, and the messy soft clips that we see in the original BAM file are gone. This confirms that HaplotypeCaller utilized the soft-clipped sequences during the realignment stage and took that sequence into account when calling variants. Incidentally, if you expand the reads in the output BAM (right-click > Expanded view), and you can see that all of the insertions are in phase with the neighboring C/T variant. Such consistency tends to be a good sign; whereas if you had inconsistent segregation between neighboring variants, you would worry that at least one of them is an artifact.

So, to summarize, this shows that HaplotypeCaller found a different alignment after performing its local graph assembly step. The reassembled region provided Haploty peCaller with enough evidence to call the indel, whereas a position-based caller like UnifiedGenotyper would have missed it.

That being said, there is still a bit more to the bamout than meets the eye—or, at least, what you can see in this view of IGV. Right-click the mother_variants.snip pet.debug.bam track to open the view options menu. Select "Color alignments by," and then choose "read group." Your gray reads should now be colored similar to the screenshot shown in Figure 5-6.

 We have described most figures in a way that makes sense whether you're reading a grayscale or color version of this book, but color coding is the feature we want to explain for figures like 5-6 and 5-7. For that reason, you can find all the color figures for the book in our online repository (*https://oreil.ly/genomics-repo*).

Figure 5-6. Bamout shows artificial haplotypes constructed by HaplotypeCaller.

Some of the first reads, those shown in one shade at the top of the pile in Figure 5-6, are not *real* reads. These represent artificial haplotypes that were constructed by `Hap lotypeCaller` and are tagged with a special read group identifier, `RG:Z:Artificial HaplotypeRG`, to differentiate them from actual reads. You can click an artificial read to see this tag under Read Group.

Interestingly, the tool seems to have considered three possible haplotypes, including two with insertions of different lengths: either two bases or three bases inserted, before ultimately choosing the three-base case as being most likely to be real. To examine this a bit more closely, let's separate these artificial reads to the top of the track. Right-click the track, and then, on the menu that opens, select "Group align-ments by" and then choose "read group." Next let's color the reads differently. Right-click and select "Color alignments by," choose "tag," and then type **HC**.

`HaplotypeCaller` labels reads that have unequivocal support for a particular haplo-type with an HC tag value that refers to the corresponding haplotype. Now the color shows us which reads in the lower track support which of the artificial haplotypes in the top track, as we can see in Figure 5-7. Gray reads are unassigned, which means there was some ambiguity about which haplotype they support best.

This shows us that `HaplotypeCaller` was presented with more reads supporting the three-base insertion unambiguously compared to the two-base insertion case.

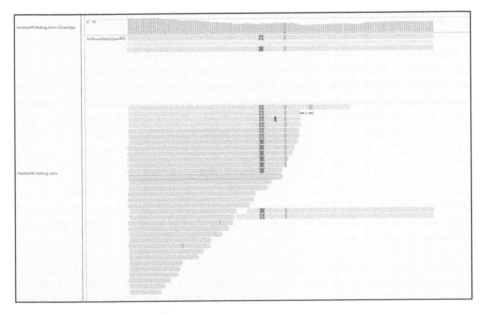

Figure 5-7. Bamout shows support per haplotype.

Filtering Based on Variant Context Annotations

At this point, we have a callset of potential variants, but we know from the earlier overview of variant filtering that this callset is likely to contain many *false-positive calls*; that is, calls that are caused by technical artifacts and do not actually correspond to real biological variation. We need to get rid of as many of those as we can without losing real variants. How do we do that?

A commonly used approach for filtering germline short variants is to use *variant context annotations*, which are statistics captured during the variant calling process that summarize the quality and quantity of evidence that was observed for each variant. For example, some variant context annotations describe what the sequence context was like around the variant site (were there a lot of repeated bases? more GC or more AT?), how many reads covered it, how many reads covered each allele, what proportion of reads were in forward versus reverse orientation, and so on. We can choose thresholds for each annotation and set a *hard filtering* policy that says, for example, "For any given variant, if the value of this annotation is greater than a threshold value of X, we consider the variant to be real; otherwise, we filter it out."

Let's now look at how annotation values are distributed in some real data, how that compares to a truth set, and see what that tells us about how we can use the annotations for filtering.

Understanding variant context annotations

We're going to look at a variant callset made from the same WGS data called `mother.bam` that we used earlier for our `HaplotypeCaller` tutorial, except here we have calls from the entire chromosome 20 instead of just a small slice. The Mother data (normally identified as sample NA12878) has been extensively studied as part of a benchmarking project carried out by a team at the National Institute of Standards and Technology (NIST), called Genome in a Bottle (GiaB). Among many other useful resources, the GiaB project has produced a highly curated truth set for the Mother data based on consensus methods involving multiple variant calling tools and pipelines. Note that the term *truth set* here is not meant as a claim of absolute truth, given that it is practically impossible to achieve absolute truth under the experimental conditions in question, but to reflect that we are very confident that *most* of the variant calls it contains are probably real and that there are probably *very few* variants missing from it.

Thanks to this resource, we can take a variant callset that we made ourselves from the original data and annotate each of our variants with information based on the GiaB truth set, if it is present there. If a variant in our callset is not present in the truth set, we'll assume that it is a false positive—in other words, that we were wrong. This will not always be correct, but it is a reasonable approximation for the purpose of this section. As a result, we'll be able to contrast the annotation value distributions for variants from our callset that we believe are real to those that we believe are false. If there is a clear difference, we should be able to derive some filtering rules.

Let's pull up a few lines of the original variant callset. We use the common utility `zcat` (a version of `cat` that can read gzipped files) to read in the variant data. We then pipe (|) the variant data to another common utility, `grep`, with the instruction to discard lines starting with ## (i.e., header lines in the VCF). Finally, we pipe (|) the remaining lines to a text-viewing utility, head, with the instruction to display only the first three lines of the file (-3). As a reminder, we're still running all of this from the */home/book/data/germline* directory.

```
# zcat vcfs/motherSNP.vcf.gz | grep -v '##' | head -3
#CHROM  POS     ID      REF     ALT     QUAL    FILTER  INFO    FORMAT  NA12878
20      61098   .       C       T       465.13  .
AC=1;AF=0.500;AN=2;BaseQRankSum=0.516;ClippingRankSum=0.00;DP=44;ExcessHet=3.0103;FS=0.000;
MQ=59.48;MQRankSum=0.803;QD=10.57;ReadPosRankSum=1.54;SOR=0.603
GT:AD:DP:GQ:PL0/1:28,16:44:99:496,0,938
20      61795   .       G       T       2034.16 .       AC=1;AF=0.500;AN=2;
BaseQRankSum=-6.330e-01;ClippingRankSum=0.00;DP=60;ExcessHet=3.9794;FS=0.000;MQ=59.81;
MQRankSum=0.00;QD=17.09;ReadPosRankSum=1.23;SOR=0.723
GT:AD:DP:GQ:PL 0/1:30,30:60:99:1003,0,1027
```

This shows you the raw VCF content, which can be a little rough to read—more challenging than when we were looking at the variant calls with IGV earlier in this chapter, right? If you need a refresher on the VCF format, feel free to pause and go back to the genomics primer in Chapter 2, in which we explain how VCFs are structured.

Basically, what we're looking for here are the strings of annotations, separated by semicolons, that were produced by `HaplotypeCaller` to summarize the properties of the data at and around the location of each variant. For example, we see that the first variant in our callset has an annotation called QD with a value of 10.57:

```
AC=1;AF=0.500;AN=2;BaseQRankSum=0.516;ClippingRankSum=0.00;DP=44;ExcessHet=
3.0103;FS=0.000;MQ=59.48;MQRankSum=0.803;QD=10.57;ReadPosRankSum=1.54;SOR=0.603
```

OK, fine, but what does that mean? Well, that's something we discuss in a little bit. For now, we just want you to think of these as various metrics that we're going to use to evaluate how much we trust each variant call.

And here are a few lines of the variant callset showing the annotations derived from the truth set, extracted by using the same command as in our previous example:

```
# zcat vcfs/motherSNP.giab.vcf.gz | grep -v '##' | head -3
#CHROM  POS     ID        REF    ALT     QUAL    FILTER  INFO    FORMAT  INTEGRATION
20      61098   rs6078030  C      T       50      PASS
callable=CS_CGnormal_callable,CS_HiSeqPE300xfreebayes_callable;callsetnames=CGnormal,
HiSeqPE300xfreebayes,HiSeqPE300xGATK;callsets=3;datasetnames=CGnormal,HiSeqPE300x;
datasets=2;datasetsmissingcall=10XChromium,IonExome,SolidPE50x50bp,SolidSE75bp;filt=
CS_HiSeqPE300xGATK_filt,CS_10XGATKhaplo_filt,CS_SolidPE50x50GATKHC_filt;platformnames=
CG,Illumina;platforms=2GT:PS:DP:ADALL:AD:GQ    0/1:.:542:132,101:30,25:604
20      61795   rs4814683  G      T       50      PASS
callable=CS_HiSeqPE300xGATK_callable,CS_CGnormal_callable,CS_HiSeqPE300xfreebayes_
callable;callsetnames=HiSeqPE300xGATK,CGnormal,HiSeqPE300xfreebayes,10XGATKhaplo,
SolidPE50x50GATKHC,SolidSE75GATKHC;callsets=6;datasetnames=HiSeqPE300x,CGnormal,
10XChromium,SolidPE50x50bp,SolidSE75bp;datasets
=5;datasetsmissingcall=IonExome;platformnames=Illumina,CG,10X,Solid;platforms=4
GT:PS:DP:ADALL:AD:GQ    0/1:.:769:172,169:218,205:1337
```

Did you notice that there are a lot more annotations in these calls, and they're quite different? The annotations you see here are not actually variant context annotations in the same sense as before. These describe information at a higher level; they refer not to the actual sequence data, but to a meta-analysis that produced the calls by combining multiple callsets derived from multiple pipelines and data types. For example, the `callsets` annotation counts how many of the multiple callsets used for making the GiaB truth set agreed on each variant call, as an indicator of confidence. You can see that the first variant record has `callsets=3`, whereas the second record has `callsets=6`, so we will trust the second call more than the first.

In the next section, we use those meta-analysis annotations to inform our evaluation of our own callset.

Plotting variant context annotation data

We prepared some plots that show the distribution of values for a few annotations that are usually very informative with regard to the quality of variant calls. We made two kinds of plots: density plots and scatter plots. Plotting the *density of values* for a single annotation enables us to see the overall range and distribution of values observed in a callset. Combining this with some basic knowledge of what each annotation represents and how it is calculated, we can make a first estimation of value thresholds that segregate false-positive calls (FPs) from true-positive calls (TPs). Plotting the *scatter of values* for two annotations, one against the other, additionally shows us the trade-offs we make when setting a threshold on annotation values individually. The full protocol for generating the plots involves a few more GATK commands and some R code; they are all documented in detail in a tutorial that we reference again in Chapter 12 when we cover interactive analysis.

The annotation that people most commonly use to filter their callsets, rightly or wrongly, is the confidence score QUAL. In Figure 5-8, we plotted the density of the distribution observed for QUAL in our sample. It's not terribly readable, because of a very long tail of extremely high values. What's going on there? Well, you wouldn't be alone in thinking that the higher the QUAL value, the better the chance that the variant is real, given that QUAL is supposed to represent our confidence in a variant call. And for the bulk of the callset, that is indeed true. However, QUAL has a dirty secret: its value becomes horribly inflated in areas of extremely high depth of coverage. This is because each read contributes a little to the QUAL score, so variants in regions with deep coverage can have artificially inflated QUAL scores, giving the impression that the call is supported by more evidence than it really is.

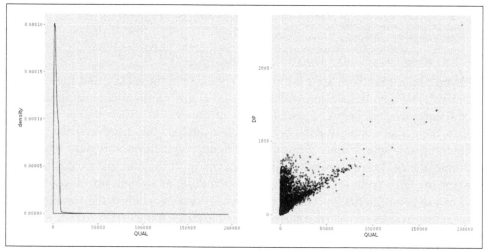

Figure 5-8. Density plot of QUAL (left); scatter plot of QUAL versus DP (right).

We can see this relationship in action by making a scatter plot of QUAL against DP, the depth of coverage, shown in Figure 5-8. That plot shows clearly that the handful of variants with extremely high QUAL values also have extremely high DP values. If you were to look one up in IGV, you would see that those calls are basically junk—they are poorly supported by the read data and located in regions that are very messy. This tells us we can fairly confidently disregard those calls. So let's zoom into the QUAL plot, this time restricting the x-axis to eliminate extremely high values. In

Figure 5-9, we're still looking at all calls in aggregate, whereas in Figure 5-9 we stratify the calls based on how well they are supported in the GiaB callset, as indicated by the number of callsets in which they were present.

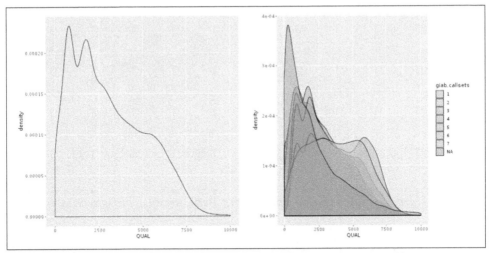

Figure 5-9. Density plot of QUAL: all calls together (left); stratified by callsets annotation (right).

You can see on the plot that most of the low-credibility calls (under the curve that reaches the highest vertical point) are clustering at the low end of the QUAL range, but you also see that the distribution is continuous; there's no clear separation between bad calls and good calls, limiting the usefulness of this annotation for filtering. It's probable that the depth-driven inflation effect is still causing us trouble even at more reasonable levels of coverage. So let's see if we can get that confounding factor out of the equation. Enter QD, for QualByDepth: the QD annotation puts the variant confidence QUAL score into perspective by normalizing for the amount of coverage available, which compensates for the inflation phenomenon that we noted a moment ago.

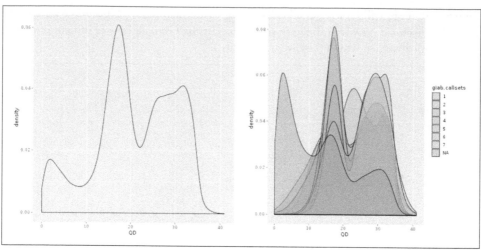

Figure 5-10. Density plot of QD: all calls together (left); stratified by callsets annotation (right).

Even without the truth set stratification option, we already see in Figure 5-10 that the shape of the density curve suggests this annotation is going to be more helpful. And indeed, when we run with the split turned on, we see a clear pattern with more separation between "bad variants" and the rest in Figure 5-10. You see the small "shoulder" on the low end of the QD distribution? Those are all the very-low-quality variant calls that are most probably junk, as confirmed by the GiaB stratification. In general, the QD value tends to be a rather reliable indicator of variant call quality.

Interpreting the Shape of the QD Distribution

Another thing that's neat about QD is that when you're looking at QD annotation values from a single sample, you can see the variants segregating into two camel humps. These correspond to heterozygous- versus homozygous-variant calls, and they cluster around QD values that are predictable relative to one another.

You can generally expect that the second hump will be centered on a value that's twice the value of the first. This is because of the way QD is calculated: it is normalized against the amount of read depth *that supports the alternate allele*! And on average, you should expect homozygous-variant calls to have twice as much supporting depth for the alternate allele compared to heterozygous calls. So, for heterozygous calls, the denominator in the QD calculation is twice as large, and therefore the final QD value is half as much as you'd see for a homozygous-variant call.

This can be useful for new experiments for which you don't have the luxury of having a truth set available. Even though you can't easily predict what the "good" values of QD should be, you can plot the QD density distribution and make some reasonable assumptions based on its shape and where the shoulder and the camel humps line up. Just keep in mind that this shape is this clear only for a single sample, because when you add more samples, you will have many sites where some samples are heterozygous and others are homozygous, causing the humps to blend together. On the bright side, the shoulder-of-crap should become more clearly defined as you add more samples.

Because the various annotations measure different aspects of the variant context, it's interesting to examine how well they might perform when combined. One way to approach this question is to combine the scatter and density plots we've been making into a single figure, like Figure 5-11, which shows both the two-way scatter of two annotations (here, QD and DP) and their respective density curves on the sides of the scatter plot. This provides us with some insight into how probable true variants seem to cluster within the annotation space. If nothing else, the blobs of color in the middle (which is where the "goodies" are clustering) are reasonably clearly demarcated from the gray smear of likely false positives on the left side of the plot.

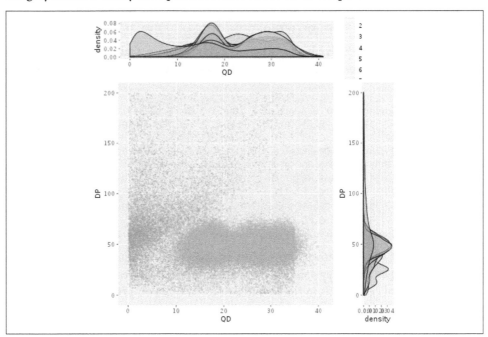

Figure 5-11. A scatter plot with marginal densities of QD versus DP.

So what do we do with this insight? We use it as a guide to set thresholds with reasonable confidence on each annotation, reflecting how much each of them contributes to allowing us to weed out different subsets of false-positive variants. The resulting combined filtering rule might be more powerful than the sum of its parts.

Applying hard filters to germline SNPs and indels

Looking at the plots in the previous section gave us some insight into what the variant context annotations represent and how we can use them to derive some rudimentary filtering rules. Now let's see how we can apply those rules to our callset.

Earlier, we looked at the QualByDepth annotation, which is captured in VCF records as QD. The plots we made showed that variants with a QD value below 2 are most likely not real, so let's filter out sites with QD < 2 using the GATK tool `VariantFil tration`, as follows. We give it a reference genome (`-R`), the VCF file of variants we want to filter (`-V`), and we specify an output filename (`-O`). In addition, we also provide a filtering expression stating the filtering logic to apply (here, `QD < 2.0` means reject any variants with a QD value below 2.0) and the name to give to the filter (here, `QD2`):

```
# gatk VariantFiltration \
    -R ref/ref.fasta \
    -V vcfs/motherSNP.vcf.gz \
    --filter-expression "QD < 2.0" \
    --filter-name "QD2" \
    -O sandbox/motherSNP.QD2.vcf.gz
```

This produces a VCF with all the original variants; those that failed the filter are annotated with the filter name in the `FILTER` column, whereas those that passed are marked `PASS`:

```
# zcat sandbox/motherSNP.QD2.vcf.gz | grep -v '##' | head -3
#CHROM POS      ID      REF      ALT      QUAL      FILTER  INFO     FORMAT  NA12878
20     61098    .       C        T        465.13    PASS
AC=1;AF=0.500;AN=2;BaseQRankSum=0.516;ClippingRankSum=0.00;DP=44;
ExcessHet=3.0103;FS=0.000;MQ=59.48;MQRankSum=0.803;QD=10.57;
ReadPosRankSum=1.54;SOR=0.603GT:AD:DP:GQ:PL  0/1:28,16:44:99:496,0,938
20     61795    .       G        T        2034.16   PASS
AC=1;AF=0.500;AN=2;BaseQRankSum=-6.330e-01;ClippingRankSum=0.00;DP=60;
ExcessHet=3.9794;FS=0.000;MQ=59.81;MQRankSum=0.00;QD=17.09;
ReadPosRankSum=1.23;SOR=0.723          GT:AD:DP:GQ:PL      0/1:30,30:60:99:1003,0,1027
```

The two variant records shown above both passed the filtering criteria, so their `FIL TER` column shows `PASS`. Try viewing more records in the file to find some that have been filtered. You can do this in the terminal or in IGV, as you prefer.

Now, what if we wanted to combine filters? We could complement the QD filter by also filtering out the calls with extremely high DP because we saw those are heavily associated with false positives. We could run `VariantFiltration` again on the previous output with the new filter, or we could go back to the original data and apply both filters in one go like this:

```
# gatk VariantFiltration \
    -R ref/ref.fasta \
    -V vcfs/motherSNP.vcf.gz \
```

```
--filter-expression "QD < 2.0 || DP > 100.0" \
--filter-name "lowQD_highDP" \
-O sandbox/motherSNP.QD2.DP100.vcf.gz
```

The double pipe (||) in the filtering expression is equivalent to a logical OR operator. You can use it any number of times, and you can also use && for the logical AND operator, but beware of building overly complex filtering queries. Sometimes, it is better to keep it simple and split your queries into separate components, both for clarity and for testing purposes.

The GATK hard-filtering tools offer various options for tuning these commands, including deciding how to handle missing values. Keep in mind, however, that hard filtering is a relatively simplistic filtering technique, and it is not considered part of the GATK Best Practices. The advantage of this approach is that it is fairly intuitive if you have a good understanding of what the annotations mean and how they are calculated. In fact, it can be very effective in the hands of an expert analyst. However, it gives you a lot of dials to tune, can be overwhelming for newcomers, and tends to require new analytical work for every new project because the filtering thresholds that are appropriate for one dataset are often not directly applicable to other datasets.

We included hard filtering here because it's an accessible way to explore what happens when you apply filters to a callset. For most "real" work, we recommend applying the appropriate machine learning approaches covered in Chapter 6.

 This concludes the hands-on portion of this chapter, so you can stop your VM now, as discussed in Chapter 4. We don't recommend deleting it, because you will need it in the next chapter.

Introducing the GATK Best Practices

It's all well and good to know how to run individual tools and work through a slice of an example as we did in this chapter, but that's very different from doing an actual end-to-end analysis. That's where the GATK Best Practices come in.

The GATK Best Practices are workflows developed by the GATK team for performing variant discovery analysis in high-throughput sequencing data. These workflows describe the data processing and analysis steps that are necessary to produce high-quality variant calls, which can then be used in a variety of downstream applications. As of this writing, there are GATK Best Practices workflows for four standard types of variant discovery use cases—germline short variants, somatic short variants, germline copy-number variants, and somatic copy-number alterations—plus a few specialized use cases—mitochondrial analysis, metagenomic analysis, and liquid biopsy analysis. All GATK workflows currently designated "Best Practices" are designed to operate on DNA sequencing data, though the GATK methods development team has recently started a new project to extend the scope of GATK to the RNAseq world.

The GATK Best Practices started out as simply a series of recommendations, with example command lines showing how the tools involved should typically be invoked. More recently, the GATK team has started providing reference implementations for each specific use case in the form of scripts written in WDL, which we cover in detail in Chapter 8. The reference implementation WDLs demonstrate how to achieve optimal scientific results given certain inputs. Keep in mind, however, that those implementations are optimized for runtime performance and cost-effectiveness on the computational platform used by Broad Institute, which we use later in this book, but other implementations might produce functionally equivalent results.

Finally, it's important to keep in mind that although the team validates its workflow implementations quite extensively, at this time it does so almost exclusively on human data produced with Illumina short read technology. So if you are working with different types of data, organisms, or experimental designs, you might need to adapt certain branches of the workflow as well as certain parameter selections, values, and accessory resources such as databases of previously identified variants. For help with these topics, we recommend consulting the GATK website and support forum, which is populated by a large and active community of researchers and bioinformaticians as well as the GATK support staff, who are responsive to questions.

Best Practices: Evolution and Deviations

The term *GATK Best Practices* carries specific meaning and corresponds to specific actionable recommendations that change over time. If someone hands you a script and tells you, "This runs the GATK Best Practices," start by asking which version of GATK it uses, when it was written, and what key steps it includes. Both the software and the usage recommendations evolve in step with the rapid pace of technological and methodological innovation in the field of genomics; so, what was Best Practice last year (let alone in 2010) might no longer be applicable.

Furthermore, any GATK workflow that has been significantly adapted or customized, whether for performance reasons or to fit a use case that the GATK team does not explicitly cover, should be referred to as "based on" or "adapted from" GATK Best Practices. When in doubt about whether a particular customization constitutes a significant divergence, you are welcome to reach out to the GATK support team, which will review the available information and help you evaluate its significance and possible consequences for your work.

Best Practices Workflows Covered in This Book

In this book, we cover the following GATK Best Practices workflows in detail, as represented in Figure 5-12:

- Germline short sequence variants (Chapter 6)
- Somatic short sequence alterations (Chapter 7)
- Somatic copy-number alterations (Chapter 7)

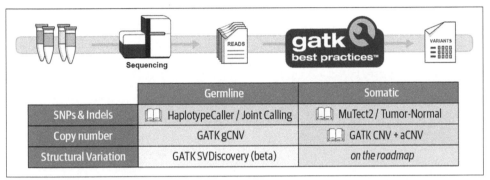

Figure 5-12. Table of standard variant discovery use cases covered by GATK Best Practices.

Other Major Use Cases

If you are interested in the following use cases, we suggest that you visit the Best Practices section of the GATK website (*https://oreil.ly/3LqiZ*), which offers documentation for all GATK Best Practices pipelines that are currently available or in development:

- Germline copy number variation
- Structural variation
- Mitochondrial variation
- Blood biopsy
- Pathogen/contaminant identification

Wrap-Up and Next Steps

In this chapter, we introduced you to GATK, the Broad Institute's popular open source software package for genomic analysis. You learned how to compose GATK commands and practiced running them in a GATK container on your VM in GCP. Then, we walked you through an introductory example of germline short variant analysis with `HaplotypeCaller` on some real data, and you got a taste of how to interpret and investigate variant calls with the open source IGV. We also discussed variant filtering and talked through some of the concepts and basic methodology involved. Finally, we introduced the GATK Best Practices workflows, which feature heavily in the next few chapters. In fact, the time has come for you to tackle your first end-to-end genomic analysis: the GATK Best Practices for germline short variant discovery, from unaligned sequence read data all the way to an expertly filtered set of variant calls that can be used for downstream analysis. Let's head over to Chapter 6 and get started!

## CHAPTER 6

GATK Best Practices for Germline Short Variant Discovery

By now, we've introduced all of the major concepts and tools, and we've walked through a few examples, calling variants with the GATK `HaplotypeCaller` and exploring its output. It's time to turn up the dial on the analytical complexity and take a closer look at the three GATK Best Practices workflows.

In this chapter, we explore the most widely used variant discovery pipeline in the genomics community: the GATK Best Practices for germline short variant discovery in multiple samples, which is designed to take advantage of the benefits of including more data in the analysis process. We begin with an overview of the initial data processing and quality control steps that are necessary to render the data suitable for analysis. Then, we'll dive into the details of the joint calling workflow. In the final part of this chapter, we also cover the highlights of an alternate workflow that supports single-sample calling for use when multisample calling is not an option, as in the context of clinical analysis.

Data Preprocessing

As we discussed in the genomics primer, the data produced by the sequencing process is not immediately suitable for performing variant discovery analysis. We need to preprocess that raw output to transform it into the appropriate format and clean up some technical issues. In this section, we outline the key steps and their rationale. However, we don't run through the entire preprocessing workflow in detail, nor do we perform any hands-on exercises. At this point, it is becoming increasingly more common for the sequencing service providers to perform this as an extension of the data-generation process, and rarer for researchers to need to do this work themselves, so we would rather focus on later stages for which a deeper exploration will likely be of more practical utility to you.

As Figure 6-1 summarizes, the preprocessing workflow consists of three main steps: mapping reads to the genome reference, marking duplicates, and recalibrating base quality scores. In practice, a full pipeline implementation will typically involve some additional "plumbing" steps. For full details, we recommend consulting the Broad Institute's reference implementations for whole-genome processing (*https://oreil.ly/ XvpGZ*) and for exome processing (*https://oreil.ly/YThVc*).

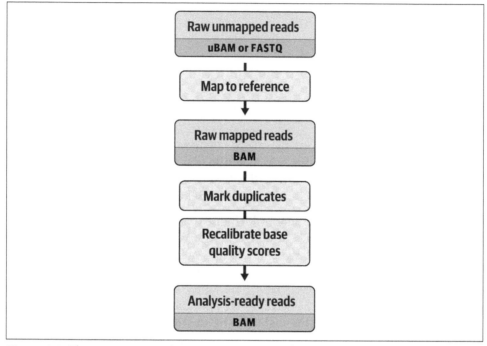

Figure 6-1. The main steps in the preprocessing workflow.

 Old versions of the preprocessing included an additional step called *Indel Realignment*, which corrected for alignment errors in the mapping step. That step is no longer needed if you use a graph-based variant caller for calling short variants, as discussed in the section on the HaplotypeCaller tool in Chapter 5.

The preprocessing workflow is designed to operate independently on data from individual samples. The per-sample data is usually further divided into distinct subsets called *read groups*. As a reminder, read groups correspond to the intersection of DNA libraries and flowcell lanes generated through multiplexing, as explained in Chapter 2.

Some sequence providers collapse multiple read groups into a single FASTQ file, with the very unfortunate consequence that you lose the ability to distinguish read groups. This is bad for two reasons. First, it forces you to process a large monolithic file through the mapping step, which is going to be a performance bottleneck in the pipeline, requiring you to do extra work in the workflow implementation. Second, it hamstrings the base recalibration algorithm, which prefers to operate per read group for reasons that we get to shortly.

In contrast, keeping read group data in separate files in the initial stages allows us to process those read groups independently through the mapping step, which is computationally more favorable. We'll bring those read groups together into a single file for each step when we run the duplicate marking step.

We apply the same preprocessing steps for almost all sequencing data that are destined for variant calling, regardless of the type of variants we're looking for, with a few exceptions. For example, we use a different mapper for RNAseq in order to deal with intronic regions appropriately. Fortunately, if we're looking for multiple types of variants in the same data, we need to do this preprocessing only once; the resulting BAM file is suitable for input to all the GATK Best Practices variant discovery workflows.

Mapping Reads to the Genome Reference

Because the sequencer spits out individual reads without any indication of where each read belongs in the genome, the first essential step in the workflow is to map reads to the reference genome and assign an exact position for each base in the read. As a result, mapping is sometimes called *alignment* instead, but we prefer to use the term *mapping* because it is more reflective of the primary value we draw from this step: identifying where the read belongs in the genome. As you'll see later in this chapter, we don't completely trust the fine local alignment provided by the mapper. It's a bit like the GPS navigation system in a car: you can usually trust it to get you to your desired street location, but you wouldn't trust it to parallel park the car along the curb.

Many mappers are available, and in theory you could choose to use whatever mapper conforms to the main requirements of the GATK Best Practices guidelines in terms of inputs and outputs, and produces high-quality results. However, in practice, the GATK development team currently recommends using a specific tool, bwa mem (*https://arxiv.org/abs/1303.3997*), in order for your data to be compatible for cross-analysis with other datasets, for the reasons we outlined in "Functional Equivalence Pipeline Specification" on page 51. It is possible that this recommendation might change in the future.

Here's an example of a simple command used to map sequence data with bwa mem on an interleaved FASTQ file, which contains both forward and reverse reads for each pair, sorted by query name (you don't need to run this on your VM; it's just for illustration):

```
bwa mem -M -t 7 -p reference.fasta unmapped_reads.fq > mapped_reads.sam
```

This command takes the FASTQ file and produces an aligned SAM format file containing the mapped read data. Each read record now includes read mapping position and alignment information as well as additional metadata. The reads are sorted in the same order in the output as they were in the input FASTQ file.

 In previous chapters, you've seen us use a prompt with $ to indicate commands to run on your VM and a prompt with # to indicate a command to run inside a Docker container. When we show you sample commands that don't start with $ or #, these commands are just for your reference, and we don't intend you to run them directly.

In the Broad Institute's production implementation of the GATK Best Practices, the unmapped read data is stored in unmapped BAM files instead of FASTQ. As we noted in Chapter 2, read data can be stored in BAM format without any mapping information. The advantage of using this format is that it allows us to store metadata associated with the sequence (such as read group) within the same file, rather than relying on additional files for that purpose. This simplifies data management operations and reduces the chance that metadata might be lost or associated with the wrong data. This does, however, add a few steps to the pipeline implementation, which you can see in action in the reference implementation workflow (*https://oreil.ly/qw1j-*).

When processing sample data that is distributed into multiple read groups, we run this step individually per read group, which speeds up processing if you are able to run the read group data in parallel on multiple machines. In addition, because mapping operates on each read independently of any others, it can be further parallelized at the CPU level. The example command we provided specifies that the process should use seven threads (-t 7) to maximize efficiency on a common type of machine with eight cores available. However, you can parallelize this step to a much higher degree if you have specialized hardware available. That is how computing platforms that offer accelerated genomics pipelines such as DRAGEN can derive major speed gains from boosting parallelization of the mapping step through specialized hardware such as FPGAs (see Chapter 3 for an explanation of these terms).

As soon as we have generated the mapped data in BAM (or equivalent) format, we could technically go straight to the variant calling step. However, we know there are some technical sources of error that we can deal with at this stage: the presence of duplicate reads and of systematic biases in the assignment of base quality scores. Let's apply the recommended mitigation strategies before we try to call variants, as prescribed by the GATK Best Practices.

Marking Duplicates

Duplicate reads are reads that are derived from the same original physical fragment in the DNA library, originating at two different stages of the preparation and sequencing processes. During library preparation itself, if the protocol involves a PCR amplification step, each cycle of PCR will produce increasing amounts of duplicate fragments. This used to be necessary to generate enough starting material for the sequencing process, but recent improvements in the technology have made PCR amplification unnecessary in most modern protocols, so that source of duplicates has been significantly reduced. The other major source of duplicates, during the sequencing process, is caused by various optical confusions—for example, when a single DNA cluster on the Illumina flowcell is called as two distinct clusters and treated as two different reads.

We consider these duplicates to be nonindependent observations, so we want to take only one representative of each group into account in downstream analyses. To that end, we'll use a Picard tool called MarkDuplicates that identifies groups of duplicates, scores them to determine which one has the highest sequence quality, and marks all the other ones with a tag that allows downstream tools (such as variant callers) to ignore them, as illustrated in Figure 6-2.

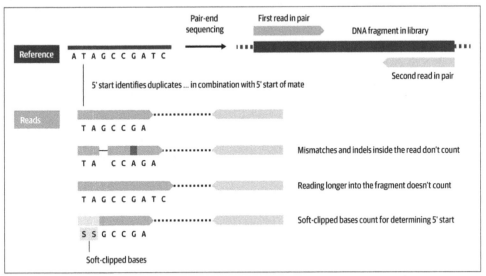

Figure 6-2. Reads marked as duplicates because they originated from the same DNA fragment in the library.

This processing step is performed per sample across read groups, if applicable. When processing sample data that is distributed into multiple read groups, this is the step where we consolidate the read group data files into a single per-sample file by feeding all read group input files to the MarkDuplicates tool. Conveniently, the tool will produce a single sorted file containing the data from all the read groups. This means that we can avoid running the consolidation in a separate step. Here's an example showing what the command looks like when run on data from three read groups belonging to the same sample:

```
gatk MarkDuplicates \
    -R reference.fasta \
    -I mapped_reads_rg1.bam \
    -I mapped_reads_rg2.bam \
    -I mapped_reads_rg3.bam \
    -O sample_markdups.bam
```

The BAM file output by the duplicate marking step still contains all of the same reads as before, but now some of them have been tagged, and downstream tools can choose to use them or ignore them, depending on what is appropriate for their purposes, as shown in Figure 6-3.

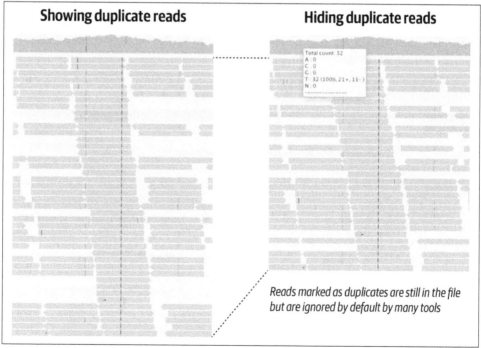

Figure 6-3. The effect of duplicate marking visualized in Integrated Genome Viewer.

This step has long constituted a major bottleneck in traditional pipelines because it involves making a large number of comparisons between all of the read pairs belonging to the sample, across all of its read groups. It also requires a sorting operation because it operates across all reads belonging to the sample. To address this issue, the GATK team ported `MarkDuplicates` in the GATK4 framework using Spark to speed up its operation, which yields major performance gains when using multicore CPUs for processing. The `MarkDuplicatesSpark` tool has not yet been integrated into the reference implementation of the Best Practices pipeline, but that should happen in the near future.

Recalibrating Base Quality Scores

The final processing step we need to apply to our data aims to detect and correct for patterns of systematic errors in the base quality scores, which you'll recall are the confidence scores emitted by the sequencer for each base. Base quality scores play an important role in weighing the evidence for or against possible variant alleles during the variant discovery process for short variants, so it's important to correct any systematic bias observed in the data. Such biases can originate from biochemical processes during library preparation and sequencing, from manufacturing defects in the flowcell chips, or instrumentation defects in the sequencer.

Base Quality Score Recalibration (BQSR) involves collecting covariate statistics from all base calls in the dataset, building a model from those statistics, and applying base quality adjustments to the dataset based on the resulting model.

Figure 6-4 shows two plots that help us understand the effect of BQSR. In panel A, we're comparing the base qualities reported by the instrument (Reported Quality) against the base qualities that we derive from gathering all the observations with the same reported quality and seeing how many errors there are (Empirical Quality). For example, before applying the BQSR correction to the data (Original series, pink/gray), we find that a base with a quality of 30 according to the instrument should in fact be given a quality of 25, according to the quality assessed by counting mismatches in this group. This tells us that the sequencer was overly confident of its own accuracy for that subset of data, and indeed we observe that this is the case for most of the bases in this particular dataset, although there are a few bins where this trend is reversed. After applying the BQSR correction, if we run the modeling procedure again on the recalibrated data (Recalibrated series, blue/black), all of the dots align on the diagonal, which tells us that there is no longer a gap between the qualities assigned to the bases and the observed quality.

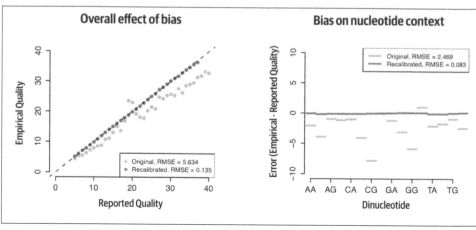

Figure 6-4. Visualizing the effect of BQSR.

Note that panel A in Figure 6-4 shows overall effect, which is based on calculations from multiple covariates. In panel B, we're looking at one of those covariates, the dinucleotide context. This takes into account the sequence context of each base (specifically, the two bases before the base of interest), which we know can affect the accuracy of sequencing. This plot shows the average error in the base quality assignment for each of the possible dinucleotides (note that the axis legend does not show all labels; use the order A, C, G, T to identify the missing labels), where Error is defined as the difference between the quality estimated by the model (Empirical) and the quality reported by the instrument (Reported). Negative values mean that the instrument was overconfident; for example, we can say based on this plot that in this dataset, a base called after CG typically had a quality score that was inflated by ~7.5 before recalibration compared to what it should have been. After recalibration, the Error goes to zero for all bins, which means there is no longer a difference between the assigned base qualities and those calculated by modeling.

The patterns of base quality biases that we find in the data are specific to each library and flowcell, so they must be modeled anew every time; it is not possible to create a model on one dataset and apply it subsequently to other data. Here is an example command for building the model:

```
gatk BaseRecalibrator \
    -R reference.fasta \
    -I sample_markdups.bam \
    --known-sites known_variation.vcf \
    -O recal_data.table
```

This command takes the reference genome (*reference.fasta*) and the BAM file containing the mapped read data that has already been processed through duplicate marking (*sample_markdups.bam*) as well as a VCF file of variants that have been previously reported for the organism of interest (*known_variation.vcf*). The latter plays an important role in the recalibration model; the algorithm uses the known sites to mask out positions that it should not use in building the model. For a complete explanation of why and how this works, see the BQSR method documentation (*https://oreil.ly/B69n1*). The output of the command is a text file containing recalibration tables produced by the modeling process, which we then use to apply the model to the data. Here is an example command for applying the recalibration:

```
gatk ApplyBQSR \
    -R reference.fasta \
    -I sample_markdups.bam \
    --bqsr-recal-file recal_data.table \
    -O sample_markdups_recal.bam
```

This produces the final output (*sample_markdups_recal.bam*), a BAM file containing the read data with recalibrated base qualities that is ready for analysis.

This processing step is performed per sample, but the algorithm itself will differentiate between read groups because many of the biases it tracks are read group specific. The initial statistics collection can be parallelized by scattering across genomic coordinates, typically by chromosome or batches of chromosomes, but this can be broken down further to boost throughput if needed. Gathering per-region statistics into a single genome-wide model of covariation cannot be parallelized, but it is computationally trivial and therefore not a bottleneck. The final application of recalibration rules is best parallelized in the same way as the initial statistics collection, over genomic regions, then followed by a final file merge operation to produce a single analysis-ready file per sample.

Joint Discovery Analysis

Finally, the real fun begins! We're going to call variants on a cohort of multiple people. But before we get into the weeds of how it works, let's take a few minutes to talk about why we do it this way.

When it comes to understanding germline variation, a single person's genome is rarely useful in isolation. Most research questions in this area will benefit from looking at data from multiple people, whether it's a small number, as in the case of the investigation of disease inheritance from parents to their children, or a much larger number, as in the case of population genetics. So, from a scientific standpoint, we're usually going to want to analyze multiple samples together. In addition, there are technical advantages to grouping data from multiple samples, mostly related to statistical power and compensation for technical noise.

Overview of the Joint Calling Workflow

In the early days of variant discovery, multisample analysis was done by running variant calling on each sample independently and then combining the resulting lists of variants. This approach had several weaknesses, including that it didn't account for the difference between "no variant call because no variation" versus "no variant call because no data." Joint analysis solves this problem because it takes only one sample with a variant of interest for a call to be made at the corresponding site across all samples. The output will then contain detailed genotyping information for all samples, allowing us to make the necessary distinctions, as demonstrated in Figure 6-5.

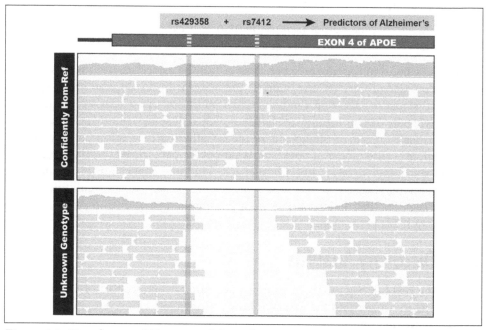

Figure 6-5. Sites that would be omitted from the VCF in a single-sample callset.

The other big advantage of joint analysis is that it allows us to boost the statistical power of the variant calling calculations by increasing the amount of data that we can take into account. This comes into play in two places: at the initial stage of calculating the confidence score of a particular variant call, and a little later, at the stage of filtering variants. The fundamental principle at play here is that when we don't have much data to work with, we are more vulnerable to technical artifacts. For example, consider the first sample shown in Figure 6-6. We have only a handful of reads covering a particular site in that sample, and we see only one or two reads that carry what looks like a variant allele, so we're going to be very skeptical because that could easily be a library preparation artifact or a sequencing error. We might emit a variant call, but with a very low score that is likely to get it thrown out in filtering. However, when we

---

look at data from multiple samples, we find that we see the same potential variant allele in another sample, so we'll be more inclined to believe there is real variation here, and we'll give the variant a better score.

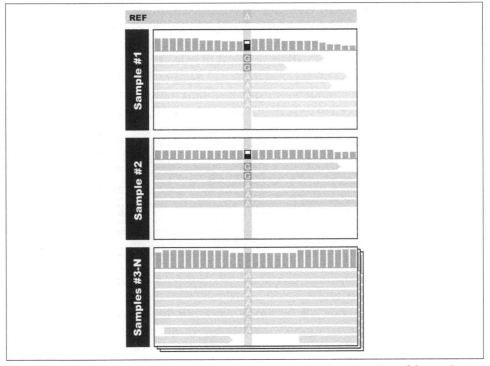

Figure 6-6. Seeing concordant evidence in multiple samples boosts our confidence that there is real variation.

 Boosting confidence when we see a variant in multiple samples does not mean we penalize those that show up in only one sample in a cohort, which are called *singletons*. The joint calling algorithm is designed to avoid penalizing singletons, so a singleton that is well supported by read data will still be called confidently.

In addition to this signal-boosting effect, which acts at the stage of calculating confidence scores, we also see important benefits later on during filtering. Having data from many samples compensates for fluctuations in data quality, reduces the effect of extreme outliers, and stabilizes the distribution of variant context annotations, a type of metadata collected during the variant calling process that we discuss in more detail in the filtering section. The benefits of joint calling for filtering are especially obvious in cohorts of up to a few hundred samples; then the gains begin to taper off, providing diminishing returns as you keep adding more samples. You can read more about this pattern in the HaplotypeCaller preprint (*https://oreil.ly/Sdcm3*).

So how does it work? In earlier versions of the variant discovery phase, you had to run the variant caller directly on multiple per-sample BAM files. However, that scaled very poorly with the number of samples, posing unacceptable limits on the size of the study cohorts that could be analyzed in that way. In addition, it was not possible to add samples incrementally to a study; you had to redo all variant calling work when you added new samples. This is called the *N + 1 problem*: imagine you've analyzed a cohort of *N* samples at considerable expense of time and money, and then you add one more sample and need to redo all that work (Figure 6-7). That's a big problem.

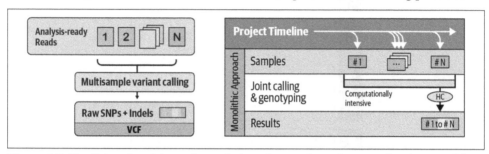

Figure 6-7. Traditional multisample analysis scales poorly and causes the N + 1 problem

Starting with GATK version 3.0, the GATK team introduced a new approach, which decoupled the two internal processes that previously composed variant calling: (1) the initial per-sample collection of variant context statistics and calculation of all possible genotype likelihoods, given each sample by itself, which requires access to the original BAM file reads and is computationally expensive; and (2) the calculation of the posterior probability of each site being variant, given the genotype likelihoods across all samples in the cohort, which is computationally cheap. The tool developers separated these two steps as described shortly, enabling incremental growth of cohorts as well as scaling to large cohort sizes.

This workflow, illustrated in Figure 6-8, is designed to operate on a set of per-sample BAM files that have been preprocessed as described in the first section of this chapter.

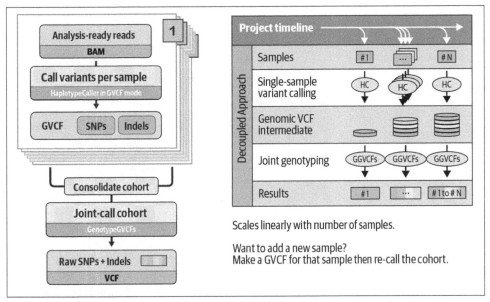

Figure 6-8. The GVCF workflow improves the scaling of joint calling and solves the N + 1 problem.

The very first step is to call variants with `HaplotypeCaller` per sample in *reference confidence mode*, also colloquially called *GVCF mode*, which generates an intermediate file called a GVCF. We cover the GVCF format in more detail in the hands-on exercise portion of this section; for now, just consider that it is an intermediate output that contains information for every single position of the genome territory covered by the analysis, including the many positions for which the tool did not see any evidence of variation.

After we have GVCFs for all the samples that we want to joint-call, we consolidate them into a `GenomicsDB` datastore, which is similar to a database that takes the form of a collection of files. We can then perform joint-calling to produce the raw cohort callset, which contains variant calls made across all samples, with detailed genotype information for each sample. This step, shown in Figure 6-9, runs quite quickly and can be rerun at any point when samples are added to the cohort, thereby solving the N + 1 problem.

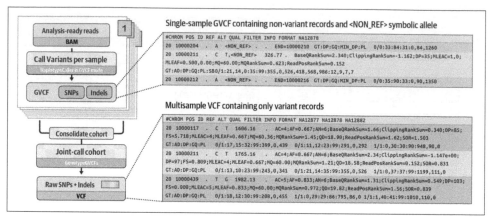

Figure 6-9. Progression from per-sample GVCFs to final cohort VCF.

This cohort-wide analysis empowers sensitive detection of variants even at difficult sites, and produces a squared-off matrix of genotypes that provides information about all sites of interest in all samples considered, which is important for many downstream analyses.

Whew, that was a lot of theory. Are you ready to get your hands back on the keyboard for a few exercises? That's right; it's time to start up your VM again and jump back into your GATK container, just as you did in Chapter 5. You're going to work from within the */home/book/data/germline* directory and use the same sandbox directory that you created in Chapter 5.

Calling Variants per Sample to Generate GVCFs

First, we're going to run `HaplotypeCaller` in GVCF mode on the mother BAM file we played with in Chapter 5. It's almost exactly the same command, except now we're adding the `-ERC GVCF` argument and naming the output with a `.g.vcf` extension (plus `.gz` for compression purposes). Make sure you run this from within the */home/book/data/germline* directory in the container on your VM:

```
# gatk HaplotypeCaller \
    -R ref/ref.fasta \
    -I bams/mother.bam \
    -O sandbox/mother_variants.200k.g.vcf.gz \
    -L 20:10,000,000-10,200,000 \
    -ERC GVCF
```

This produces a GVCF file that contains likelihoods for each possible genotype for the variant alleles, including a symbolic allele, `<NON-REF>` in the `ALT` column. This symbolic allele is present in all records, whether they are variant records or nonvariant records. You can view what the calls look like the same way as before, using the zcat tool, `grep`, to get rid of the header lines, and `head` to limit the number of lines you want to look at:

```
# zcat sandbox/mother_variants.200k.g.vcf.gz | grep -v '##' | head -3
#CHROM  POS      ID     REF    ALT     QUAL     FILTER  INFO     FORMAT   NA12878
20      10000000  .     T      <NON_REF>        .       .
END=10000008    GT:DP:GQ:MIN_DP:PL     0/0:16:48:16:0,48,497
20      10000009  .     A      <NON_REF>        .       .
END=10000009    GT:DP:GQ:MIN_DP:PL     0/0:16:32:16:0,32,485
```

These are nonvariant sites for which `HaplotypeCaller` found no evidence of variation. The first is a block that covers positions 20:10,000,000 to 20:10,000,008, whereas the second covers only position 20:10,000,009. If you look a little further in the file, you'll also find some actual variant calls like this one:

```
20      10000117  .     C      T,<NON_REF>      283.60  .
BaseQRankSum=-0.779;DP=23;ExcessHet=3.0103;MLEAC=1,0;MLEAF=0.500,0.00;
MQRankSum=1.044;RAW_MQandDP=84100,23;ReadPosRankSum=0.154
GT:AD:DP:GQ:PL:SB       0/1:11,12,0:23:99:291,0,292,324,327,652:9,2,9,3
```

Now see what GVCFs look like in the IGV. We supply precomputed GVCF files for all three members of the CEU Trio (Mother, Father, and Son), but you could generate them yourself in the same way as the first from the BAMs we also provide.

Start a new session to clear your IGV screen (File > New Session) and then load the GVCF for each CEU Trio member (*gvcfs/mother.g.vcf*, *gvcfs/father.g.vcf*, *gvcfs/son.g.vcf*) using the paths to the file locations in the book bucket.[1] This time, be sure to provide the index file paths to IGV because it is unable to get them automatically for compressed files. Zoom in on the genomic coordinates 20:10,002,371-10,002,546. You should see something like Figure 6-10.

Notice anything different in the GVCF compared to the VCFs that you saw in Chapter 5? Instead of just seeing a few variant sites, you see many gray blocks. If you click one to examine the details of the call, you'll see that the genotype assignment is homozygous reference, and the position information includes an END coordinate, which indicates that the block covers an interval of one or more bases. These blocks represent intervals where the sample appears to be nonvariant and for which the contiguous sites it covers have similar genotype likelihoods. `HaplotypeCaller` groups nonvariant sites into blocks during the calling process based on their likelihoods as a way of compressing the information in the GVCF without losing useful information in areas where there is more variability in the quality of the available information.

---

1 The full file paths are *gs://genomics-in-the-cloud/v1/data/germline/gvcfs/mother.g.vcf.gz*, *gs://genomics-in-the-cloud/v1/data/germline/gvcfs/mother.g.vcf.gz.tbi*, *gs://genomics-in-the-cloud/v1/data/germline/gvcfs/father.g.vcf.gz*, *gs://genomics-in-the-cloud/v1/data/germline/gvcfs/father.g.vcf.gz.tbi*, *gs://genomics-in-the-cloud/v1/data/germline/gvcfs/son.g.vcf.gz*, and *gs://genomics-in-the-cloud/v1/data/germline/gvcfs/son.g.vcf.gz.tbi*.

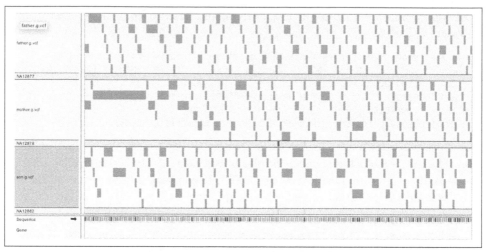

Figure 6-10. GVCFs viewed in IGV show tiled nonvariant blocks.

The HaplotypeCaller calculates the likelihoods for these GVCF blocks using an abstract nonreference allele that represents all the possible variation that it did not observe; they serve as an estimate of how confident we are that these sites are truly nonvariant. Later, the joint calling step will retain only sites that are likely enough to be variant compared to the reference in at least one sample in the cohort. For variant sites, the nonreference allele serves to estimate the likelihood that a new allele seen in one sample was missed in the other samples.

Consolidating GVCFs

For the next step, we need to consolidate the GVCFs into a GenomicsDB datastore in order to improve scalability and speed in the next step, joint genotyping. That might sound complicated, but it's actually very straightforward. Let's try running this on two of the samples we have available:

```
# gatk GenomicsDBImport \
    -V gvcfs/mother.g.vcf.gz \
    -V gvcfs/father.g.vcf.gz \
    --genomicsdb-workspace-path sandbox/trio-gdb \
    --intervals 20:10,000,000-10,200,000
```

This will create a new directory called *trio-gdb* containing the datastore contents. We named our datastore trio-gdb aspirationally, because, ultimately, we want to include all three samples from the family in the analysis, but we're pretending that the son's sample hasn't been completed yet—some kind of delay happened in the lab.

The way the consolidated GVCF information is represented in the database makes it difficult to examine directly, so if we want to visualize it, we need to extract the combined data from the `GenomicsDB` database using `SelectVariants`:

```
# gatk SelectVariants \
    -R ref/ref.fasta \
    -V gendb://sandbox/trio-gdb \
    -O sandbox/duo_selectvariants.g.vcf.gz
```

This time, when you run `zcat` to view the contents, you'll see that you now have two columns with genotype information, one for each of the samples that we originally included:

```
$ zcat sandbox/duo_selectvariants.g.vcf.gz | grep -v '##' | head -3
#CHROM POS     ID      REF     ALT     QUAL    FILTER INFO     FORMAT  NA12877 NA12878
20      10000000        .       T       <NON_REF>       .       .
END=10000001    GT:DP:GQ:MIN_DP:PL      ./.:15:39:15:0,39,585 ./.:16:48:16:0,48,497
20      10000002        .       G       <NON_REF>       .       .       .
GT:DP:GQ:MIN_DP:PL      ./.:15:39:15:0,39,585 ./.:16:48:16:0,48,497
```

This extraction step is not part of the routine pipeline, but it can be helpful for understanding what happens under the hood when we combine GVCFs from multiple samples. It also comes in handy when we need to troubleshoot unexpected or problematic results.

OK, so we just combined the GVCFs from the mother and father samples. Now, let's imagine that the son's sample was just delivered, and we want to add it to our datastore without having to rerun everything. We're going to use the `GenomicsDBImport` tool again, but this time we're going to make it update the existing datastore (rather than creating a new one) by using `--genomicsdb-update-workspace-path` instead of `--genomicsdb-workspace-path` to specify the datastore:

```
# gatk GenomicsDBImport \
    -V gvcfs/son.g.vcf.gz \
    --genomicsdb-update-workspace-path sandbox/trio-gdb
```

You can check that the datastore was updated with the son's samples by running the same basic extraction command as before:

```
# gatk SelectVariants \
    -R ref/ref.fasta \
    -V gendb://sandbox/trio-gdb \
    -O sandbox/trio_selectvariants.g.vcf.gz
```

Run `zcat` again to view the contents of the new file (which we named slightly differently), and you should now see three columns with genotype information:

```
$ zcat sandbox/trio_selectvariants.g.vcf.gz | grep -v '##' | head -3
#CHROM POS     ID      REF     ALT     QUAL    FILTER INFO     FORMAT  NA12877 NA12878 NA12882
20      10000000        .       T       <NON_REF>       .       .
END=10000001    GT:DP:GQ:MIN_DP:PL      ./.:15:39:15:0,39,585 ./.:16:48:16:0,48,497
./.:20:60:20:0,60,573
20      10000002        .       G       <NON_REF>       .       .       .
GT:DP:GQ:MIN_DP:PL      ./.:15:39:15:0,39,585 ./.:16:48:16:0,48,497    ./.:20:57:20:0,57,855
```

Success! You just defeated the $N + 1$ problem by adding samples incrementally to a GenomicsDB datastore.

This ability to add samples incrementally to a GenomicsDB datastore is subject to a few limitations. The tool documentation (*https://oreil.ly/kT8-J*) recommends making a backup of the datastore before proceeding to update it, because any interruptions of the process could corrupt the data and leave it in an unusable state. In addition, it's not possible to specify new intervals when updating an existing datastore; the tool will automatically use the intervals that were originally used to create it. That's why our second GenomicsDBImport command didn't include any intervals.

It's important to understand that the process of consolidating GVCFs that you just worked through is not equivalent to the joint genotyping step, which we tackle in a minute. We can't consider variants in the resulting merged GVCF as having been called jointly. No analysis was involved; this is really just a matter of organizing the information differently for logistical reasons.

 Prior to GATK4, this step was done through hierarchical merges with a tool called CombineGVCFs. That tool is included in GATK4 for legacy purposes, but performance is far superior when using GenomicsDBImport, as in the previous example.

Applying Joint Genotyping to Multiple Samples

It's time to apply the joint calling tool, GenotypeGVCFs, to the consolidated GVCF or GenomicsDB datastore. The tool will look at the available information for each site from both variant and nonvariant samples across all the samples in the cohort and will produce a VCF containing only the sites that it found to be variant in at least one sample:

```
# gatk GenotypeGVCFs \
    -R ref/ref.fasta \
    -V gendb://sandbox/trio-gdb \
    -O sandbox/trio-jointcalls.vcf.gz \
    -L 20:10,000,000-10,200,000
```

Finally, we have our multisample callset! If we compare that final VCF to the original GVCFs, we've now removed sites that were nonvariant in all the samples that we joint-called, and the symbolic <NON_REF> allele is gone, as well, because it has served its purpose:

```
# zcat sandbox/trio-jointcalls.vcf.gz | grep -v '##' | head -3
#CHROM  POS    ID    REF   ALT   QUAL  FILTER  INFO    FORMAT NA12877 NA12878 NA12882
20      10000117    .     C     T     1624.94 .
AC=4;AF=0.667;AN=6;BaseQRankSum=1.91;DP=85;ExcessHet=3.9794;FS=5.718;MLEAC=4;MLEAF=0.667;
MQ=60.25;MQRankSum=1.04;QD=19.12;ReadPosRankSum=2.21;SOR=1.503
GT:AD:DP:GQ:PL     0/1:17,15:32:99:399,0,439
0/1:11,12:23:99:291,0,292       1/1:0,30:30:90:948,90,0
```

```
20      10000211      .      C      T      1783.94 .
AC=4;AF=0.667;AN=6;BaseQRankSum=2.63;DP=96;ExcessHet=3.9794;FS=0.809;MLEAC=4;MLEAF=0.667;
MQ=60.00;MQRankSum=0.00;QD=18.78;ReadPosRankSum=-4.210e-01;
SOR=0.831 GT:AD:DP:GQ:PL      0/1:13,10:23:99:243,0,341
0/1:21,14:35:99:355,0,526          1/1:0,37:37:99:1199,111,0
```

The calls made by `GenotypeGVCFs` and `HaplotypeCaller` run in multisample mode should mostly be equivalent, especially as cohort sizes increase. However, borderline calls can have marginal differences; for instance, low-quality variant sites, in particular for small cohorts with low coverage. Since the release of GATK 4.1, this should be less of a problem, thanks to the introduction of a new QUAL score model (available to some earlier versions via the `-new-qual` flag). Alternatively, it is possible to perform joint genotyping directly with `HaplotypeCaller` on small cohorts.

This is what the command to run `HaplotypeCaller` jointly on the three samples would look like:

```
# gatk HaplotypeCaller \
    -R ref/ref.fasta \
    -I bams/mother.bam \
    -I bams/father.bam \
    -I bams/son.bam \
    -O sandbox/trio_jointcalls_hc.vcf.gz \
    -L 20:10,000,000-10,200,000
```

You can run it if you want to compare the calls to the joint-called results we generated earlier.

Now let's circle back to the locus we examined at the start. Load *sandbox/trioGGVCF.vcf* into IGV and navigate to the genomic coordinates 20:10,002,376-10,002,550. As shown in Figure 6-11, you should see the variant call on the top track and the genotypes for each of the three samples below.

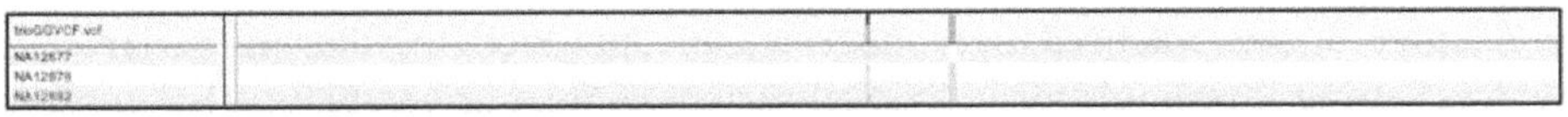

Figure 6-11. Variant call with genotype assignment for the three samples.

Now we have genotypes at this site for both the father (NA12877) and son (NA12882) in addition to the mother. If you click each one, you'll see both the mother and the son were called homozygous variant, yet the father was called homozygous reference; that is, nonvariant. Given the familial relationship for the three samples, what do you think about this set of genotype calls?

If you're thinking it doesn't make sense, you're right! The evidence we have for both the mother and the child being truly homozygous variant is fairly strong, but then the father is extremely unlikely to be homozygous reference because that's a flagrant Mendelian violation (i.e., it breaks Mendel's rules of genetic inheritance). That would mean the son received one variant allele from his mother and one reference allele from his father, and then experienced an early mutation that produced a matching *de novo* variant—not impossible, but very unlikely. So two possibilities exist here; either the daddy is not the daddy (it happens more often than you want to know), or the tool made an error in one or more of the genotypes. We happen to know from other sources that NA12877's paternity is not in question, so we need to accept that the tool made a mistake. This is not so unusual when there is some ambiguity in the data, especially when homopolymers are involved, as is the case here.

In fact, we ran a few tests with other pipelines and found that other variant callers also stumbled on the genotype assignments for this variant in the same data. Some of them also called the father homozygous reference, whereas others called it heterozygous. So this is a site in the genome where joint calling allows us to make a good variant call, in that we are confident that one or more samples in the group show some variation there, but some of the individual genotypes might be questionable. How do we deal with this? Well, we're going to put this problem in the fridge for now, because our first order of business after calling is filtering out the bad variant calls, which does not involve taking individual genotypes into account. However, when that's done, we'll circle back to deal with the genotypes using a process called *Genotype Refinement*.

Filtering the Joint Callset with Variant Quality Score Recalibration

At this point in the variant discovery process, we have in hand a set of raw variant calls, but we know that some of our raw calls are most probably artifacts, given that the caller is explicitly designed to be very permissive. That is why we now need to filter the variant calls to distinguish those that are most likely to be real variants in the biological material from those that are most probably just artifacts introduced by biochemical processes or technical errors.

In Chapter 5, we practiced hard filtering, an approach to filtering variants that involves setting thresholds for individual variant context annotations. That approach can produce reasonable results in the hands of an experienced analyst, but it is labor intensive and difficult to extend consistently across different studies and datasets. For a more scalable and standardizable solution, we look to machine learning approaches that enable us to build models of what real variants and artifacts look like in our data, respectively.

Machine Learning in the GATK

Machine learning is a subfield of data science related to statistics. GATK analytical tools all use some kind of statistical technique; in some cases, the key algorithms belong to the machine learning family. Classic GATK machine learning methods that have been around since the early days of GATK include the aforementioned BQSR and Variant Quality Score Recalibration. In Variant Quality Score Recalibration (VQSR), the core algorithm is a Gaussian mixture model that aims to classify variants based on how their annotation values cluster, given a training set of high-confidence variants.

Meanwhile, an increasingly popular subfield of machine learning called *deep learning* uses techniques based on *neural networks*, a type of algorithm that mimics neural pathways in animal brains. Deep learning has been around for a long time, but until recently, neural networks were too computationally intensive to tackle anything more than contrived problems. Now, thanks to recent technological developments, they can tackle much bigger problems, and have become intensely popular as a way to pursue artificial intelligence. Later in this chapter, you'll have the opportunity to try out GATK tools that apply neural networks to classify and filter variants in a single sample.

That being said, new advances in machine learning approaches in genomics are not limited to deep learning. The GATK Best Practices pipeline for germline copy-number variant (gCNV) discovery uses a type of machine learning algorithm called a Probabilistic Graphical Model (PGM) to deliver gCNV calling on either a single sample or a cohort of samples. It outperforms established methods in the field significantly, both in terms of accuracy and of computational scalability. Unfortunately we don't cover it in this book, but you can learn more about it in the relevant GATK documentation (*https://oreil.ly/xJOhO*).

When filtering a multisample callset, the GATK Best Practices recommendation is to use the Variant Quality Score Recalibration (VQSR) method. This method involves using variant context annotations from variants that are present in a training set to model how variants that are likely to be real cluster together. Then, it's a matter of applying the model to all of the variants in our callset of interest. You can think of this as a process in which we're measuring where each variant lies relative to the clusters of real variants across all of the annotation dimensions. The closer a variant is to a cluster, the more likely it is to be real. Of course, this can be difficult to visualize when we take all the annotations into account because there are more than the three that our puny human brains are generally limited to handling at a time. So we usually represent subsets of the model by making pairwise scatter plots of two annotations at a time to visualize the clustering process, as shown in Figure 6-12.

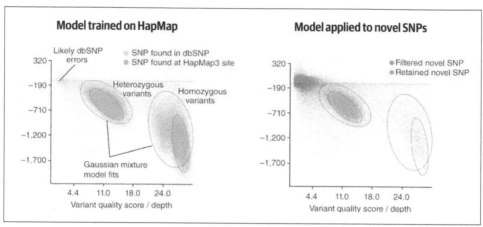

Figure 6-12. Gaussian clusters learned from a training set are applied to novel variant calls.

In practice, the VQSR tools use the model to assign a new confidence score to each variant, called VQSLOD, which is more reliable than the QUAL confidence score originally calculated by the caller because it takes more information into account. Specifically, this new score takes into account all the variant context annotations that would be informative enough for hard filtering, and the resulting score encapsulates the information contributed by every one of those annotations in a single value.

In Figure 6-13, each dot at the bottom is an individual variant, plotted where its annotation value falls on the x-axis. The variants clustering to the left are considered good based on preliminary scoring, whereas those on the right are considered bad. The density curves above them show the positive model (left) and negative model (right), respectively, representing the proportion of variants with corresponding annotation values. The VQSLOD for a given variant is the log ratio of the variant's probabilities of belonging to the positive model and to the negative model, respectively. Here we're showing how this applies to a single annotation; in reality, we'll calculate the VQSLOD score based on multiple annotations, so the tools will do some additional mathematical integration that we don't show here to produce a single unified score.

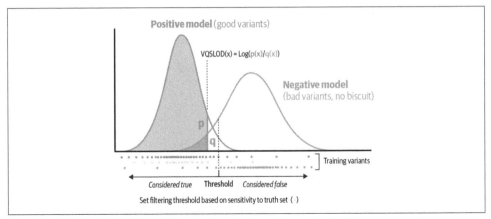

Figure 6-13. How the VQSLOD score is calculated for an individual annotation.

This single unified score is very convenient because it means that all that remains for us to do is to set a filtering threshold for that single score instead of trying to set thresholds for all the annotations separately. Contrast that to what we covered on the topic of hard filtering; you can see this is much handier.

So how do we set a threshold on that one score? Indirectly. There is no absolute scale that tells us what is a good VQSLOD. Instead, given a truth set that overlaps substantially with our callset, we can use a truth sensitivity threshold as a more meaningful proxy for finding the "right" VQSLOD score. This means that we can specify a target sensitivity—for example, 99.9%—and have the tools internally determine whatever VQSLOD would be appropriate for the filtering process to guarantee that 99.9% of the variants we know are true will still be considered real in the resulting filtered callset. In other words, this comes down to saying, "Filter my callset with whatever VQSLOD would give me back 99.9% of the variants that I know are true in this callset." That target sensitivity value itself is not an absolute recommendation, given that different projects might require different levels of sensitivity and precision, but at least it provides a principled basis for the analyst to choose their threshold.

We don't provide hands-on exercises for this step, because the VQSR algorithm is greedy for data—it needs to see all the variant data at once—so it's difficult to create an example case that will run in a short amount of time. Instead, here are example commands for the two steps involved in VQSR.

To build the model and apply it to a joint-called dataset and calculate VQSLOD scores for the variants it contains, we would use VariantRecalibrator, as follows:

```
gatk VariantRecalibrator \
    -R reference.fasta \
    -V jointcalls_hc.vcf.gz \
    --resource:hapmap,known=false,training=true,truth=true,prior=15.0 \
    hapmap_sites.vcf.gz \
```

```
--resource:omni,known=false,training=true,truth=false,prior=12.0 \
1000G_omni2.5.sites.vcf.gz \
--resource:1000G,known=false,training=true,truth=false,prior=10.0 \
1000G_phase1.snps.high_conf.vcf.gz \
--resource:dbsnp,known=true,training=false,truth=false,prior=2.0 dbsnp.vcf.gz \
-an QD -an MQ -an MQRankSum -an ReadPosRankSum -an FS -an SOR \
-mode SNP \
-O output.recal \
--tranches-file output.tranches
```

This is one of the longer GATK commands you're likely to encounter, so it's worth taking a minute to go over what it involves. In addition to the reference genome of the VCF of joint calls, which are its main input, the command takes in a set of resource files (via the `--resource` parameter) that play an essential role in the modeling process. These should be sets of variant calls that have been previously reported and curated to some degree of confidence expressed by the `prior` metric. Here, we show the resources that are used for human genome analysis; for other organisms, you would have to adapt the resources you use accordingly. The `-an` parameters specify the variant context annotations that we want to use in the recalibration process, and here we show the default recommendations provided by the GATK team. The `-mode` parameter controls which kind of variants the tool should recalibrate. Because of some key differences between how SNPs and indels show up in the sequence, we need to recalibrate them in separate passes, so we would run this once in SNP mode and then once again in INDEL mode. Finally, the command will produce two output files: the main recalibration file (*output.recal*), and a tranches file (*output.tranches*) that maps several confidence levels to specific VQSLOD cutoffs.

After we've run that, we still need to choose a confidence level to use as sensitivity threshold and actually generate a filtered VCF for our callset, which we would do by using `ApplyVQSR`, as follows:

```
gatk ApplyVQSR \
  -R reference.fasta \
  -V jointcalls_hc.vcf.gz \
  -O jointcalls_filtered.vcf.gz \
  --truth-sensitivity-filter-level 99.9 \
  --tranches-file output.tranches \
  --recal-file output.recal \
  -mode SNP
```

In this command, we're mostly just feeding in the reference, original data and the two files output by `VariantRecalibrator`, the tranches files, and the recalibration file. We also pass in the confidence level we want to use with the `--truth-sensitivity-filter-level` argument, using 99.9% because it is the recommended default for SNPs in human genomic analysis, but keep in mind that this is something you would need to decide based on your experimental goals. The final output of the procedure will be the filtered callset, which will still contain all the original variant sites, but the ones that fail to pass the VQSLOD threshold will be marked as filtered and ignored in downstream analyses.

The downside of how variant recalibration works is that the algorithm requires high-quality sets of known variants to use as training and truth resources, which for many organisms are not yet available. It also requires quite a lot of data in order to learn the profiles of good versus bad variants, so it can be difficult or even impossible to use on small datasets containing only one or a few samples, on targeted sequencing data, on RNAseq, and on nonmodel organisms. If for any of these reasons you cannot perform variant recalibration on your data, you will need to use either the neural networks–based technique referenced in "Single-Sample Calling with CNN Filtering" on page 173, or hard filtering, which consists of setting flat thresholds for specific annotations and applying them to all variants equally, as described in Chapter 5.

You can learn more about the variant recalibration algorithm from its documentation (*https://oreil.ly/wCLrk*) and see these tools in action in the reference implementation (*https://oreil.ly/Uk7-I*) of the joint genotyping workflow.

Refining Genotype Assignments and Adjusting Genotype Confidence

Now that we have filtered our callset at the site level, we can be reasonably confident that we have the right variant calls, but as we observed earlier, some unresolved ambiguity might remain at the genotype level. For studies that require high-quality genotype information (sooo...most of them) we're going to add on an optional step called *Genotype Refinement* that uses supplemental sources of information like pedigree (i.e., familial relationships) and allele frequencies observed in relevant populations to refine genotype assignments. Where applicable, this will lead us to adjust our confidence in specific genotypes or even change what genotype we assign.

Let's apply this to the two "problem" calls we produced for the mother-father-son trio earlier—and yes, we're back to working hands-on in the GATK container on your VM. In the command that follows, we provide the genotype refinement tool, `Calcula teGenotypePosteriors`, with a pedigree file that describes how the three individuals are related (mother, father, and child) and a sites-only VCF file that contains allele frequencies observed in a large database of known variation called gnomAD (*https://oreil.ly/_t0kq*):

```
# gatk CalculateGenotypePosteriors \
    -V sandbox/trio-jointcalls.vcf.gz \
    -ped resources/trio-pedigree.ped \
    --supporting-callsets resources/af-only-gnomad.vcf.gz \
    -O sandbox/trio-refined.vcf.gz
```

Run that and then load the resulting VCF into IGV and check out the genotype assignment we now have for the father. Zoom in on the genomic coordinates 20:10,002,438-10,002,479. What happened here? Between the trio pedigree and the population file, the tool decided that the father's genotype was highly questionable, so it downgraded the genotype quality score, as depicted in Figure 6-14, to the point that this genotype would be filtered out as being unreliable for downstream analysis.

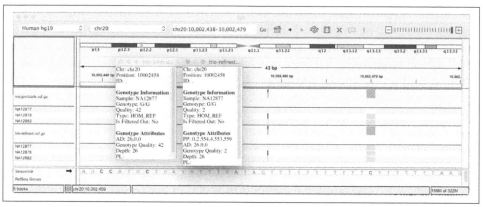

Figure 6-14. Genotype assignments corrected on the basis of pedigree and population priors.

It's important to understand that the tool does not "brute-force" the genotype assignment; if we had a confident genotype call to begin with, the external information would not be sufficient to overturn the caller's initial judgment. However, when we have a low-confidence call, where the caller is essentially telling us "here's my best guess but I'm probably wrong," the supplemental information allows us to resolve the ambiguity and rescue a variant call that might otherwise be unusable for downstream analysis, as in the case of the SNP. Or, as in the case of the indel, it recalibrates the confidence we have in the genotype call.

Next Steps and Further Reading

This concludes the whistle-stop tour of the joint discovery workflow, at least as far as this book is concerned. There is a lot more to learn about how the algorithms and tools that we presented here operate, and about how you can customize these commands to suit your particular data and experimental goals. We recommend that you check out the relevant documentation (*https://oreil.ly/SBnZc*) on the GATK website as well as the reference implementations (*https://oreil.ly/WDqVr*) of the workflows, though you might benefit from waiting until you've worked through the rest of this book to tackle those. If you're not familiar with WDL workflows, you'll especially want to read Chapter 8 through Chapter 11 first, in which we teach you what you need to know to decipher the workflows and use them effectively for your own work.

Before we move on to somatic variant calling, however, let's consider the case of single-sample calling, which is a bit of a niche use case compared to joint calling but is of great importance in clinical genomics.

Single-Sample Calling with CNN Filtering

As discussed earlier, sometimes we are constrained to analyzing a single sample at a time. When that's the case, we don't need to follow the GVCF workflow; instead, we can simply apply the basic command for `HaplotypeCaller`, as shown in the very first exercise in this chapter. However, that leads us to having too little data (specifically, too few variants) for filtering with VQSR, which, as we have noted, is a very greedy algorithm that requires lots of data.

Over the years, the GATK development team tried a variety of algorithms to address this issue. Then, a breakthrough: the advent of "cheap and easy" machine learning tools, and the rise of deep learning in particular, empowered the team to apply a technique called *Convolutional Neural Networks*, or CNN. This new approach proved to be impressively effective for single-sample germline short variant calling, performing well even on indels, which are traditionally the more difficult problem in this space compared to SNVs.

CNNs Applied to Variant Calling

If you're not familiar with CNNs, you're not alone—and the good news is, you don't really need to understand every detail of the theory to use them, just like you don't need to master the intricacies of internal combustion engines in order to drive a car. But it can be helpful to get a sense of what's under the hood, so let's go over the main points together.

First, what's with the name? The *neural networks* part refers to a type of machine learning techniques that are loosely modeled after neural connections in the human brain and specialize in recognizing patterns in numerical data. The *convolution* bit in the CNN name refers to the specific type of mathematical operation used to process data. From the Latin *convolvere*, meaning *roll together*, convolution involves combining signals from different sources of information or from different scales. In the case of the GATK CNN algorithm, this allows the tool to capture features like 10 basepair sequence motifs.

A commonly described application of neural networks is for classifying images. The internet has many examples, and, of course, it's the internet, so the most popular ones compare various kinds of dogs to items that are most decidedly not dogs. This leads to classic memes like "sheepdog or mop," "chihuahua or muffin," "puppy or bagel," and, our personal favorite, "labradoodle or fried chicken," as shown in Figure 6-15, taken from an NPR article (*https://oreil.ly/SAPBd*). It turns out that although it's fairly trivial for a minimally focused human brain to make these distinctions on sight, describing the relevant rules to a computer is quite a bit more challenging.

Figure 6-15. Labradoodle or fried chicken? (Source: Karen Zack, @teenybiscuit)

Doing so involves detecting edges in the picture to identify the minimal silhouette, divested of any unhelpful visual noise, that will allow us to make the distinction we're after. In image analysis, this is done separately for the red, green, and blue channels that images are composed of. The resulting edge maps are then combined into a *tensor*, a data structure that provides a way to stack information from different sources. Then, the tensors representing various images are fed into the neural network and grouped together based on similarities. Long story short, if we provide a training set with images that are labeled one way or another, the algorithm will be able to cross-reference how the labeled images were grouped with the rest and therefore produce labels for all images.

So how does this relate to variant calls? The basic idea is that we have different types of information available about our variant calls: variant context annotation values, reference sequence context, and optionally read data, that we want to use to make a judgment about the veracity of each call. We encode all that information into tensors just like the edge maps in the image analysis case, and feed the tensors to a modeling architecture that can detect predictive features in the bundle of information provided.

This is a deep topic about which there is a lot to learn, but it can be quite…convoluted, so to speak. In the interest of not turning this book into a treatise on machine learning, let's skip straight to how it differs from previous approaches and how we apply the relevant tools in practice.

Overview of the CNN Single-Sample Workflow

At a very high level, the procedure for applying CNN to variant calling appears similar to VQSR, which we covered earlier: the tools build a model of what real variation looks like based on training sets, and when we apply that model to our callset, we get a new confidence score for each of our variants. Under the hood, however, the modeling architecture is wildly different—and quite a bit more complicated. Again, the good news is that you don't need to care too much about how it works at that level, as long as you understand its key benefit: we can include more information than just the variant context annotations in the modeling process. So far, we were limited to using annotations, which are essentially summaries of the original sequence data, for both hard filtering and VQSR.

Now, with CNN, we can also take advantage of what we know about the reference sequence context, and we can even take into account the raw read data itself. That gives us more power to discriminate between real variants and artifacts, especially for indels in difficult areas of the genome where the sequence context is littered with *short tandem repeats* (repetitions of the same few bases, like TATATATATA) and *homopolymers* (runs of the same one base, like AAAAAAAAAA) that are classic confounding factors for variant calling. In practice, we'll refer to the basic approach that uses annotations supplemented with just reference context information as *1-D CNN*, and the somewhat more sophisticated approach that also uses read data as *2-D CNN*.

Another important benefit of how the CNN modeling works is that it allows us to create models that can be applied out of the box to new datasets as long as they are similar enough to the original training data in terms of experimental design and technology (i.e., how the sequencing was done). This is also a big difference compared to the VQSR method, which requires that you train a new model for every new callset that you want to filter. The GATK package includes a set of precomputed models trained on highly validated callsets produced from reference datasets such as SynDip (*https://oreil.ly/WyPhN*), GiaB (*https://oreil.ly/XJlBQ*), and Platinum Genomes (*https://oreil.ly/aW1on*). For convenience, the GATK CNN tools are set to use the appropriate default model (1D or 2D modeling) based on which inputs are provided (or not). That being said, you can train your own models if you prefer.

Let's run through the main two options for running with a precomputed model. You'll see that this method is easy to use even if you don't understand the underlying algorithm in detail. Note that because the CNN tools are very new as of this writing, the examples use a separate set of resource files, so you now need to change our working directory within the GATK container: you were previously in */home/book/data/germline*, so now you need to change your working directory to */home/book/data/cnn*. To be clear, the CNN tools are intended for germline variant discovery; the naming and file bundle structure simply reflects the fact that these tools have a somewhat special status for now:

```
# cd ../cnn
# mkdir sandbox
# ls
bams  ref  resources  sandbox  vcfs
```

As previously, you create a sandbox for collecting the outputs of the commands that you're going to run.

Applying 1D CNN to Filter a Single-Sample WGS Callset

The simplest form, or architecture, of CNN modeling available in GATK is called *1D* for one-dimensional. That term may seem a little misleading because it takes into account variant context annotations and reference sequence context, all of which most of us would probably not think of as representing a single dimension. Nevertheless, as with much of the machine learning jargon, it might be best to just accept it and move on with your life. If it helps, you can think of the dimensions as if you were visualizing the read data in IGV. 1D is a line, which is the reference. 2D is that reference line and also the depth of all the reads.

Conveniently, the GATK package includes a precomputed 1D model that is compatible with our tutorial data. To apply it to our callset of interest, we simply feed our VCF to the CNNScoreVariants tool, which will calculate a new confidence score called CNN_1D for each variant. The precomputed model is used by default unless we explicitly specify a model file on the command line (we demonstrate that usage later in this section):

```
# gatk CNNScoreVariants \
    -R ref/Homo_sapiens_assembly19.fasta \
    -V vcfs/g94982_b37_chr20_1m_15871.vcf.gz \
    -O sandbox/my_1d_cnn_scored.vcf
```

This produces a new VCF file in which our variants have not yet been filtered, just annotated with the new score as well as sensitivity tranche information. To actually apply a filtering rule, we use the same sensitivity-based logic as VQSR, providing known variant callsets as resources and the desired thresholds to the generalized filtering tool FilterVariantTranches:

```
# gatk FilterVariantTranches \
    -V sandbox/my_1d_cnn_scored.vcf \
    -O sandbox/my_1d_cnn_filtered.vcf \
    --resource resources/1000G_omni2.5.b37.vcf.gz \
    --resource resources/hapmap_3.3.b37.vcf.gz \
    --info-key CNN_1D \
    --snp-tranche 99.9 \
    --indel-tranche 95.0
```

We choose 99.9% and 95.0% for the SNP and indel sensitivity thresholds because those are the same recommended defaults as we encountered in the VQSR filtering section, but remember that the same admonition applies: you should choose these settings based on your experimental goals.

The tool helpfully includes a filtering summary at the end of its run:

```
13:47:51.613 INFO  ProgressMeter - Traversal complete. Processed 31742 total
variants in 0.0 minutes.
13:47:51.614 INFO  FilterVariantTranches - Filtered 303 SNPs out of 12929 and
filtered 669 indels out of 2932 with INFO score: CNN_1D.
```

This command produces the filtered VCF in which variants are annotated with the modeling verdict in the FILTER field: PASS for variants classified as real and the name of the tranche for variants classified as artifacts. The tool accepts an optional --invalidate-previous-filters flag (not used here) that allows us to clear any previously annotated filtering information if desired. This is what the output looks like for the preceding command:

```
# cat sandbox/my_1d_cnn_filtered.vcf | grep -v '##' | head -3
#CHROM  POS     ID      REF     ALT     QUAL    FILTER  INFO    FORMAT  NA12878
20      1000072 rs6056638       A       G       998.77  PASS
AC=2;AF=1.00;AN=2;CNN_1D=3.256;DB;DP=32;ExcessHet=3.0103;FS=0.000;MLEAC=2;MLEAF
=1.00;MQ=60.00;POSITIVE_TRAIN_SITE;QD=31.21;SOR=0.818;VQSLOD=20.79;culprit=MQ
GT:AD:DP:GQ:PL  1/1:0,32:32:96:1027,96,0
20      1000152 rs6056639       C       T       678.77  PASS
AC=2;AF=1.00;AN=2;CNN_1D=2.313;DB;DP=28;ExcessHet=3.0103;FS=0.000;MLEAC=2;MLEAF
=1.00;MQ=60.00;POSITIVE_TRAIN_SITE;QD=24.24;SOR=0.693;VQSLOD=18.18;culprit=QD
GT:AD:DP:GQ:PL  1/1:0,28:28:81:707,81,0
```

If you look at more lines of output, you can find sites like this candidate indel call that has been filtered out by the 1D CNN modeling:

```
20      1012919 rs34579666      CT      C       439.73
CNN_1D_INDEL_Tranche_95.00_100.00       AC=2;AF=1.00;AN=2;CNN_1D=-2.925;DB;DP=31;
ExcessHet=3.0103;FS=0.000;MLEAC=2;MLEAF=1.00;MQ=59.62;POSITIVE_TRAIN_SITE;
QD=19.12;SOR=1.708;VQSLOD=3.87;culprit=SOR  GT:AD:DP:GQ:PL1/1:0,23:23:67:477,67,0
```

You might have noticed that, unlike for the VQSR protocol, we did not need to run the CNN model application and filtering on SNPs and indels separately. Algorithmically, these tools still treat the two variant classes differently, just like the VQSR tools, but they handle this internally for us. This is a deliberate decision on the part of the developers to make the tools easier to use.

Applying 2D CNN to Include Read Data in the Modeling

As noted earlier, the CNN tools are capable of taking in the read data underlying the variant calls that we want to filter. This might seem redundant with the variant context annotations, which are, after all, intended to summarize the salient aspects of the read data in a more compact form. Yet that is one of their main flaws, given that the act of summarizing the information causes us to lose some of the more fine-grained detail in the reads.

That puts us at the mercy of two limiting factors: the fidelity of the mathematical formula chosen to summarize the data, and the imagination of the developers who decided which aspects of the data would be useful for filtering purposes. Even though it is arguably not difficult to identify a few heuristics that are bound to play a big role in predicting variant quality for the majority of calls—some form of coverage evaluation, strand orientation bias, and base quality, perhaps—it is much more challenging to identify the subtle effects that will tip the scales in difficult cases. That is a big part of what makes deep learning so appealing for variant filtering; it promises to identify (and quantify!) those effects for us without a ton of up-front analysis.

To include read data in the scoring process (while still using a precomputed model), we provide the `CNNScoreVariants` tool with a bamout produced by `HaplotypeCaller`, and we add the `--tensor-type read_tensor` argument to switch on the 2D model architecture. This will automatically activate the default precomputed 2D model, unless you specify your own on the command line:

```
# gatk CNNScoreVariants \
    -R ref/Homo_sapiens_assembly19.fasta \
    -I bams/g94982_chr20_1m_10m_bamout.bam \
    -V vcfs/g94982_b37_chr20_1m_895.vcf \
    -O sandbox/my_2d_cnn_scored.vcf \
    --tensor-type read_tensor \
    --transfer-batch-size 8 \
    --inference-batch-size 8
```

Notice that we feed a bamout file (created by `HaplotypeCaller`) to `CNNScoreVariants` when we run it the 2D mode, not just the original read data. We do this because the read alignments must match what `HaplotypeCaller` saw when it made the variant calls. This does mean that you need to add the `-bamout` argument to your initial `HaplotypeCaller` command when you do the variant calling step, which we don't normally do, so it's something you should plan for ahead of time.

 If you're using a different variant caller than `HaplotypeCaller` that doesn't generate a bamout equivalent, this doesn't apply, and you simply provide the original read data. Keep in mind that if you use another local alignment caller like Platypus, you could run into the problem of calls that don't quite match the original read mapping.

When you run the command, it produces a VCF with a new confidence score called CNN_2D annotated to each variant along with tranche information. To apply filtering, we run the FilterVariantTranches as before, switching the filtering info-key appropriately:

```
# gatk FilterVariantTranches \
    -V sandbox/my_2d_cnn_scored.vcf \
    -O sandbox/my_2d_cnn_filtered.vcf \
    --resource resources/1000G_omni2.5.b37.vcf.gz \
    --resource resources/hapmap_3.3.b37.vcf.gz \
    --info-key CNN_2D \
    --snp-tranche 99.9 \
    --indel-tranche 95.0
```

The output is similar but not identical to what we produced using the 1D model architecture:

```
# cat sandbox/my_2d_cnn_filtered.vcf | grep -v '##' | head -3
#CHROM  POS      ID         REF       ALT     QUAL    FILTER  INFO    FORMAT   NA12878
20      1000072  rs6056638  A         G       998.77  PASS
AC=2;AF=1.00;AN=2;CNN_2D=7.687;DB;DP=32;ExcessHet=3.0103;FS=0.000;MLEAC=2;MLEAF
=1.00;MQ=60.00;POSITIVE_TRAIN_SITE;QD=31.21;SOR=0.818;VQSLOD=20.79;culprit=MQ
GT:AD:DP:GQ:PL  1/1:0,32:32:96:1027,96,0
20      1000152  rs6056639  C         T       678.77  PASS
AC=2;AF=1.00;AN=2;CNN_2D=5.349;DB;DP=28;ExcessHet=3.0103;FS=0.000;MLEAC=2;MLEAF
=1.00;MQ=60.00;POSITIVE_TRAIN_SITE;QD=24.24;SOR=0.693;VQSLOD=18.18;culprit=QD
GT:AD:DP:GQ:PL  1/1:0,28:28:81:707,81,0
```

And a little farther down, that indel we looked at earlier is still filtered out:

```
20      1012919  rs34579666  CT       C       439.73
CNN_2D_INDEL_Tranche_95.00_100.00
AC=2;AF=1.00;AN=2;CNN_2D=5.576;DB;DP=31;ExcessHet=3.0103;FS=0.000;MLEAC=2;
MLEAF=1.00;MQ=59.62;POSITIVE_TRAIN_SITE;QD=19.12;SOR=1.708;VQSLOD=3.87;
culprit=SOR
GT:AD:DP:GQ:PL1/1:0,23:23:67:477,67,0
```

If you'd like to inspect these in more detail to compare calls filtered by one but not the other, an easy way is to load the VCFs in IGV as described previously and scroll until you find a difference like the one shown in Figure 6-16. Note the lighter colors for one of the variant call markers in the 2D_CNN track; that indicates the call was filtered.

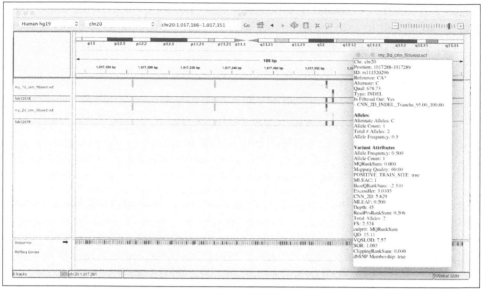

Figure 6-16. Different calls made by 1D and 2D CNN models.

Finally, keep in mind that the model provided by default is tailored to Illumina short read sequencing data from humans. It should be widely applicable also to other organisms sequenced with the same or similar technology, but the further out you go, the less fidelity you'll get because the precomputed model will not be as good a fit for the reality of your data. At some point, you might be better off generating your own custom model. There is a Python script for training custom models that is available in the GATK repository on GitHub, but the GATK team has not yet developed formal recommendations for how to use it because it is an area of ongoing development. For additional resources, see the Best Practices documentation (*https://oreil.ly/NHY4e*).

This concludes the exercises for this chapter, so unless you're moving on to the next chapter right away (what stamina!), remember to stop your VM so you don't end up paying while you're not playing.

Wrap-Up and Next Steps

You have now successfully worked through the most widely used germline joint variant discovery pipeline and its alternate single-sample form. You have a solid understanding of the overall logic of the pipeline, you've applied most of the key steps to real data, and you've observed the results for yourself. Next up, we tackle somatic variant discovery, mainly as it applies to cancer research.

First, we go over a few important challenges that affect the design of analytical approaches in cancer genomics. Then, we dive into the pipeline for somatic short variant analysis and work through some exercises to get a firm grasp of the key steps. After that, we branch out into the exploration of somatic copy-number alterations, which involves a very different approach compared to short variant analysis.

## CHAPTER 7

GATK Best Practices for Somatic Variant Discovery

In this chapter, we tackle somatic variant discovery, mainly as it applies to cancer research. This is going to introduce new challenges and, correspondingly, new experimental and computational designs. We begin with somatic short variants, which still have a lot in common with germline short variants, with enough new challenges to keep things interesting. Then we expand our horizons to include copy-number variation, which involves a very different approach from what we've taken so far.

Challenges in Cancer Genomics

Before we get into the weeds, let's make one thing clear: everything is more difficult with respect to cancer. Consider what is happening in the body before a tumor even begins to develop. Living cells in our bodies are not static; they are metabolically active, taking in nutrients, doing work, and pumping out waste. Some of them are dividing at various rates. Enzymes are unraveling DNA to transcribe it and/or copy and repackage it. At the scale of a single cell, very few of those molecular transactions ever go wrong, but because this is happening all the time in millions of cells, the numbers add up to a lot of little things going wrong, causing mutations across the board.

Most of the time, these mutations have no discernible effect whatsoever. But from time to time, a mutation arises that destabilizes the affected cell's metabolism in a way that causes it to begin dividing more rapidly and produce a tumor. That's still not cancer yet, mind you; most tumors in our bodies remain relatively contained and harmless. However, some then experience other destabilizing mutations that kick off a chain reaction, leading to cancerous development, including aggressive growth and in the worst-case scenario, *metastasis*: dissemination to other parts of the body. The mutation events that drive this progression are appropriately called *driver mutations*, in contrast to the others, which are called *passenger mutations*. Driver mutations are typically the main focus of somatic analysis in cancer research.

The difficulty here is that as the tumor progresses, various metabolic functions begin to break down and the cells lose their ability to repair damaged DNA. As a result, the frequency of mutation events increases, and variants accumulate in a variety of divergent cell lineages, producing a highly heterogeneous environment. When a pathologist samples the tumor tissue, they end up extracting a mix of multiple cell lines instead of the well-behaved population of largely clonal cells that you expect from healthy tissue.

This is a huge problem for detecting somatic variation, because depending on the internal state of the tumor tissue at the time we sample it, the variants caused by driver mutations might be represented at only a low fraction of the total genomic material. The most direct consequence is that we can't allow the variant caller to make any assumptions about the fraction of variant allele that we expect to find. This is a big difference compared to the germline case, in which variant allele fractions should roughly follow the ploidy of the organism. That is why we use different variant callers for somatic versus germline variant discovery analysis, which, under the hood, use different algorithms to model genomic variation in the data.

In addition, the sampling process itself introduces confounding factors. Depending on the skill of the pathologist, the stage of progression of the tumor, and the type of biological tissue affected, we can end up with a tumor sample that is contaminated with healthy "normal" cells. Conversely, if we are able to obtain a *matched normal sample*, a sample taken from supposed healthy tissue from the same person—that sample can, in fact, be contaminated with low levels of tumor cells, as depicted in Figure 7-1.

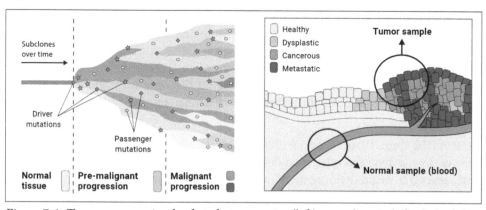

Figure 7-1. Tumor progression leads to heterogeneity (left); sampling is difficult (right).

The practical consequence of all this for somatic variant analysis is that our work will necessarily be deeply vulnerable to sources of contamination and error because the signal we are trying to detect is much lower compared to the noise that is present in the data, particularly in the case of advanced cancers.

Somatic Short Variants (SNVs and Indels)

Let's commence our exploration of somatic variant calling with short variants—SNVs and indels. Some of the core concepts we cover here should feel familiar if you've already worked through Chapter 6, and indeed some of the tools and underlying algorithms are the same. As we've done previously, we'll be looking at piles of reads and their exact sequences to identify substitutions and small insertions and deletions.

However, the basic experimental design that we follow is actually very different. In germline variant discovery, we were mainly interested in identifying variants that were present in some people but not in others, which led us to favor the joint calling approach. In somatic variant discovery, we're primarily looking for differences between samples coming from the same person, so we will shift our focus to a different paradigm: the *Tumor-Normal* pair analysis.

The most useful source of information for identifying somatic short variants in the context of cancer is the comparison of the tumor against a sampling of healthy tissue from the same individual, which, as we said previously, is called a *matched normal*, or often *normal* for short. The idea here is that any variation observed in the normal sample is part of the germline genome of the individual and can be discounted as background, whereas any variant observed in the tumor sample that is not present in the normal sample is more likely to be a somatic mutation, as demonstrated in Figure 7-2.

Unfortunately, some confounding factors muddy the waters. For example, biopsy samples preserved with formalin and paraffin, a widely used laboratory technique (FFPE, for formalin-fixed paraffin-embedded), are subject to biochemical alterations that happen during sample preparation and cause patterns of false-positive variant calls that are specific to the tumor sample. Conversely, the presence of tumor cells in normal tissue (either due to contamination during sampling or metastatic dispersion) can cause real somatic variants to be erroneously discounted as being part of the germline of the individual. Finally, as we discussed earlier, the various sources of technical error and contamination that affect all high-throughput sequencing techniques to some degree will have a proportionately higher impact on somatic calling. So, we will need to apply some robust processing and filtering techniques to mitigate all these problems. (The overall workflow is illustrated in Figure 7-3.)

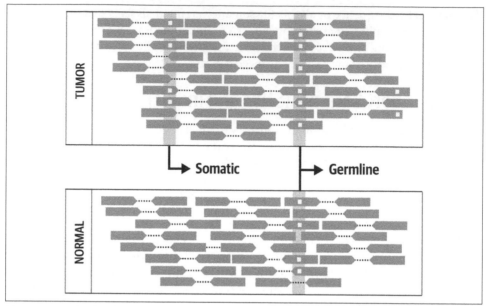

Figure 7-2. The fundamental concept of Tumor-Normal comparison.

Before we get into the details of the Tumor-Normal workflow in the next subsection, it's important to note that sometimes it's not possible to apply the full Tumor-Normal pair analysis to a sample or set of samples, usually because we simply don't have matched normals for them. This is particularly frustrating when that lack comes from an investigator's failure to collect a matched normal from the participant or patient. Such cases highlight the importance of making sure that researchers map out the full analysis protocol before engaging in data collection. On the other hand, in some cases it's biologically impossible to obtain a matched normal; for example, blood-borne cancers typically reach every tissue in the body, so we can't assume that any tissue sample would constitute an appropriate normal.

Whatever the reason for the absence of a matched normal, the workflow that we present in the next subsection can be run on a tumor sample with only a few minor modifications, which are detailed in the online GATK documentation (*https://oreil.ly/ ZgIrB*). However, you should be aware that the results will necessarily be noisier because you will be relying on the germline population resource to exclude the person's own germline genome.

Overview of the Tumor-Normal Pair Analysis Workflow

This workflow, illustrated in Figure 7-3, is designed to be applied to a tumor sample and its matched normal. We can apply it equally to whole genome data as well as exome data, but both the tumor and the normal must be of the same data type and have been run through the same preprocessing workflow, as we described for germ-line variant discovery.

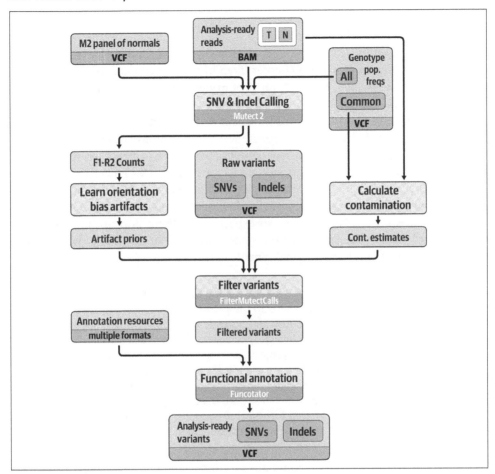

Figure 7-3. Best Practices for somatic short variant discovery.

The main variant calling step, which involves a tool called Mutect2, is applied on both the tumor and the normal sample, with the addition of two important resources: a Panel of Normals (PoN) and a database of germline population frequencies of all germline variants that have been previously reported. The PoN is a resource that allows us to eliminate common technical artifacts; we dedicate the next subsection to explaining how it works and how to create one, although you won't run it yourself for practical reasons. We talk about the germline population frequencies resource in "Filtering Mutect2 Calls" on page 193. The filtering process will take into account read orientation statistics, an estimate of cross-sample contamination, germline background, and concordance with the PoN. Finally, we apply functional annotation to the variant callset to predict the effect of candidate variants.

Creating a Mutect2 PoN

The PoN is a callset in VCF format that aims to capture common recurring artifacts, so we can filter those out if we encounter them in our tumor data. We generate it by running the somatic caller, Mutect2, on a set of normal samples that were produced with the same sequencing and library preparation technologies as our tumor sample. The idea is that we're asking the caller to show us everything it could consider a somatic variant, excluding obvious germline variants (unless otherwise specified). Because these are normals, not tumor samples, we can reasonably assume that any variant calls that are produced this way are either germline variants or artifacts. For those that are more likely to be artifacts, we can therefore further assume that if we see the same artifacts in multiple normals, they might have been caused by the technological processes involved, so we can expect to see them pop up in some of our tumor samples that were processed the same way. If we do see them, we'll want to discount their likelihood of being real somatic variants.

 The normal samples used for making the PoN usually come from people who are unrelated to the person whose tumor we are analyzing. They can be matched normals from other cancer research participants or germline samples collected from healthy individuals in unrelated research projects, as long as the appropriate consents have been provided. The background genetics of the participants or their population of origin don't really matter; it's all about the technical profile of the data. Ideally, we want to include at least 40 samples; most PoNs used for research purposes at the Broad Institute are composed of several hundred samples.

We won't run the PoN creation commands in this chapter, because they would take too long to run. Instead, we demonstrate usage with some generic command lines. You can find links to full implementations in the GATK Best Practices documentation (*https://oreil.ly/ZgIrB*).

If we were to do this, we would first run Mutect2 in tumor-only mode on the normals selected to be in our PoN. We would do this separately on all the normals in our panel. The following command (and subsequent ones in this section) are just examples; we'll have you execute commands on your VM a little later:

```
gatk Mutect2 \
    -R reference.fasta \
    -I normal_1.bam \
    -O normal_1.vcf.gz \
    --max-mnp-distance 0
```

This command applies the somatic variant calling model to the normal tissue sample's BAM file and produces a VCF file of calls that are most likely all either artifacts or germline variants.

 The `--max-mnp-distance 0` setting disables the tool's ability to combine neighboring SNVs into multinucleotide variants in order to maximize comparability across all the normals in the PoN that we're putting together.

Next, we would consolidate all the per-normal VCFs using `GenomicsDB`, which you might recognize from the germline joint calling workflow. At that time, we were consolidating GVCFs, whereas here we're working with regular VCFs, but it doesn't make any difference: the `GenomicsDB` tools are quite happy to work with either type of VCF:

```
gatk GenomicsDBImport \
    -R reference.fasta \
    -L intervals.interval_list \
    -V normal_1.vcf.gz \
    -V normal_2.vcf.gz \
    -V normal_3.vcf.gz \
    --genomicsdb-workspace-path pon_db
```

Finally, we would extract the calls that are relevant to the goals of the PoN. By default, the rule is to keep any variant call observed in two or more samples but below a certain frequency threshold in the germline population resource:

```
gatk CreateSomaticPanelOfNormals \
    -R reference.fasta \
    -V gendb://pon_db \
    --germline-resource af-only-gnomad.vcf.gz \
    -O pon.vcf.gz
```

You might recognize gnomAD (*https://oreil.ly/gNICj*) as the population resource we used in Chapter 6, in the single-sample filtering protocol. The gnomAD database is a public resource that includes variant calls from more than 100,000 people, and for each, it provides the frequency at which each allele is found in the population of the database. We use that information here to gauge the likelihood that a variant call that shows up in several normals in the panel might in fact be a germline variant rather than a recurring artifact.

This generates the final VCF that we will use as a PoN. It does not include any of the sample genotypes; only site-level information is retained. We provide an example PoN that you can examine with good old `zcat`, `grep`, and `head`, as we've done previously, if you want to see the structure of the information within the file (execute this command in the */home/book/data/somatic* directory inside the GATK container):

```
# zcat resources/chr17_m2pon.vcf.gz | grep -v '##' | head -3
#CHROM  POS     ID      REF     ALT     QUAL    FILTER  INFO
```

```
chr6    29941027        .       G       A       .       .       .
chr6    29941061        .       G       C       .       .       .
```

Note that this example PoN file includes data only in a subset of regions of interest that are located within chromosome 6, chromosome 11, and chromosome 17.

Running Mutect2 on the Tumor-Normal Pair

Now that you know how to create a PoN, let's move to the main event of this section, which consists of running `Mutect2` on the Tumor-Normal pair using a PoN that we created for you from 40 publicly available normal samples from the 1000 Genomes Project. Just as in Chapter 6, we're going to do this work within the GATK container on our trusty VM, so turn on your VM if it's not already running, and spin up the GATK container as we showed you in Chapter 5. This time, however, our working directory is */home/book/data/somatic*, and we'll need to make another sandbox:

```
# cd /home/book/data/somatic
# mkdir sandbox
```

We run `Mutect2` on the Tumor-Normal pair together, specifying which sample name (as encoded in the `SM` tag in the BAM file) should be used for the normal with the `-normal` argument. We provide the PoN and the gnomAD germline resource with the `-pon` and `--germline-resource` arguments, respectively. Plus, we specify intervals (using `-L` and an interval list file) because we're running on exome data:

```
# gatk Mutect2 \
    -R ref/Homo_sapiens_assembly38.fasta \
    -I bams/tumor.bam \
    -I bams/normal.bam \
    -normal HCC1143_normal \
    -L resources/chr17plus.interval_list \
    -pon resources/chr17_m2pon.vcf.gz \
    --germline-resource resources/chr17_af-only-gnomad_grch38.vcf.gz \
    -bamout sandbox/m2_tumor_normal.bam \
    -O sandbox/m2_somatic_calls.vcf.gz
```

This will run for a few minutes, so it's a good time to take a stretch break. After about five minutes, you should see the completion message from `Mutect2`:

```
15:07:19.715 INFO  ProgressMeter - Traversal complete. Processed 285005 total
regions in 5.0 minutes.
```

This `Mutect2` command produces a VCF file containing possible variant calls as well as a bamout file. You might remember the bamout file from the exercises we ran with `HaplotypeCaller` in Chapter 5; this is the same thing. `Mutect2` is built on most of the same code as `HaplotypeCaller`, and pulls the same graph-based realignment trick to improve sensitivity to indels. The big difference between them is the genotyping model. For more information on their respective algorithms, see the `Haplotype Caller` and `Mutect2` documentation on GATK's website (*https://oreil.ly/Bits1*). For

---

now, just keep in mind that `Mutect2` also realigns the reads, so if we want to look at the read data as part of the variant review process later on, we'll need the bamout. Because this is something analysts do very frequently for somatic calls (much more, relatively speaking, than for germline calls) we just set the `Mutect2` command to produce the bamout systematically.

Estimating Cross-Sample Contamination

As we outlined earlier, identification of somatic mutations is more heavily affected by sources of low-grade noise than the germline case. Even small amounts of contamination between samples (which happens more often than researchers like to think) can confuse the caller and lead to a lot of FPs. Indeed, we have no direct way to distinguish between an allele that shows up in the data because there is a real mutation in the biological tissue, and one that shows up because our sample is contaminated with DNA from someone else who has that allele in their germline.

We can't directly correct for such contamination, but we can estimate how much it affects any given sample. From there, we can flag any mutation calls that are observed at an allelic fraction equal to or smaller than the contamination rate. It doesn't mean we rule out those calls as being necessarily false, but it does mean we'll be extra skeptical of those calls when we move on to the next phase of analysis.

To estimate the contamination rate in our tumor sample, we first identify homozygous-variant sites in the Normal to choose what sites to look at in the Tumor and then evaluate how much contamination we might have in the tumor based on the fraction of reads supporting the reference allele at those sites. The idea is that because those sites should be homozygous variant, any reference reads present must have come from someone else through a contamination event.

To select the sites that we use in the calculation, we look at biallelic sites that are found in a catalog of common variation (so that there's a good chance they will be relevant) but have an alternate allele frequency that is rather low (so there won't be too much noise from alternate alleles). Then, for every site in that subset, we do a quick pileup-based genotyping run to identify those that are homozygous-variant in that sample. We do this on both the tumor sample and the matched normal:

```
# gatk GetPileupSummaries \
    -I bams/normal.bam \
    -V resources/chr17_small_exac_common_3_grch38.vcf.gz \
    -L resources/chr17_small_exac_common_3_grch38.vcf.gz \
    -O sandbox/normal_getpileupsummaries.table

# gatk GetPileupSummaries \
    -I bams/tumor.bam \
    -V resources/chr17_small_exac_common_3_grch38.vcf.gz \
    -L resources/chr17_small_exac_common_3_grch38.vcf.gz \
    -O sandbox/tumor_getpileupsummaries.table
```

These two commands should run very quickly. If you see a warning about the GetPi
leupSummaries tool being in a beta evaluation stage and not yet ready for production,
you can ignore it; this is a leftover of the development process and will be removed in
the near future.

 GATK developers tend to be very cautious and label their tools as
alpha or beta when they start working on them, to make sure that
researchers don't mistake an experimental tool for something that
has been fully vetted. However, they sometimes forget to remove or
update those labels when the tools have been fully vetted, so the
warnings remain longer than necessary. If you're ever in doubt
about the state of a particular tool, don't hesitate to ask the support
team by posting in the community support forum (*https://oreil.ly/
KmHzX*).

Each run produces a table that summarizes the allelic counts at each site that was
selected as well as the allele frequency annotated in the population resource:

```
# head -5 sandbox/normal_getpileupsummaries.table
#<METADATA>SAMPLE=HCC1143_normal
contig  position     ref_count    alt_count     other_alt_count allele_frequency
chr6    29942512     7      4      0     0.063
chr6    29942517     12     4      0     0.062
chr6    29942525     13     7      0     0.063

# head -5 sandbox/tumor_getpileupsummaries.table
#<METADATA>SAMPLE=HCC1143_tumor
contig  position     ref_count    alt_count     other_alt_count allele_frequency
chr6    29942512     9      0      0     0.063
chr6    29942517     13     1      0     0.062
chr6    29942525     13     7      0     0.063
```

Now we can do the actual estimation by providing both tables to the calculator tool:

```
# gatk CalculateContamination \
    -I sandbox/tumor_getpileupsummaries.table \
    -matched sandbox/normal_getpileupsummaries.table \
    -tumor-segmentation sandbox/segments.table \
    -O sandbox/pair_calculatecontamination.table
```

This command produces a final table that tells us the likely rate of cross-sample contamination in the Tumor callset. Viewing the contents of the file shows us that the
level of contamination of the tumor is estimated at 1.15%, with an error of 0.19%:

```
$ cat sandbox/pair_calculatecontamination.table
sample  contamination    error
HCC1143_tumor    0.011485364960150258     0.0019180421331441303
```

This means that one read out of every hundred is likely to come from someone else.
This percentage will effectively be our floor for detecting low-frequency somatic
events; for any potential variant called at or below that AF, we would have zero power
to judge whether they were real or created through contamination. And even a larger

---

AF could be due entirely to contamination. We'll feed the table into the filtering tool so that it can take the contamination estimate properly into account.

Filtering Mutect2 Calls

Much like `HaplotypeCaller`, `Mutect2` is designed to be very sensitive and capture as many true positives as possible. As a result, we know that the raw callset will be rife with false positives of various types. To filter these calls, we're going to use a tool called `FilterMutectCalls`, which filters calls based on a single quantity: the probability that a variant is a somatic mutation. Internally the tool takes into account multiple factors, including the germline population frequencies provided during the variant-calling step, the contamination estimates, and the read orientation statistics computed earlier. After it has calculated that probability for each variant call, the tool then determines the threshold that optimizes the *F score*, the harmonic mean of sensitivity and precision across the entire callset. As a result, we can simply run this one command to apply default filtering to our callset:

```
# gatk FilterMutectCalls \
    -R ref/Homo_sapiens_assembly38.fasta \
    -V sandbox/m2_somatic_calls.vcf.gz \
    --contamination-table sandbox/pair_calculatecontamination.table \
    -O sandbox/m2_somatic_calls.filtered.vcf.gz \
    --stats sandbox/m2_somatic_calls.vcf.gz.stats \
    --tumor-segmentation sandbox/segments.table
```

The main output of this operation is the VCF output file containing all variants with filter annotations applied as appropriate:

```
15:25:14.742 INFO  ProgressMeter - Traversal complete. Processed 333 total
variants in 0.0 minutes.
```

 If you want to tweak the balance point one way or the other, you can adjust the relative weight of sensitivity versus precision in the harmonic mean. Setting the `-f-score-beta` parameter to a value greater than its default of 1 increases sensitivity, whereas setting it lower favors greater precision.

Alright, it's time to have a look at the calls we produced in IGV. You might have noticed that we're now working with the hg38 reference build, so we'll need to adjust the reference setting in IGV before we can try to load any files.

You're going to load the following files: the original BAM files for the Tumor and the Normal, the bamout generated by `Mutect2`, and the filtered callset. Identify the files based on the commands you ran a moment ago and then follow the same procedure as in Chapter 4 through Chapter 6 to copy the files to your Google Cloud Storage bucket and then load them in IGV using the File > URL pathway.

In IGV, set the view coordinates to **chr17:7,666,402-7,689,550** after the files are loaded, as shown in Figure 7-4, which are centered on the TP53 locus. We care about this gene because it is known as a tumor suppressor; it is activated by DNA damage and produces a protein called p53 that can either repair the damage or trigger killing off the cell. Any mutations to this gene that interfere with this very important function could contribute to allowing unchecked tumor development.

To review this call in detail, zoom into the somatic call in the output BAM track, using coordinates **chr17:7,673,333-7,675,077**. Hover over or click the gray call in the variant track to view annotations, and try scrolling through the data to evaluate the amount of coverage for each sample. In the sequence tracks, we see a C to T variant light up in red for the Tumor but not the Normal (Figure 7-4).

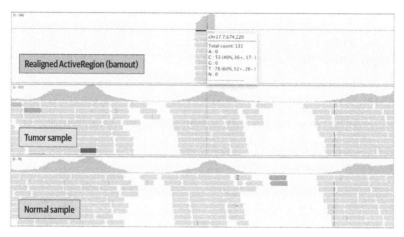

Figure 7-4. Zooming in on TP53 in IGV.

What do you think of this somatic variant call?

Tweaking IGV Visualization Settings

Tweaking some IGV settings to make the review easier can be helpful. First, you can make room to focus on certain tracks by minimizing the space taken up by others or removing them entirely. You can resize each track panel by clicking the gray separator lines and dragging them up or down. You can also switch the data view of a specific track to a more compact display by right-clicking or Ctrl-clicking the track and then, on the menu that opens, selecting the Collapsed or Squished view options. And, of course, you can remove individual tracks by right-clicking or Ctrl-clicking (Shift-clicking several to select them at the same time) and then selecting Remove Track.

Another neat feature set available in the tracks' contextual menu is the ability to order, group, and/or color the data based on specific properties. For example, we often use

---

the following combination for examining output from both `HaplotypeCaller` and `Mutect2`: group by sample, color alignments by tag, using HC as tag, and sort by base.

It's also worth exploring the IGV Preferences pane, particularly the Alignments tab, to get a good handle on the options available there. For example, we used it to toggle soft clips for exercises in Chapter 5; you can also toggle downsampling, the display of duplicates, and other metadata-based display options.

Annotating Predicted Functional Effects with Funcotator

Filtering somatic mutation calls can be an especially daunting task compared to their germline equivalent. It can therefore be particularly helpful to annotate the calls with predicted functional effects and their significance. For example, a stop codon in the middle of a protein coding region is likely to be more significant than a mutation in the middle of an intron. Knowing this about your calls can help you prioritize which ones you spend time investigating in depth.

To gauge functional impact, we need to know which regions of the genome code for protein sequence and which correspond to elements important to gene expression. Transcript annotation resources such as GENCODE (*https://www.gencodegenes.org*) capture such information in the standardized Gene Transfer Format (*https://oreil.ly/SDmFX*) (GTF). The GATK team provides a set of resources (*https://oreil.ly/LDGdM*) containing compiled functional annotation resources, or datasources, that are suitable for many common functional annotation needs as well as a tool called `Funcotator` to annotate your callset with information from these datasources. You can also create your own custom data sources. In addition, although `Funcotator` outputs results in VCF by default, you can make it output Mutation Annotation Format (MAF), instead, which the cancer genomics community uses extensively. See the Funcotator documentation (*https://oreil.ly/6qw0X*) for details on both options.

For now, though, let's apply `Funcotator` to our filtered `Mutect2` callset:

```
# gatk Funcotator \
    --data-sources-path resources/funcotator_dataSources_GATK_Workshop_20181205/ \
    --ref-version hg38 \
    -R ref/Homo_sapiens_assembly38.fasta \
    -V sandbox/m2_somatic_calls.filtered.vcf.gz \
    -O sandbox/m2_somatic_calls.funcotated.vcf.gz \
    --output-file-format VCF
```

This runs very quickly and produces a VCF in which our somatic callset has now been lovingly annotated based on the GENCODE functional data. Here's a sneak peek of what the annotations look like in their natural habitat:

```
# zcat sandbox/m2_somatic_calls.funcotated.vcf.gz | grep -v '##' | head -3
#CHROM POS    ID     REF    ALT    QUAL   FILTER INFO    FORMAT HCC1143_normal
HCC1143_tumor
chr17  1677390 .      A      T      .      weak_evidence
CONTQ=19;DP=23;ECNT=1;FUNCOTATION=[PRPF8|hg38|chr17|1677390|1677390|INTRON||SNP|A|A|T|
```

```
g.chr17:1677390A>T|ENST00000572621.5|-|||c.e13-
175T>A|||0.4463840399002494|CTGCCTCTCAAGGCCCCAGAA|PRPF8_ENST00000304992.10_INTRON];
GERMQ=32;MBQ=33,31;MFRL=301,311;MMQ=60,60;MPOS=22;NALOD=1.06;NLOD=3.01;POPAF=6.00;SEQQ
=1;STRANDQ=17;TLOD=4.22 GT:AD:AF:DP:F1R2:F2R1:SB      0/0:10,0:0.081:10:6,0:4,0:6,4,0,
0      0/1:9,2:0.229:11:5,1:3,1:6,3,2,0
chr17   2394409 .       G       T       .       PASS
CONTQ=93;DP=106;ECNT=1;FUNCOTATION=[MNT|hg38|chr17|2394409|2394409|INTRON||SNP|G|G|T|
g.chr17:2394409G>T|ENST00000575394.1|-|||c.e1-
6231C>A|||0.6209476309226932|TTCCTGACCAGCGCCGCCACC|MNT_ENST00000174618.4_INTRON];
GERMQ=93;MBQ=32,31;MFRL=148,177;MMQ=60,60;MPOS=14;NALOD=1.56;NLOD=10.52;POPAF=6.00;SEQQ
=93;STRANDQ=93;TLOD=30.05  GT:AD:AF:DP:F1R2:F2R1:SB      0/0:35,0:0.027:35:15,0:19,0:4,
31,0,0  0/1:53,13:0.206:66:23,5:29,8:13,40,4,9
```

Now let's have a look at the annotations specifically for the TP53 mutation that we viewed earlier in IGV, at chr17:7674220:

```
# zcat sandbox/m2_somatic_calls.funcotated.vcf.gz | grep 7674220
chr17   7674220 .       C       T       .       PASS
CONTQ=93;DP=134;ECNT=1;FUNCOTATION=[TP53|hg38|chr17|7674220|7674220|MISSENSE||SNP|C|C|T|
g.chr17:7674220C>T|ENST00000269305.8|-|7|933|c.743G>A|c.(742-
744)cGg>cAg|p.R248Q|0.5660847880299252|GATGGGCCTCCGGTTCATGCC|TP53_ENST00000445888.6_MISSENSE_
p.R248Q/TP53_ENST
00000420246.6_MISSENSE_p.R248Q/TP53_ENST00000622645.4_MISSENSE_p.R209Q/
TP53_ENST00000610292.4_MISSENSE_p.R209Q/TP53_ENST00000455263.6_MISSENSE_p.R248Q/
TP53_ENST00000610538.4_MISSENSE_p.R209Q/TP53_ENST00000620739.4_MISSENSE_p.R209Q/
TP53_ENST00000619485.4_MISSENSE_p.R209Q/TP53_ENST00000510385.5_MISSENSE_p.R116Q/
TP53_ENST00000618944.4_MISSENSE_p.R89Q/TP53_ENST000005
04290.5_MISSENSE_p.R116Q/TP53_ENST00000610623.4_MISSENSE_p.R89Q/
TP53_ENST00000504937.5_MISSENSE_p.R116Q/TP53_ENST00000619186.4_MISSENSE_p.R89Q/
TP53_ENST00000359597.8_MISSENSE_p.R248Q/TP53_ENST00000413465.6_MISSENSE_p.R248Q/
TP53_ENST00000615910.4_MISSENSE_p.R237Q/TP53_ENST00000617185.4_MISSENSE_p.R248Q];
GERMQ=93;MBQ=31,32;MFRL=146,140;MMQ=60,60;MPOS=21;NALOD=1.73;NLOD=15.33;POPA
F=6.00;SEQQ=93;STRANDQ=93;TLOD=264.54    GT:AD:AF:DP:F1R2:F2R1:SB
0/0:51,0:0.018:51:22,0:29,0:36,15,0,0  0/1:0,76:0.987:76:0,38:0,38:0,0,52,24
```

This SNV is a C-to-T substitution that is predicted to cause an arginine-to-glutamine *missense mutation*. Missense mutations change the protein that is produced from the gene, which can be functionally neutral, but can also have deleterious effects. If we look at the rest of the file, out of our 124 mutation records, 21 were annotated as MISSENSE; and of these, 10 passed all filters. That information should help us prioritize what to investigate in the next stage of our analysis.

With that, we're done with the Tumor-Normal pair analysis workflow for discovery of somatic SNVs and indels! Next, we're going to go over the process we apply for investigating somatic copy-number alterations.

 If you're taking a break before continuing on with the next section, consider stopping your VM to avoid paying for it while you're away living your best life.

Somatic Copy-Number Alterations

So far, we've been looking exclusively at small variants, SNVs, and indels that are characterized based on the presence or absence of specific nucleotides compared to the reference sequence. Moving to the copy number involves a big mental (and methodological!) shift given that now we're going to be looking at the *relative amounts of sequence* in the sample of interest, measured relative to itself (Figure 7-5 and Figure 7-6). In other words, we're not actually going to measure the copy number that is physically present in the biological material. What we measure is the copy ratio, which we use as a proxy because it is a *consequence* of copy-number alterations (CNAs). As part of that process, we're going to use the reference genome primarily as a coordinate system for defining segments that display CNAs.

As shown in Figure 7-5, the *copy number* is the absolute, integer-valued number of copies of each locus (i.e., gene or other meaningful segment of DNA). The copy ratio is the relative, real-valued ratio of the number of copies of each locus to the average ploidy.

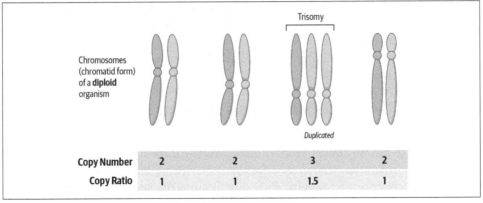

Figure 7-5. Difference between copy number and copy ratio.

Figure 7-6 shows the striking contrast in genomic composition between a normal cell line that has the expected copy number for all chromosomes (left) and a cancer cell line that has suffered multiple major CNAs (right), not to mention a variety of structural alterations. For the normal cell line in this figure (and indeed, most healthy human cells), any given chromsome's copy number is 2 and its copy ratio is 1. Meanwhile, in the cancer cell line, there is a lot of variation between chromosomes; for example, it looks like the copy number for chromosome 7 is 5, whereas for chromosome 12 it is 3, and we can't really guess their copy ratio because the alterations are so extensive that we don't have a good sense of the average ploidy of this genome.

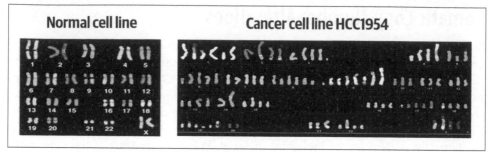

Figure 7-6. Spectral karyotyping paints each chromosome pair with a color, showing various chromosomal segments that are amplified or missing (colors in left and right panels are not expected to match).

The good news is that copy-number analysis is less sensitive than short variant analysis to some of the challenges that we enumerated earlier because we don't care as much about exact sequence. However, it is more sensitive to technical biases that affect the evenness of sequence coverage distribution. The consequence of these two points for the analysis design is that it is less important to have a matched normal available, whereas it is even more important to have a PoN composed of normals that have a closely matching technical profile. Note, however, that the copy-number PoN is completely different from the PoN used for short variant analysis. We go over the specifics in the next section.

Overview of the Tumor-Only Analysis Workflow

The standard workflow for somatic copy-number analysis involves four main steps (Figure 7-7). First, we collect proportional coverage statistics for the case sample as well as any normals that will be used for the PoN. Setting the case sample aside for a moment, we create a copy-number PoN that captures the amount of systematic skew in coverage observed across the exome or genome. Then, we use that information to normalize the coverage data we collected for the case sample, which eliminates technical noise and reveals actual copy ratio differences in the data. Finally, we run a segmentation algorithm that identifies contiguous segments of sequence that presents the same copy ratio, and emit CNA calls for those segments whose copy ratio diverges significantly from the mean. Optionally, we can produce plots to visualize the data at several stages, which is helpful for interpretation.

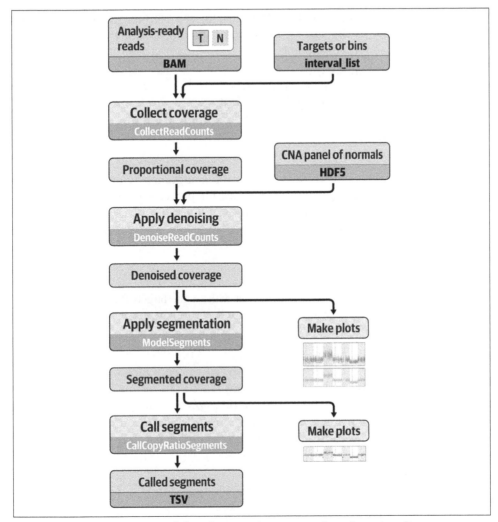

Figure 7-7. Best Practices workflow for somatic copy-number alteration discovery.

Again, you're going to do this work from within the GATK container on your VM, so make sure everything is turned on and you're in the */home/book/data/somatic* directory. The exercises in this section use the same reference genome and some of the same resource files as the first section, and we're going to use the same sandbox for writing outputs.

Collecting Coverage Counts

We count *read depth*, or coverage, over a series of intervals. Here we use exome capture targets because we're analyzing exomes, but if we were analyzing whole genomes, we would switch to using bins of arbitrary length. For exomes, we take the interval list that corresponds to the capture targets (specific to every manufacturer's exome prep kit) and first add some padding:

```
# gatk PreprocessIntervals \
    -R ref/Homo_sapiens_assembly38.fasta \
    -L resources/targets_chr17.interval_list \
    -O sandbox/targets_chr17.preprocessed.interval_list \
    --padding 250 \
    --bin-length 0 \
    --interval-merging-rule OVERLAPPING_ONLY
```

In this command, the `--interval-merging-rule` argument controls whether the GATK engine should merge adjacent intervals for processing. In some contexts, it makes sense to merge them, but in this context, we prefer to keep them separate in order to have more fine-grained resolution, and we'll merge only intervals that actually overlap.

 If we wanted to generate an intervals list for whole genome data, we would run the same tool but without the targets file, and with the `--bin-length` argument set to the length of bin we want to use (set to 100 bases by default).

When we have the analysis-ready intervals list, we can run the read count collection tool, which operates by counting how many reads start within each interval. By default, the tool writes data in HDF5 format (*https://oreil.ly/uP_J4*), which is handled more efficiently by downstream tools, but here we change the output format to tab-separated values (TSV) because it is easier to peek into:

```
# gatk CollectReadCounts \
    -I bams/tumor.bam \
    -L sandbox/targets_chr17.preprocessed.interval_list \
    -R ref/Homo_sapiens_assembly38.fasta \
    -O sandbox/tumor.counts.tsv \
    --format TSV \
    -imr OVERLAPPING_ONLY
```

The output file includes a header that recapitulates the sequence dictionary, followed by a table of counts for each target:

```
# head -5 sandbox/tumor.counts.tsv
@HD     VN:1.6
@SQ     SN:chr1 LN:248956422
@SQ     SN:chr2 LN:242193529
@SQ     SN:chr3 LN:198295559
@SQ     SN:chr4 LN:190214555
```

Annoyingly, the sequence dictionary for the hg38 reference build is extremely long because of the presence of all the *ALT contigs* (explained in the genomics primer in Chapter 2). So, to peek into our file, it's better to use `tail` instead of `head`:

```
# tail -5 sandbox/tumor.counts.tsv
chr17   83051485        83052048        1
chr17   83079564        83080237        0
chr17   83084686        83085575        1010
chr17   83092915        83093478        118
chr17   83094004        83094827        484
```

Ah, that's more helpful, isn't it? The first column is the chromosome or contig, the second and third are the start and stop of the target region we're looking at, and the fourth is the count of reads overlapping the target. Figure 7-8 provides an example of what this looks like visualized.

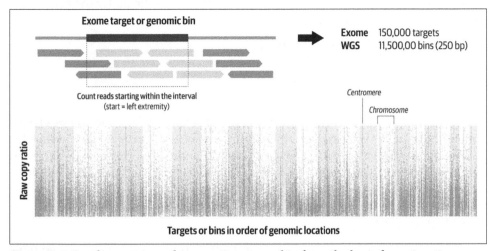

Figure 7-8. Read counts in each genomic target or bin form the basis for estimating segmented copy ratio, and each dot is the value for a single target or bin.

Each dot in the figure represents one of the intervals for which we collected coverage, arranged in the order of their position on the genome. The gray and white stripes in the background represent alternating chromosomes, and the dotted lines represent the position of their centromeres. Notice how the dots form a wavy smear because of the huge variability in the amount of coverage present from one to the next. At this point, it's impossible to identify any CNAs with any kind of confidence.

Creating a Somatic CNA PoN

As we discussed in Chapter 2, a lot of technical variability affects the sequencing process, and some of it is fairly systematic—some regions will produce more sequence data than others, mainly because of chemistry. We can establish a baseline of what normal variability looks like by using an algorithm called singular value decomposition, which is conceptually similar to principal component analysis and aims to capture systematic noise from a panel of normals that were sequenced and processed the same way. The minimum number of samples in a CNA PoN should be at least 10 normal samples for the tools to work properly, but the Best Practices recommendation is to use 40 or more normal samples if possible.

Creating the PoN is a lot of work, so we provide one for you made from 40 samples from the 1000 Genomes Project. Each of the 40 samples was run individually through the `CollectReadCounts` tool, as described earlier (but leaving the default output format set to HDF5; hence, the *.hdf5* extension in the command line, shown in the next code snippet). Then the panel was created by running the command that follows on all the read count files. To be clear, we're just showing this as an example, so don't try to run it; it won't work with the data we've provided for the book. for the book; we're just showing it as an example.

```
gatk CreateReadCountPanelOfNormals \
    -I file1_clean.counts.hdf5 \
    …
    -I file40_clean.counts.hdf5 \
    -O cnaponC.pon.hdf5
```

The PoN produced by this command is completely different from the one we created for the short variant section earlier in this chapter. For the short variants pipeline, the PoN was just a kind of VCF file with annotated variant calls. In this case, however, the file produced by aggregating the read count data from the 40 normals is not meaningfully readable by the naked eye.

Applying Denoising

With our sample read counts in hand and the PoN ready to go, we have everything we need to run the most important step in this workflow: denoising the case sample read counts:

```
# gatk DenoiseReadCounts \
    -I cna_inputs/hcc1143_T_clean.counts.hdf5 \
    --count-panel-of-normals cna_inputs/cnaponC.pon.hdf5 \
    --standardized-copy-ratios sandbox/hcc1143_T_clean.standardizedCR.tsv \
    --denoised-copy-ratios sandbox/hcc1143_T_clean.denoisedCR.tsv
```

Even though we're running only a single command, internally the tool is performing two successive data transformations. First, it standardizes the read counts by the median counts recorded in the PoN, which involves a base-2 log transformation and normalization of the distribution to center around one, to produce *copy ratios*. Then, it applies the denoising algorithm to these standardized copy ratios using the principal components of the PoN.

You can tune the denoising tool to be more or less aggressive by adjusting the --number-of-eigensamples parameter, which affects the resolution of results; that is, how smooth the resulting segments will be. Using a larger number will produce a higher level of denoising but can reduce the sensitivity of the analysis.

Again, the output files are not really human readable, so let's plot their contents to get a sense of what the data shows at this point:

```
# gatk PlotDenoisedCopyRatios \
    --sequence-dictionary ref/Homo_sapiens_assembly38.dict \
    --standardized-copy-ratios sandbox/hcc1143_T_clean.standardizedCR.tsv \
    --denoised-copy-ratios sandbox/hcc1143_T_clean.denoisedCR.tsv \
    --minimum-contig-length 46709983 \
    --output sandbox/cna_plots \
    --output-prefix hcc1143_T_clean
```

The resulting plot includes both the standardized copy ratios produced by the first internal step and the final denoised copy ratios (Figure 7-9).

You can use gsutil to copy the output plots from your VM to your Google bucket and then view them (or download them) from the GCP console. If you need a refresher on this, see Chapter 4. The following is done from your VM (which has the gsutil tool installed and configured in Chapter 4) rather than the Docker container you've run the gatk commands in, and uploads to the bucket that you specify (my-bucket) in this example:

```
$ export BUCKET="gs://my-bucket"
$ gsutil -m cp -R sandbox/cna_plots $BUCKET/somatic-sandbox/
```

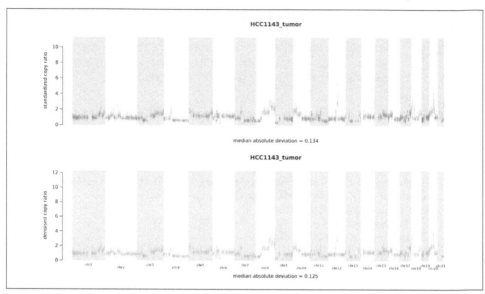

Figure 7-9. Copy-number alteration analysis plots showing the standardized copy ratios after the first step of denoising (top) and the fully denoised copy ratios after the second round (bottom).

It can be difficult to see very clearly in the book version of these images, but if you zoom in on the plots you generated yourself, you should be able to see that the range of values in the second-round image is tighter than in the first round, which shows the improvement from one step to the next. In any case, we can agree that these show much better resolution compared to the raw read counts shown in Figure 7-8, and we can already discern some probable CNAs.

Performing Segmentation and Call CNAs

Now that we have our lovingly denoised copy ratios, all we need to do is identify exactly which regions demonstrate evidence of CNA. To do that, we're going to group neighboring intervals that have similar copy ratios into segments. Then, we should be able to determine which segments have an overall copy ratio that supports the presence of an amplification or deletion relative to the rest.

We do this using the ModelSegments tool, which uses a Gaussian-kernel binary-segmentation algorithm to perform multidimensional kernel segmentation.

```
# gatk ModelSegments \
    --denoised-copy-ratios sandbox/hcc1143_T_clean.denoisedCR.tsv \
    --output sandbox  \
    --output-prefix hcc1143_T_clean
```

As for the previous plots, the easiest way to view these is to transfer them to your Google bucket and open (or download) them from the GCP console. This command produces multiple files, so the output name we provide is simply the name that we want to give to the new directory that will contain them. Again, the outputs are not particularly user friendly, so we're going to make some plots to visualize the results. We provide the plotting tool with the denoised copy ratios (from DenoiseRead Counts), the segments (from ModelSegments), and the reference sequence dictionary:

```
# gatk PlotModeledSegments \
    --denoised-copy-ratios sandbox/hcc1143_T_clean.denoisedCR.tsv \
    --segments sandbox/hcc1143_T_clean.modelFinal.seg \
    --sequence-dictionary ref/Homo_sapiens_assembly38.dict \
    --minimum-contig-length 46709983 \
    --output sandbox/cna_plots \
    --output-prefix hcc1143_T_clean
```

Figure 7-10 shows the segments identified by the PlotModelSegments tool. The dots representing targets or bins on alternate segments are colored in either blue and orange, whereas segment medians are drawn in black. Most segments have a copy ratio of approximately 1, which constitutes the baseline. Against that backdrop, we can observe multiple amplifications and deletions of different sizes and different copy ratios. You might notice that most of these don't correspond to the numbers that you would expect if you had, for example, four copies instead of two for a given segment. That is because some of the CNAs occur in subclonal populations, meaning that only part of the sampled tissue is affected. As a result, the effect is diluted and you might see a copy ratio of 1.8 instead of 2.

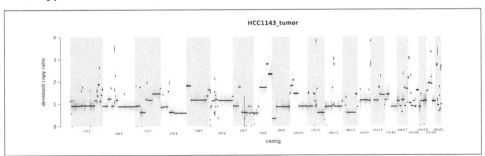

Figure 7-10. Plot of segments modeled based on denoised copy ratios.

More generally, it seems like a lot is going on in this sample; in fact, we can look in the segmentation file in the output directory and see that we've predicted 235 segments for this sample. This isn't typical of regular tumor samples, although some can be pretty messed up; what's going on here is that we're looking at a sample taken from a cell line. Cancer cell lines are convenient for teaching and testing software because the data can easily be made open access, but biologically they tend to have accumulated very high numbers of alterations. So this represents an extreme situation. On the bright side, most samples that you'll encounter in the real world are likely to be cleaner (and easier to interpret) than this. So you have something to look forward to!

That being said, the software can still help us get a bit more clarity even on a messy sample like this. We can get final CNA calls—that is, a determination of which segments we can consider to have significant evidence of amplification or deletion—by running the last tool in this workflow, as follows:

```
# gatk CallCopyRatioSegments \
    -I sandbox/hcc1143_T_clean.cr.seg \
    -O sandbox/hcc1143_T_clean.called.seg
```

This tool adds a column to the segmented copy-ratio *.cr.seg* file from ModelSegments, marking amplifications (+), deletions (-), and neutral segments (0):

```
# tail -5 sandbox/hcc1143_T_clean.called.seg
chrX    118974529    139746109    864    0.475183     +
chrX    139749773    139965748    28    -0.925385     -
chrX    140503468    153058699    277    -0.366860     -
chrX    153182138    153580550    47    0.658197     +
chrX    153588113    156010661    544    0.075279     0
```

And there you have it; this shows you the output of the somatic copy-number workflow. If we look at the full progression, you can see that we went from raw coverage counts all the way to having identified specific regions of deletion or amplification, as illustrated in Figure 7-11 (using different data and only a subset of chromosomes).

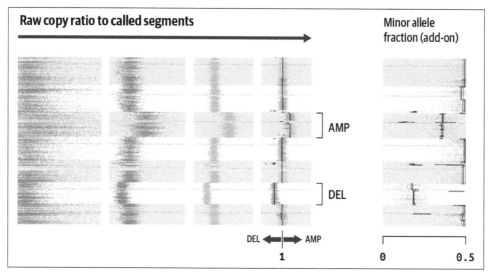

Figure 7-11. Full progression from raw data to results.

Copy ratio counts in Figure 7-11 are calculated and processed as shown in previous figures, rotated 90° to the right. The leftmost panel shows raw copy ratio counts calculated per target or genomic bin. The second and third panels show standardized and denoised counts, respectively. The fourth panel shows segmentation and calling results, including one whole chromosome amplification (AMP) and one deletion (DEL). The rightmost panel shows a minor allele fraction calculated using the procedure outlines in the next section.

Ironically, these results don't include an actual copy number, just the boundaries of each segment, its copy ratio, and the final call. To get the copy number, you will need to do some additional work that is beyond the scope of this chapter. See the documentation (*https://oreil.ly/NXTuz*) for more information and next steps.

Additional Analysis Options

We've just covered the standard workflow that is most commonly used for performing copy-number analysis with GATK tools. This involved collecting coverage counts, using a PoN to apply denoising, and performing segmentation in order to model and finally call copy-number–altered segments on the tumor case sample. However, you can apply two additional options to further beef up the quality of the results.

Tumor-Normal pair analysis

The Tumor-only workflow gets you decent, usable results for somatic CNA discovery. But if you have a matched normal available, you can gain additional benefits! For one thing, you can run the workflow, as described earlier on the normal sample, which is helpful for understanding the baseline that you're working with. In many cases, having the matched normal can also help the denoising process.

Feel free to run through the commands on your own. In the final segmentation results, notice the probable CNAs on chromosome 2 and chromosome 6. This matched normal is itself from a cell line, albeit derived from healthy tissue instead of a tumor. It's not uncommon to see some background alterations occur even in normal tissue when it's been propagated as a cell line.

 This concludes the exercises for this chapter, so once again, remember to stop your VM if you're taking a break before moving on to the next chapter.

Allelic copy ratio analysis

So far, we've used only read coverage as evidence for CNAs, but we can also include *allelic copy ratio*: the copy ratio for different alleles at sites with heterozygous SNVs. This can reveal interesting things about the heterozygosity of our case sample, which can have important biological consequences—especially when there is a *loss of heterozygosity*.

We can detect imbalances in the heterozygosity of segments by looking at the allelic counts (i.e., the read counts for different alleles at each site) in the case sample at sites that are commonly variant in the population. For a diploid organism, we expect to see any heterozygous SNV manifest as two different alleles present in equal proportions. Suppose that we have an A/T variant; we expect to see approximately 50% A and 50% T in the sequence reads (give or take some technical variability). We can detect those sites on the fly when we're counting the reads in the first step of the workflow. Then, we can use those sites as markers; if we look at those sites in the tumor and find that their allelic ratios are significantly skewed, that gives us additional information about the degree of CNA that affects the segments of which these sites are part.

If you're interested in exploring this additional analysis mode, have a look at the somatic copy-number documentation (*https://oreil.ly/NXTuz*) on the GATK website.

Wrap-Up and Next Steps

Well then, between the germline workflow and these two somatic analysis workflows, we've already run a lot of genomics commands. So, does that mean we're ready to run full-scale genomic analyses?

Hah, no. At this point you can run individual tools one by one—which is going to be useful at various stages of your work, especially for initial testing/evaluation as well as troubleshooting failures—but you also know that the variant discovery Best Practices are composed of multiple tasks involving various tools. Trust us, you don't want to be running each task manually for each sample or set of samples that you need to process. You're going to want to chain these tasks into scripts that automate the bulk of the work to cut down on the mindless tedium as well as to reduce the chance of human error. So in Chapter 8, we show you how to write workflow scripts using WDL, which automates calling each step in the overall analysis pipeline. Then, we run them using a workflow management system called Cromwell that is equally comfortable running on your laptop, cluster, or cloud platform.

Automating Analysis Execution with Workflows

So far, we've been running individual commands manually in the terminal. However, the overwhelming majority of genomics work proper—the secondary analysis, in which we go from raw data to distilled information that we'll then feed into downstream analysis to ultimately produce biological insight—involves running the same commands in the same order on all the data as it rolls off the sequencer. Look again at the GATK Best Practices workflows described in Chapters 6 and 7; imagine how tedious it would be to do all that manually for every sample. Newcomers to the field often find it beneficial to run through the commands involved step by step on test data, to gain an appreciation of the key steps, their requirements, and their quirks—that's why we walked you through all that in Chapters 6 and 7—but at the end of the day, you're going to want to automate all of that as much as possible. Automation not only reduces the amount of tedious manual work you need to do, but also increases the throughput of your analysis and reduces the opportunity for human error.

In this chapter, we cover that shift from individual commands or one-off scripts to repeatable workflows. We show you how to use a workflow management system (*Cromwell*), and language, WDL, that we selected primarily for their portability. We walk you through writing your first example workflow, executing it, and interpreting Cromwell's output. Then, we do the same thing with a couple of more realistic workflows that run GATK and use scatter-gather parallelism.

Introducing WDL and Cromwell

You might remember that we introduced the concept of workflows in the technology primer (Chapter 3). As we noted at the time, many tooling options are available for writing and running workflows, with a variety of features, strengths, and weaknesses. We can't tell you the *best option overall*, because much of that determination would depend on your particular backgrounds and needs. For the purposes of this book, we've selected an option, the combination of Cromwell and WDL, that is most suitable for the breadth of audience and scope of needs that we are targeting.

We already briefly introduced WDL at that time, as well, but let's recap the main points given that it was half a book ago and a lot of water under the bridge. As just mentioned, its full name is *Workflow Description Language*, but it's generally referred to as WDL (pronounced "widdle") for short. It's a DSL commonly used for genomics workflows that was originally developed at the Broad Institute and has since then evolved into a community-driven open source project under the auspices of a public group named OpenWDL (*https://openwdl.org*). As a workflow language, it's designed to be very accessible to bioinformatics newcomers who do not have any formal training in software programming while maximizing portability across different systems.

The workflow management system we're going to use to actually *run* workflows written in WDL is Cromwell, an open source application, also developed at the Broad Institute. Cromwell is designed to be runnable in just about any modern Java-enabled computing environment, including popular HPC systems like Slurm and SGE, commercial clouds like GCP and AWS, and any laptop or desktop computer running on some flavor of Linux. This portability is a core feature of Cromwell, which is intended to facilitate running the same workflows at multiple institutions in order to maximize scientific collaboration and computational reproducibility in research. Another major portability-oriented feature of Cromwell is that it supports (but does not require) the use of containers for providing the code that you want to run within each component task of a given workflow. You'll have the opportunity to try running both ways in the exercises in this chapter.

Finally, continuing in the spirit of collaboration and interoperability, Cromwell is also designed to support multiple workflow languages. Currently, it supports WDL as well as the CWL, another popular workflow language designed for portability and computational reproducibility.

Cromwell offers two modes of operation: the one-shot run mode and the server mode. The *one-shot run mode* is a simple way of running Cromwell that involves a single command line: you give it a workflow and a file that lists workflow inputs, and it will start running, execute the workflow, and finally shut down when the workflow is done. This is convenient for running workflows episodically with minimal hassle, and it's what we use in the exercises in this chapter. The *server mode* of operation involves setting up a persistent server that is always running, and submitting workflow execution requests to that server via a REST API (a type of programmatic interface).

Starting the Cromwell server is quite easy. After it's running, it offers functionality that is not available in the single-run mode, some of which we cover in Chapter 11. However, managing its operation securely on an ongoing basis requires a specialized skill set that most individual researchers or small groups without dedicated support staff do not possess. In Chapter 11, we introduce you to Terra, a managed system operated by the Broad Institute that provides access to a persistent Cromwell server through a GUI as well as an API. That will give you the opportunity to try out Cromwell in server mode without having to administer a server yourself.

> We won't cover Cromwell server administration in this book, so see the Cromwell documentation (*https://oreil.ly/8T7j8*) if you're interested in learning more.

Jamie Cromwell, the Warp Pig

Wildly popular in sticker form, Jamie the Warp Pig is the Cromwell development team's homage to James Cromwell, the American movie actor who starred in classics including nerd favorite *Star Trek 8: First Contact* and the children's live-action farm tale *Babe*.

In *First Contact*, Mr. Cromwell played Zefram Cochrane, the brilliant but frequently inebriated inventor of warp drive. In *Babe*, he played the gruff but ultimately kindhearted farmer who wins the movie's eponymous piglet at a country fair and changes his Christmas dinner plans when he realizes the young pig has the makings of a great sheepdog (er, sheep-pig?). The development team at the Broad Institute recognized in Mr. Cromwell's dramatic range the flexibility that is frequently required from Cromwell the workflow engine, which is expected to run workloads ranging from trivial to extreme in a variety of computing environments. That'll do, pig. That'll do.

Whether you run it as a one-shot or in server mode, Cromwell has interesting features that aim to promote efficiency and scalability—but no one wants to read a laundry list of features in a book like this, so let's move on to the exercises and we'll bring up those key features as they become relevant along the way.

Installing and Setting Up Cromwell

In this chapter, we examine and execute some workflows written in WDL to become acquainted with the basic structure of the language and learn how Cromwell manages inputs and outputs, logs, and so on. For continuity with the previous chapters, we run Cromwell on the GCP Compute Engine VM that we used earlier, in Chapters 4 and 5. However, we're no longer running anything from within the GATK container. Instead, we install and run Cromwell directly in the VM environment.

You'll need to run a few installation commands because Cromwell requires Java, which is not preinstalled on the VM we're using. To do so, log back in to your VM via SSH, just as you did in earlier chapters. Remember that you can always find your list of VM instances in the GCP console by going directly to the Compute Engine (*https://oreil.ly/sGeug*), or, in the menu of GCP services on the left side of the console, click Compute Engine, if you've forgotten the URL.

In your VM, type **java -version** at the prompt. You should get the following output:

```
$ java -version
Command 'java' not found, but can be installed with:
apt install openjdk-11-jre-headless   # version 11.0.3+7-1ubuntu2~19.04.1, or
apt install default-jre               # version 2:1.11-71
apt install openjdk-8-jre-headless    # version 8u212-b03-0ubuntu1.19.04.2
apt install openjdk-12-jre-headless   # version 12.0.1+12-1
apt install openjdk-13-jre-headless   # version 13~13-0ubunt1
Ask your administrator to install one of them.
```

Cromwell requires Java version 8, so let's install the openjdk-8-jre-headless option, which is a lightweight environment sufficient for our needs:

```
$ sudo apt install openjdk-8-jre-headless
Reading package lists... Done
Building dependency tree
Reading state information... Done
[...]
done.
```

This triggers the installation process, which should run to completion without error. You might see a few notifications but as long as you see that final done output, you should be fine. You can run the Java version check again to satisfy yourself that the installation was successful:

```
$ java -version
openjdk version "1.8.0_222"
OpenJDK Runtime Environment (build 1.8.0_222-8u222-b10-1ubuntu1~19.04.1-b10)
OpenJDK 64-Bit Server VM (build 25.222-b10, mixed mode)
```

With Java installed, let's set up Cromwell itself, which comes with a companion utility called `Womtool` that we use for syntax validation and creating input files. They are both distributed as compiled *.jar* files, and we've included a copy in the book bundle, so you don't need to do anything fancy except point to where they are located. To keep our commands as short as possible, let's set up an environment variable pointing to their location. Let's call it *BIN* for binary, a term often used to refer to the compiled form of a program:

```
$ export BIN=~/book/bin
```

Updating Cromwell and Womtool on Your VM

In this book, we use the latest release of Cromwell and `Womtool` available as of this writing, version 53.1. Because you live in the future, newer versions might be available from the Cromwell GitHub repository (*https://oreil.ly/FJQwF*). If you want or need to use a more recent version on your VM, you can copy the newer *.jars* from the repository to your VM using `curl`, a classic command-line tool that retrieves files via URL, as shown in the following example:

```
$ curl -L -o ~/book/bin/cromwell-53.1.jar \
https://github.com/broadinstitute/cromwell/releases/download/53/cromwell-53.jar
  % Total    % Received % Xferd  Average Speed   Time    Time     Time  Current
                                 Dload  Upload   Total   Spent    Left  Speed
100   641    0   641    0     0   2774      0 --:--:-- --:--:-- --:--:--  2774
100  205M  100  205M    0     0  66.2M      0  0:00:03  0:00:03 --:--:--  80.6M
```

In this command, the `-o` specifies the destination and filename of the copy, and the `-L` flag tells `curl` to follow redirect links, which is important because GitHub uses redirects to serve the actual file (hence, two lines appearing in `curl`'s log output).

To use this for a different version of Cromwell, simply change the version number accordingly. It shows up in three places: once in the destination filename, and twice in the URL. You can run the same command for `Womtool` as well; just substitute **womtool** for `cromwell` in the destination filename and in the URL.

Let's check that we can run Cromwell by asking for its `help` output, which presents a summary of the three commands you can give it: `server` and `submit` are part of the server mode we discussed earlier, and `run` is the one-shot mode that we use shortly:

```
$ java -jar $BIN/cromwell-53.1.jar --help
cromwell 53.1
Usage: java -jar /path/to/cromwell.jar [server|run|submit] [options] <args>...
  --help                  Cromwell - Workflow Execution Engine
  --version
Command: server
Starts a web server on port 8000.  See the web server documentation for more
details about the API endpoints.
Command: run [options] workflow-source
```

```
Run the workflow and print out the outputs in JSON format.
 workflow-source            Workflow source file or workflow url.
 --workflow-root <value>    Workflow root.
 -i, --inputs <value>       Workflow inputs file.
 -o, --options <value>      Workflow options file.
 -t, --type <value>         Workflow type.
 -v, --type-version <value>
                            Workflow type version.
 -l, --labels <value>       Workflow labels file.
 -p, --imports <value>      A zip file to search for workflow imports.
 -m, --metadata-output <value>
                            An optional JSON file path to output metadata.
Command: submit [options] workflow-source
Submit the workflow to a Cromwell server.
 workflow-source            Workflow source file or workflow url.
 --workflow-root <value>    Workflow root.
 -i, --inputs <value>       Workflow inputs file.
 -o, --options <value>      Workflow options file.
 -t, --type <value>         Workflow type.
 -v, --type-version <value>
                            Workflow type version.
 -l, --labels <value>       Workflow labels file.
 -p, --imports <value>      A zip file to search for workflow imports.
 -h, --host <value>         Cromwell server URL.
```

It's also worth doing the same thing for Womtool, to get a sense of the various utility commands available:

```
$ java -jar $BIN/womtool-53.1.jar --help
Womtool 53.1
Usage: java -jar Womtool.jar
[validate|inputs|parse|highlight|graph|upgrade|womgraph] [options]
workflow-source
 workflow-source            Path to workflow file.
 -i, --inputs <value>       Workflow inputs file.
 -h, --highlight-mode <value>
                            Highlighting mode, one of 'html', 'console'
(used only with 'highlight' command)
 -o, --optional-inputs <value>
                            If set, optional inputs are also included in the
inputs set. Default is 'true' (used only with the inputs command)
 --help
 --version
Command: validate
Validate a workflow source file. If inputs are provided then 'validate'
also checks that the inputs file is a valid set of inputs for the
workflow.
Command: inputs
Generate and output a new inputs JSON for this workflow.
Command: parse
(Deprecated; WDL draft 2 only) Print out the Hermes parser's abstract
syntax tree for the source file.
Command: highlight
```

```
(Deprecated; WDL draft 2 only) Print out the Hermes parser's abstract
syntax tree for the source file. Requires at least one of 'html' or 'console'
Command: graph
Generate and output a graph visualization of the workflow in .dot format
Command: upgrade
Automatically upgrade the WDL to version 1.0 and output the result.
Command: womgraph
(Advanced) Generate and output a graph visualization of Cromwell's
internal Workflow Object Model structure for this workflow in .dot format
```

Of the functions just listed, you'll have the opportunity to use inputs, validate, and graph in this chapter.

Now let's check that you have all the workflow files that we provide for this chapter. If you followed the setup instructions in Chapter 4, you should have a code directory that you cloned from GitHub. Under *~/book/code*, you'll see a directory called *workflows* that contains all of the code and related files that you'll use in this chapter (aside from the data, which came from the bucket). You're going to run commands from the home directory (instead of moving into a subdirectory as we did in earlier chapters), so to keep paths short in the various commands, let's set up an environment variable to point to where the workflow files reside:

```
$ export WF=~/book/code/workflows
```

Finally, let's talk about text editors. In all but one of the exercises that follow, you're simply going to view and run prewritten scripts that we provide, so for viewing you could just download or clone the files to your laptop and open them in your preferred text editor. In one exception, we suggest you modify a WDL to break it in order to see what Cromwell's error messaging and handling behavior looks like, so you'll need to actually edit the file. We show you how to do this using one of the shell's built-in text editors, called nano, which is considered one of the most accessible for people who aren't used to command-line text editors. You are of course welcome to use another shell editor like vi or emacs if you prefer; if so, it will be up to you to adapt the commands we provide accordingly.

Choosing a Text Editor for WDL Development

Moving beyond the scope of the book, if you want to write or edit your own WDLs, you might find the shell command-line text editors too limited. Whether you prefer to use a full integrated development environment (IDE) application like IntelliJ, or a simpler text editor with a GUI like Sublime, you can find a few recommended options that accept WDL syntax highlighting plug-ins in this online documentation (*https:// oreil.ly/JSNo1*). Note that you'll need to transfer any code that you edit on your local machine back to your VM every time you want to try it out. Alternatively, you could test your code locally (Cromwell can run on your desktop) and run in the cloud only when you want to run at scale.

> Another alternative is to use Google Cloud Shell, which we introduced in Chapter 4, for your WDL development purposes. Cloud Shell provides a point-and-click code editor interface (currently in beta) which is a solid option for working entirely in the cloud. Unfortunately, it does not yet accept WDL syntax highlighting.

Whatever you decide to use as a text editor, just make sure not to use a *word processor* like Microsoft Word or Google Docs. Those applications can introduce hidden characters and are therefore not appropriate for editing code files. With that all sorted out, let's buckle up and tackle your very first WDL workflow.

Your First WDL: Hello World

We begin with the simplest possible working example of a WDL script: the quintessential HelloWorld. If you're not familiar with this, it's a common trope in the documentation of programming languages; in a nutshell, the idea is to provide an introductory example with the minimum amount of code that produces the phrase HelloWorld!. We're actually going to run through three basic WDL workflows to demonstrate this level of functionality, starting with the absolute minimum example, and then adding on just enough code to show core functionality that is technically not required yet needed for realistic use.

Learning Basic WDL Syntax Through a Minimalist Example

Let's pull up the simplest example by loading the *hello-world.wdl* workflow file into the nano editor:

```
$ nano $WF/hello-world/hello-world.wdl
```

As noted earlier, nano is a basic editor. You can use the arrow keys on your keyboard to move through the file. To exit the editor, press Ctrl+X.

This is what the minimalistic Hello World for WDL looks like:

```
version 1.0

workflow HelloWorld {
  call WriteGreeting
}

task WriteGreeting {
  command {
      echo "Hello World"
  }
  output {
      File output_greeting = stdout()
  }
}
```

First, let's ignore everything except the one line that has the phrase HelloWorld in it, the one in which it's in quotes. Do you recognize the command on that line? That's right, it's a simple echo command; you can run that line by itself right now in your terminal:

```
$ echo "Hello World"
Hello World
```

So that's the command at the heart of our script that performs the desired action, and everything else is wrapping to make it runnable in scripted form through our workflow management system.

Now let's unpack that wrapping. At the highest level, we have just two distinct stanzas, or blocks of code: the one starting with workflow HelloWorld, and the one starting with task WriteGreeting, with several lines of code between the curly braces in each case (the original designer of WDL really liked curly braces; you'll see a lot more of them). We can summarize them like this:

```
workflow HelloWorld {...}

task WriteGreeting {...}
```

This makes it really clear that our script is structured in two parts: the workflow block, which is where we call out the actions that we want the workflow to perform, and a task block, where we define the action details. Here we have only one task, which is not really typical given that most workflows consist of two or more tasks; we cover workflows with multiple tasks further in this section.

Let's take a closer look at how the action—that is, the command—is defined in the WriteGreeting task:

```
task WriteGreeting {
  command {
    echo "Hello World"
  }
  output {
    File output_greeting = stdout()
  }
}
```

In the first line, we're declaring that this is a task called WriteGreeting. Within the outermost curly braces, we can break the structure of the code into another two blocks of code: command {...} and output {...}. The command block is quite straightforward: it contains the echo "Hello World" command. So that's pretty self-explanatory, right? In general, you can stick just about anything in there that you would run in your terminal shell, including pipes, multiline commands, and even blocks of "foreign" code like Python or R, provided that you wrap it in heredoc syntax (*https://oreil.ly/VK1F8*). We provide examples of what that looks like in Chapter 9.

Meanwhile the `output` block is perhaps a bit less obvious. The goal here is to define the output of the `command` block we plan to run. We're declaring that we expect the output will be a `File`, which we choose to call `output_greeting` (this name can be anything you want, except one of the reserved keywords, which are defined in the WDL specification). Then, in the slightly tricky bit, we're stating that the output content itself will be whatever is emitted to `stdout`. If you're not that familiar with command-line terminology, `stdout` is short for *standard out*, and refers to the text output to the terminal window, meaning it's what you see displayed in the terminal when you run a command. By default, this content is also saved to a text file in the execution directory (which we examine shortly), so here we're saying that we designate that text file as the output of our command. It's not a terribly realistic thing to do in a genomics workflow (although you might be surprised...we've seen stranger things), but then that's what a Hello World is like!

Anyway, that's our `task` block explained away. Now, let's look at the `workflow` block:

```
workflow HelloWorld {
  call WriteGreeting
}
```

Well, that's pretty simple. First, we declare that our workflow is called `HelloWorld`, and then, within the braces, we make a `call` statement to invoke the `WriteGreeting` task. This means that when we actually run the workflow through Cromwell, it will attempt to execute the `WriteGreeting` task. Let's try that out.

Running a Simple WDL with Cromwell on Your Google VM

Exit the `nano` editor by pressing Ctrl+X and return to the shell of your VM. You're going to launch the *hello-world.wdl* workflow using the Cromwell *.jar* file that resides in the *~/book/bin* directory, which we aliased as `$BIN` during the setup part of this chapter. The command is straightforward Java:

```
$ java -jar $BIN/cromwell-53.1.jar run $WF/hello-world/hello-world.wdl
```

This command invokes Java to run Cromwell using its one-off (`run`) workflow execution mode, which we contrasted with the persistent `server` mode earlier in this chapter. So it's just going to start up, run the workflow we provided as an input to the `run` command, and shut down when that's done. For the moment, there is nothing else involved because our workflow is entirely self-contained; we cover how to parameterize the workflow to accept input files next.

Go ahead and run that command. If you have everything set up correctly, you should now see Cromwell begin to spit out a lot of output to the terminal. We show the most relevant parts of the output here, but we've omitted some blocks (indicated by [...]) that are of no interest for our immediate purposes:

```
[...]
[2018-09-08 10:40:34,69] [info] SingleWorkflowRunnerActor: Workflow submitted
b6d224b0-ccee-468f-83fa-ab2ce7e62ab7
[...]
Call-to-Backend assignments: HelloWorld.WriteGreeting -> Local
[2018-09-08 10:40:37,15] [info] WorkflowExecutionActor-b6d224b0-ccee-468f-83fa-
ab2ce7e62ab7 [b6d224b0]: Starting HelloWorld.WriteGreeting
[2018-09-08 10:40:38,08] [info] BackgroundConfigAsyncJobExecutionActor
[b6d224b0HelloWorld.WriteGreeting:NA:1]: echo "Hello World"
[2018-09-08 10:40:38,14] [info] BackgroundConfigAsyncJobExecutionActor
[...]
[2018-09-08 10:40:40,24] [info] WorkflowExecutionActor-b6d224b0-ccee-468f-83fa-
ab2ce7e62ab7 [b6d224b0]: Workflow HelloWorld complete. Final Outputs:
{
   "HelloWorld.WriteGreeting.output_greeting": "/home/username/cromwell-
executions/HelloWorld/b6d224b0-ccee-468f-83fa-ab2ce7e62ab7/call-
WriteGreeting/execution/stdout"
}
[2018-09-08 10:40:40,28] [info] WorkflowManagerActor WorkflowActor-b6d224b0-ccee-
468f-83fa-ab2ce7e62ab7 is in a terminal state: WorkflowSucceededState
[2018-09-08 10:40:45,96] [info] SingleWorkflowRunnerActor workflow finished with
status 'Succeeded'.
[...]
[2018-09-08 10:40:48,85] [info] Shutdown finished.
```

As you can see, Cromwell's standard output is a tad…well, verbose. Cromwell has been designed primarily for use as part of a suite of interconnected services, which we discuss in Chapter 11, where there is a dedicated interface for monitoring progress and output during routine use. The single-run mode is more commonly used for troubleshooting, so the development team has chosen to make the local execution mode very chatty to help with debugging. This can feel a bit overwhelming at first, but don't worry: we're here to show you how to decipher it all—or at least the parts that we care about.

Interpreting the Important Parts of Cromwell's Logging Output

First, let's check that the output of our workflow is what we expected. Find this set of lines in the terminal output:

```
WorkflowExecutionActor-b6d224b0-ccee-468f-83fa-ab2ce7e62ab7 [b6d224b0]: Workflow
HelloWorld complete. Final Outputs:
{
   "HelloWorld.WriteGreeting.output_greeting": "/home/username/cromwell-
executions/HelloWorld/b6d224b0-ccee-468f-83fa-ab2ce7e62ab7/call-
WriteGreeting/execution/stdout"
}
```

Without going into the details just yet, we see that this provides a list in JSON format of the output files that were produced; in this case, just the one file that captured the stdout of our one echo "Hello World" command. Cromwell gives us the fully qualified path, meaning it includes the directory structure above the working directory, which is really convenient because it allows us to use it in any command with a quick copy and paste. You can do that right now to look at the contents of the output file and verify that it contains what we expect:

 Keep in mind that in the command we show here, you need to replace the username and the execution directory hash. It might be easier to look for the equivalent line in your output than to customize our command.

```
$ cat ~/cromwell-executions/HelloWorld/b6d224b0-ccee-468f-83fa-
ab2ce7e62ab7/call-WriteGreeting/execution/stdout
Hello World
```

And there it is! So we know it worked.

Now let's take a few minutes to walk through the information that Cromwell is giving us in all that log output to identify the most relevant nuggets:

```
SingleWorkflowRunnerActor: Workflow submitted b6d224b0-ccee-468f-83fa-ab2ce7e62ab7
```

= *I'm looking at this one workflow and assigning it this unique identifier.*

Cromwell assigns a randomly generated unique identifier to every run of every workflow and creates a directory with that identifier, within which all of the intermediate and final files will be written. We go over the details of the output directory structure in a little bit. For now, all you really need to know is that this is designed to ensure that you will never overwrite the results of a previous run of the same workflow or experience collisions between different workflows that have the same name:

```
Call-to-Backend assignments: HelloWorld.WriteGreeting -> Local
```

= *I'm planning to send this to the local machine for execution (as opposed to a remote server).*

By default, Cromwell runs workflows directly on your local machine; for example, your laptop. As we mentioned earlier, you can configure it to send jobs to a remote server or cloud service, instead; that's what in Cromwell lingo is called a *backend assignment* (not to be confused with diaper duty):

```
Starting HelloWorld.WriteGreeting
```

= *I'm executing the WriteGreeting task call from the HelloWorld workflow now.*

Cromwell treats each task call in the workflow as a separate job to execute, and will give you individual updates about each one accordingly. If the workflow involves multiple task calls, Cromwell will organize them in a queue and send each out for execution when appropriate. We discuss some aspects of how that works a little later. With regard to the status reporting aspect, you can imagine that as soon as we move to running more complex workflows, getting these reports through the standard out is rather impractical. This is where frontend software that provides an interface to parse and organize all of this information can really come in handy; you'll have an opportunity to experience that in Chapter 11:

```
[b6d224b0HelloWorld.WriteGreeting:NA:1]: echo "Hello World"
```

= This is the actual command I'm running for this call.

It's not obvious from this particular call because we didn't include any variables in our minimal Hello World example, but what Cromwell outputs here is the real command that will be executed. In the parameterized example that comes next, you can see that if we include a variable in the script, the log output will show the form of the command in which the variable has been replaced by the input value that we provide. This fully interpreted command also is output to the execution directory for the record:

```
[b6d224b0]: Workflow HelloWorld complete. Final Outputs:
{
"HelloWorld.WriteGreeting.output_greeting": "/home/username/cromwell-
executions/HelloWorld/b6d224b0-ccee-468f-83fa-ab2ce7e62ab7/call-
WriteGreeting/execution/stdout"
}
```

= I'm done running this workflow. This is the full path to that output file(s) you wanted.

As noted earlier, this provides a list in JSON format of all the output files that were produced, identified by their full namespace. The namespace

```
HelloWorld.WriteGreeting.output_greeting
```

tells us that we are looking at the `output_greeting` output by the call to the `Write Greeting` task belonging to the `HelloWorld` workflow.

The fully qualified path to the output file shows the entire directory structure; let's unroll that and examine what each segment corresponds to:

```
~                                            (working directory)
cromwell-executions/                         (Cromwell master directory)
 HelloWorld                                  (name of our workflow)
  b6d224b0-ccee-468f-83fa-ab2ce7e62ab7       (unique identifier of the run)
   call-WriteGreeting                        (name of our task call)
    execution                                (directory of execution files)
```

The important piece in this structure is the nesting of workflow/identifier/calls. As you'll see in the next exercise, any runs of a workflow with the same name will be added under the *HelloWorld* workflow directory, in a new directory with another unique identifier.

```
SingleWorkflowRunnerActor workflow finished with status 'Succeeded'.
```

= Yo, everything worked!

After that, the program log repeats the list of outputs along with the workflow identifier, and ends with a stack of shutdown status messages, which you typically don't need to worry about. And that's really all you need to care about at this point, concluding your first Cromwell workflow execution. Well done!

Adding a Variable and Providing Inputs via JSON

OK, but running a completely self-contained WDL is unrealistic, so let's look at how we add variables to bring in some external input that can change from run to run. In the nano editor, go ahead and open the *hello-world-var.wdl* from the code directory:

```
$ nano $WF/hello-world/hello-world-var.wdl
```

What's different? The workflow block is exactly the same, but now there's a bit more going on in the WriteGreeting task block:

```
task WriteGreeting {

  input {
      String greeting
  }

  command {
      echo "${greeting}"
  }

  output {
      File output_greeting = stdout()
  }
}
```

The Hello World input to the echo command has been replaced by ${greeting}, and we now have a new input block before the command block that contains the line String greeting. This line declares the variable called greeting and states that its value should be of type String; in other words, an alphanumeric sequence. This means that we have parameterized the greeting that will be echoed to the terminal; we're going to be able to instruct Cromwell what to insert into the command on a run-by-run basis.

This leads to the next question: how do we provide Cromwell with that value? We definitely don't want to have to give it directly on the command line, because although this particular case is simple, in the future we might need to run workflows that expect dozens of values, many of them more complex than a simple String.

Cromwell expects you to provide inputs in JavaScript Object Notation (*https://www.json.org*) (JSON) text format. JSON has a key:value pair structure that allows us to assign a value to each variable. You can see an example of this in the *$WF/hello-world/hello-world.inputs.json* file that we provide:

```
{
  "HelloWorld.WriteGreeting.greeting": "Hello Variable World"
}
```

In this simple *inputs* JSON file, we have defined the `greeting` variable from our `Hel loWorld` workflow by its fully qualified name, which includes the name of the workflow itself (`HelloWorld`) and then the name of the task (`WriteGreeting`) because we declared the variable at the task level and then the name of the variable itself.

To provide the *inputs* JSON file to Cromwell, simply add it by using the `-i` argument (short for `--input`) to your Cromwell command, as follows:

```
$ java -jar $BIN/cromwell-53.1.jar run $WF/hello-world/hello-world-var.wdl \
-i $WF/hello-world/hello-world.inputs.json
```

Look for the output the same way you did earlier; you should see the message in the file output by the workflow match the text in the JSON file.

Cromwell enforces the use of fully qualified names at all levels, which makes it impossible to declare global variables. Although this might feel like a burdensome constraint, it is much safer than the alternative, because it means that you can have variables with the same name in different parts of a workflow without causing collisions. In simple workflows, it's easy enough to keep track of variables and prevent such problems, but in more complex workflows with dozens of more variables, that can become quite difficult. That is especially the case when you use imports and subworkflows to facilitate code reuse, which we cover in Chapter 9 (ooh, spoilers). Note that you can declare a variable at the workflow level (and use the input naming syntax *WorkflowName.variable* in the *inputs* JSON file), but you'll need to pass it explicitly to any task calls in which you want to use it. You'll see an example of this in action later in this chapter.

Adding Another Task to Make It a Proper Workflow

Real-world workflows usually have more than one task, and some of their tasks are dependent on the outputs of others. In the `nano` editor, open *hello-world-again.wdl*:

```
$ nano $WF/hello-world/hello-world-again.wdl
```

Here's our third iteration attempt at a Hello World example, showing two tasks chained into a proper workflow:

```
version 1.0

workflow HelloWorldAgain {

  call WriteGreeting

  call ReadItBackToMe {
    input:
```

```
      written_greeting = WriteGreeting.output_greeting
  }

  output {
     File outfile = ReadItBackToMe.repeated_greeting
  }
}

task WriteGreeting {

  input {
     String greeting
  }

  command {
     echo "${greeting}"
  }
  output {
     File output_greeting = stdout()
  }
}

task ReadItBackToMe {

  input {
     File written_greeting
     String original_greeting = read_string(written_greeting)
  }

  command {
     echo "${original_greeting} to you too"
  }
  output {
     File repeated_greeting = stdout()
  }
}
```

You can see that the `workflow` block has quite a bit more going on now; it has an additional call statement pointing to a new task, `ReadItBackToMe`, and that call statement has some code in curly braces attached to it, which we'll call the `input` block:

```
call ReadItBackToMe {
    input:
        written_greeting = WriteGreeting.output_greeting
}
```

The `input` block allows us to pass values from the workflow level to a particular task call. In this case, we're referencing the output of the `WriteGreeting` task and assigning it to a variable called `written_greeting` for use within the `ReadItBackToMe` call.

Let's have a look at that new task definition:

```
task ReadItBackToMe {

  input {
    File written_greeting
  }

  String greeting = read_string(written_greeting)

  command {
    echo "${original_greeting} to you too"
  }
  output {
    File repeated_greeting = stdout()
  }
```

The `read_string()` bit is a function from the WDL standard library that reads in the contents of a text file and returns them in the form of a single string. So this task is meant to read in the contents of a file into a `String` variable and then use that variable to compose a new greeting and echo it to `stdout`.

In light of that, the extra code attached to the `ReadItBackToMe` call statement makes perfect sense. We're calling the `ReadItBackToMe` task and specifying that the input file we used to compose the new greeting should be the output of the call to the `Write Greeting` task.

Finally, let's look at the last block of code we haven't examined in this new version of the workflow:

```
output {
    File outfile = ReadItBackToMe.repeated_greeting
}
```

This workflow has an `output` block defined at the workflow level in addition to the individual task-level `output` blocks. This workflow-level output definition is entirely optional when the workflow is intended to be run by itself; it's more a matter of convention than function. By defining a workflow-level output, we communicate which of the outputs produced by the workflow we care about. That being said, you can use this output definition for functional purposes; for example, when the workflow is going to be used as a nested subworkflow and we need to pass its output to further calls. You'll see that in action in Chapter 9. For now, try running this workflow with the same input JSON as the previous workflow, then poke around the execution directories to see how the task directories relate to each other and where the outputs are located.

Your First GATK Workflow: Hello HaplotypeCaller

Now that you have a firm grasp of basic WDL syntax, let's turn to a more realistic set of examples: actual GATK pipelines! We begin with a very simple workflow in order to build your familiarity with the language gradually. We want a workflow that runs GATK `HaplotypeCaller` linearly (no parallelization) in GVCF mode on a single sample BAM file, as illustrated in Figure 8-1.

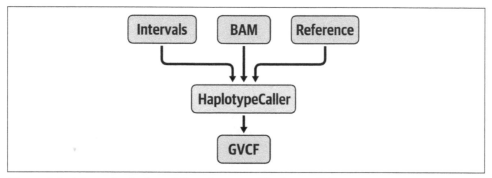

Figure 8-1. A hypothetical workflow that runs HaplotypeCaller.

The workflow should take the usual required files—the genome reference, input reads, and a file of intervals to analyze (technically optional as far as GATK is concerned, but here we made it required by the WDL)—and output a GVCF file named based on the input file.

Exploring the WDL

To illustrate all of that, we put together a WDL workflow that fulfills these requirements through a single task, `HaplotypeCallerGVCF`. Open it now in the nano editor:

```
$ nano $WF/hello-hc/hello-haplotypecaller.wdl
```

Let's walk through the main sections of the script, recalling the structure of our Hello-World example:

```
version 1.0

workflow HelloHaplotypeCaller {

    call HaplotypeCallerGVCF
}

task HaplotypeCallerGVCF {

 input {
        String docker_image
        String java_opt
```

```
        File ref_fasta
        File ref_index
        File ref_dict
        File input_bam
        File input_bam_index
        File intervals
    }

    String gvcf_name = basename(input_bam, ".bam") + ".g.vcf"

    command {
        gatk --java-options ${java_opt} HaplotypeCaller \
            -R ${ref_fasta} \
            -I ${input_bam} \
            -O ${gvcf_name} \
            -L ${intervals} \
            -ERC GVCF
    }

    output {
        File output_gvcf = "${gvcf_name}"
    }

    runtime {
        docker: docker_image
    }

}
```

Collapsing the task for clarity, you can see that it is indeed a single-task workflow, with only that single call and nothing else going on in the workflow block:

```
workflow HelloHaplotypeCaller {

    call HaplotypeCallerGVCF
}

task HaplotypeCallerGVCF { … }
```

So let's look at that HaplotypeCallerGVCF task in more detail, starting with the command block, because ultimately that's where we'll glean the most information about what the task actually does:

```
command {
    gatk --java-options ${java_opt} HaplotypeCaller \
        -R ${ref_fasta} \
        -I ${input_bam} \
        -O ${gvcf_name} \
        -L ${intervals} \
        -ERC GVCF
}
```

We see a classic GATK command that invokes `HaplotypeCaller` in GVCF mode. It uses placeholder variables for the expected input files as well as the output file. It also includes a placeholder variable for passing in Java options such as memory heap size, as described in Chapter 5. So far, that's pretty straightforward.

Those variables should all be defined somewhere, so let's look for them. Conventionally, we do this at the start of the task description, before the `command` block. This is what we see there:

```
input {
        String docker_image
        String java_opt

        File ref_fasta
        File ref_index
        File ref_dict
        File input_bam
        File input_bam_index
        File intervals
}

        String gvcf_name = basename(input_bam, ".bam") + ".g.vcf"
```

Ignoring the `String docker_image` line for now, this shows that we declared all of the variables for the input files as well as the Java options, but we didn't assign a value to them. So the task will expect to receive values for all of them at runtime. Not only that, but it will also expect values for all the accessory files we often take for granted: `refIndex`, `refDict`, and `inputBamIndex`, which refer to the indices and sequence dictionary. We don't include those files in the command itself because GATK detects their presence automatically (as long as their names conform with their master file's respective format conventions), but we do need to inform Cromwell that they exist so that it can make them available for execution at runtime.

There is one exception, though; for the output file, we see this line:

```
        String gvcf_name = basename(input_bam, ".bam") + ".g.vcf"
```

The `basename(input_bam, ".bam")` piece is a convenience function from the WDL standard library that allows us to create a name for our output file based on the name of the input file. The `basename()` function takes the full path of the input file, strips off the part of the path that's in front of the filename, and, optionally, strips off a given string from the end of the filename. In this case, we're stripping off the expected *.bam* extension and then we're using the `+ ".g.vcf"` part of the line to add the new extension that will be appropriate for the output file.

Speaking of the output file, let's now skip over to the task-level `output` block:

```
output {
    File output_gvcf = "${gvcf_name}"
}
```

This is also straightforward; we're stating that the command will produce an output file that we care about, giving it a name for handling it within the workflow, and providing the corresponding placeholder variable so that Cromwell can identify the correct file after the command has run to completion.

Technically, that is all you need to have in the workflow and task definition if you're planning to run this on a system with a local installation of the program that you want to run. However, in this chapter you're working in your VM but outside the GATK container, and as a result, GATK is not available directly to your workflow. Fortunately, Cromwell is capable of utilizing Docker containers, so we just need to add a `runtime` block to the workflow in order to specify a container image:

```
runtime {
    docker: docker_image
}
```

That's why we had that `String docker_image` line in our task variables: we're also using a placeholder variable for the container image. When we fill out the input JSON in the next step, we'll specify the `us.gcr.io/broad-gatk/gatk:4.1.3.0` image. Then, when we launch the workflow, Cromwell will spin up a new container from the image we specified and run GATK inside of it.

Technically we could hardcode the image name here, using double quotes (e.g., `docker: "us.gcr.io/broad-gatk/gatk:4.1.3.0"`) but we don't recommend doing that unless you really want to peg a particular script to a particular version, which reduces flexibility significantly. Some people use the `latest` tag to make their workflows always run with the latest available version of the program, but we consider that to be a bad practice with more downsides than upsides because you never know what might change in the latest version and break your workflow.

Generating the Inputs JSON

Alright, we've gone through all of the code in the workflow; now we need to determine how we're going to provide the inputs to the workflow when we run it. In "Your First WDL: Hello World" on page 216, we were running an extremely simple workflow, first without any variable inputs and then with a single one. In the single-input case, we created a JSON file specifying that one input. Now we have eight inputs that we need to specify. We could proceed in the same way as before—create a JSON file and write in the name of every input expected by the `HaplotypeCallerGVCF` task—but there's an easier way: we're going to use the `Womtool inputs` command to create a template JSON.

First, because we're going to be writing files that we care about for the first time in the chapter, let's make a *sandbox* directory to keep our outputs organized:

```
$ mkdir ~/sandbox-8
```

Now, you can run the Womtool command that generates the *inputs* JSON template file:

```
$ java -jar $BIN/womtool-53.1.jar \
    inputs $WF/hello-hc/hello-haplotypecaller.wdl \
    > ~/sandbox-8/hello-haplotypecaller.inputs.json
```

Because we specify an output file on the last line of this command (which is actually optional), the command writes its output to that file. If everything goes smoothly, you shouldn't see any output in the terminal. Let's look at the contents of the file that we just created:

```
$ cat ~/sandbox-8/hello-haplotypecaller.inputs.json

{
  "HelloHaplotypeCaller.HaplotypeCallerGVCF.input_bam_index": "File",
  "HelloHaplotypeCaller.HaplotypeCallerGVCF.input_bam": "File",
  "HelloHaplotypeCaller.HaplotypeCallerGVCF.ref_fasta": "File",
  "HelloHaplotypeCaller.HaplotypeCallerGVCF.ref_index": "File",
  "HelloHaplotypeCaller.HaplotypeCallerGVCF.ref_dict": "File",
  "HelloHaplotypeCaller.HaplotypeCallerGVCF.intervals": "File",
  "HelloHaplotypeCaller.HaplotypeCallerGVCF.docker_image": "String",
  "HelloHaplotypeCaller.HaplotypeCallerGVCF.java_opt": "String"
}
```

There you go: all of the inputs that the HaplotypeCallerGVCF task expects are listed appropriately, with a placeholder value stating their type, in JSON format (although they might be in a different order in yours). Now we just need to fill in the values; those are the paths to the relevant files for the first six, and the runtime parameters (container image and Java options) for the last two. In the spirit of laziness, we provide a filled-out version that uses the snippet data that we used in Chapter 5, but you could also go through the exercise of filling in the *inputs* JSON with other inputs from the data bundle that you downloaded in Chapter 4 if you like. This is what the prefilled JSON looks like (paths are relative to the home directory):

```
$ cat $WF/hello-hc/hello-haplotypecaller.inputs.json

{
"HelloHaplotypeCaller.HaplotypeCallerGVCF.input_bam_index":
"book/data/germline/bams/mother.bai",
"HelloHaplotypeCaller.HaplotypeCallerGVCF.input_bam":
"book/data/germline/bams/mother.bam",
"HelloHaplotypeCaller.HaplotypeCallerGVCF.ref_fasta":
"book/data/germline/ref/ref.fasta",
"HelloHaplotypeCaller.HaplotypeCallerGVCF.ref_index":
"book/data/germline/ref/ref.fasta.fai",
"HelloHaplotypeCaller.HaplotypeCallerGVCF.ref_dict":
"book/data/germline/ref/ref.dict",
```

```
"HelloHaplotypeCaller.HaplotypeCallerGVCF.intervals":
"book/data/germline/intervals/snippet-intervals-min.list",
"HelloHaplotypeCaller.HaplotypeCallerGVCF.docker_image": "us.gcr.io/broad-
gatk/gatk:4.1.3.0",
"HelloHaplotypeCaller.HaplotypeCallerGVCF.java_opt": "-Xmx8G"
}
```

Notice that all of the values are shown in double quotes, but this is a bit of an artifact because these values are all of `String` type. For other types such as numbers, Booleans, and arrays, you should not use double quotes. Except, of course, for arrays of strings, for which you should use double quotes around the strings, though not around the array itself, `["like","this"]`.

Running the Workflow

To run the workflow, we're going to use the same command-line syntax as earlier. Make sure to execute this in your home directory so that the relative paths in the *inputs* JSON file match the location of the data files:

```
$ java -jar $BIN/cromwell-53.1.jar \
    run $WF/hello-hc/hello-haplotypecaller.wdl \
    -i $WF/hello-hc/hello-haplotypecaller.inputs.json
```

As mentioned earlier, Cromwell output is quite verbose. For this exercise, you're looking for lines that look like these in the terminal output:

```
[2019-08-14 06:27:14,15] [info] BackgroundConfigAsyncJobExecutionActor
[9a6a9c97HelloHaplotypeCaller.HaplotypeCallerGVCF:NA:1]: Status change from
WaitingForReturnCode to Done
[2019-08-14 06:27:15,46] [info] WorkflowExecutionActor-9a6a9c97-7453-455c-8cd8-
be8af8cb6f7c [9a6a9c97]: Workflow HelloHaplotypeCaller complete. Final Outputs:
{
  "HelloHaplotypeCaller.HaplotypeCallerGVCF.output_gvcf": "/home/username/cromwell-
executions/HelloHaplotypeCaller/9a6a9c97-7453-455c-8cd8-be8af8cb6f7c/call-
HaplotypeCallerGVCF/execution/mother.g.vcf"
}
[2019-08-14 06:27:15,51] [info] WorkflowManagerActor WorkflowActor-9a6a9c97-7453-
455c-8cd8-be8af8cb6f7c is in a terminal state: WorkflowSucceededState
[2019-08-14 06:27:21,31] [info] SingleWorkflowRunnerActor workflow
status 'Succeeded'.
{
  "outputs": {
    "HelloHaplotypeCaller.HaplotypeCallerGVCF.output_gvcf":
"/home/username/cromwell-executions/HelloHaplotypeCaller/9a6a9c97-7453-455c-8cd8-
be8af8cb6f7c/call-HaplotypeCallerGVCF/execution/mother.g.vcf"
workflow HelloHaplotypeCaller {
  },
  "id": "9a6a9c97-7453-455c-8cd8-be8af8cb6f7c"
}
```

The most exciting snippets here are `Status change from WaitingForReturnCode to Done` and `finished with status 'Succeeded'`, which together mean that your workflow is done running and that all commands that were run stated that they were successful.

The other exciting part of the output is the path to the outputs. In the next section, we talk a bit about why they're listed twice; for now, let's just be happy that Cromwell tells us precisely where to find our output file so that we can easily peek into it. Of course, the output of this particular workflow is a GVCF file, so it's not exactly pleasant to read through, but the point is that the file is there and its contents are what you expect to see.

We use the `head` utility for this purpose; keep in mind that you'll need to substitute the execution directory hash in the file path shown in the following example (9a6a9c97-7453-455c-8cd8-be8af8cb6f7c) to the one displayed in your output:

```
$ head ~/cromwell-executions/HelloHaplotypeCaller/9a6a9c97-7453-455c
-8cd8-be8af8cb6f7c/call-HaplotypeCallerGVCF/execution/mother.g.vcf
##fileformat=VCFv4.2
##ALT=<ID=NON_REF,Description="Represents any possible alternative allele at this
location">
##FILTER=<ID=LowQual,Description="Low quality">
##FORMAT=<ID=AD,Number=R,Type=Integer,Description="Allelic depths for the ref and
alt alleles in the order listed">
##FORMAT=<ID=DP,Number=1,Type=Integer,Description="Approximate read depth (reads
with MQ=255 or with bad mates are filtered)">
##FORMAT=<ID=GQ,Number=1,Type=Integer,Description="Genotype Quality">
##FORMAT=<ID=GT,Number=1,Type=String,Description="Genotype">
##FORMAT=<ID=MIN_DP,Number=1,Type=Integer,Description="Minimum DP observed within
the GVCF block">
##FORMAT=<ID=PGT,Number=1,Type=String,Description="Physical phasing haplotype
information, describing how the alternate alleles are phased in relation to one
another">
##FORMAT=<ID=PID,Number=1,Type=String,Description="Physical phasing ID information,
where each unique ID within a given sample (but not across samples) connects
records
within a phasing group">
```

This run should complete quickly because we're using an intervals list that spans only a short region. Most of the time spent here is Cromwell getting started and spinning up the container, whereas GATK `HaplotypeCaller` itself runs for only the briefest of moments. As with the Hello World example, you might feel that this is an awful lot of work for such a small workload, and you'd be right; it's like swatting a fly with a bazooka. For toy examples, the overhead of getting Cromwell going dwarfs the analysis itself. It's when we get into proper full-scale analyses that a workflow management system like Cromwell really shows its value, which you'll experience in Chapters 10 and 11.

Breaking the Workflow to Test Syntax Validation and Error Messaging

Hopefully, so far everything ran as expected for you, so you know what success looks like. But in reality, things occasionally go wrong, so now let's look at what failure looks like. Specifically, we're going to look at how Cromwell handles two common types of scripting error, WDL syntax and command block syntax errors, by introducing some errors in the WDL. First, let's make a copy of the workflow file so that we can play freely:

```
$ cp $WF/hello-hc/hello-haplotypecaller.wdl ~/sandbox-8/hc-break1.wdl
```

Now, in your preferred text editor, open the new file and introduce an error in the WDL syntax. For example, you could mangle one of the variable names, one of the reserved keywords, or delete a curly brace to mess with the block structure. Here, we'll be a bit sadistic and delete the second parenthesis in the basename() function call. It's the kind of small error that is fatal yet really easy to overlook. Let's see what happens when we run this:

```
$ java -jar $BIN/cromwell-53.1.jar \
    run ~/sandbox-8/hc-break1.wdl \
    -i $WF/hello-hc/hello-haplotypecaller.inputs.json
```

As expected, the workflow execution fails, and Cromwell serves us with some verbose error messaging:

```
[2019-08-14 07:30:49,55] [error] WorkflowManagerActor Workflow 0891bf2c-4539-498c-
a082-bab457150baf failed (during MaterializingWorkflowDescriptorState):
cromwell.engine.workflow.lifecycle.materialization.MaterializeWorkflowDescriptorAct
or$$anon$1: Workflow input processing failed:
ERROR: Unexpected symbol (line 29, col 2) when parsing '_gen23'.

Expected rparen, got command .
        command {
^
$string_piece = :string

[stack trace]
[2019-08-14 07:30:49,57] [info] WorkflowManagerActor WorkflowActor-0891bf2c-4539-
498c-a082-bab457150baf is in a terminal state: WorkflowFailedState
```

There's a lot in there that we don't care about, such as the stack trace, which we're not showing here. The really important piece is Workflow input processing failed: ERROR: Unexpected symbol. That is a dead giveaway that you have a syntax issue. Cromwell will try to give you more specifics indicating where the syntax error might lie; in this case, it's pretty accurate—it expected but didn't find the closing parenthesis (rparen for right parenthesis) on line 20—but be aware that sometimes it's not as obvious.

When you're actively developing a new workflow, you probably won't want to have to launch the workflow through Cromwell each time you need to test the syntax of some new code. Good news: you can save time by using Womtool's validate command instead. That's what Cromwell actually runs under the hood, and it's a very light-weight way to test your syntax. Try it now on your broken workflow:

```
$ java -jar $BIN/womtool-53.1.jar \
    validate ~/sandbox-8/hc-break1.wdl

ERROR: Unexpected symbol (line 29, col 2) when parsing '_gen23'.

Expected rparen, got "command".
      command {
^
$string_piece = :string
```

See? You get the important part of Cromwell's output in a much shorter time frame—and you don't even need to provide valid inputs to the workflow. As an additional exercise, try introducing other WDL syntax errors; for example, try deleting a variable declaration, changing a variable name in only one of its appearances, and misspelling a reserved keyword like workflow, command, or outputs. This will help you to recognize validation errors and interpret how they are reported by Womtool. In general, we heartily recommend using Womtool validate systematically on any new or updated WDLs (and yes, it works on CWL too).

That being said, it's important to understand that Womtool's WDL syntax validation will get you only so far: it can't do anything about any other errors you might make that are outside its scope of expertise; for example, if you mess up the command syntax for the tool you want to run. To see what happens in that case, make another copy of the original workflow in your sandbox (call it *hc-break2.wdl*) and open it up to introduce an error in the GATK command this time; for example, by mangling the name of the tool, changing it to HaploCaller:

```
command {
        gatk --java-options ${java_opt} HaploCaller \
              -R ${refFasta} \
              -I ${inputBam} \
              -O ${gvcfName} \
              -L ${intervals} \
              -ERC GVCF
}
```

If you run Womtool validate, you'll see this workflow sails right through validation; Womtool cheerfully reports Success! Yet if you actually run it through Cromwell, the workflow will most definitely fail:

```
$ java -jar $BIN/cromwell-53.1.jar \
    run ~/sandbox-8/hc-break2.wdl \
    -i $WF/hello-hc/hello-haplotypecaller.inputs.json
```

Scroll through the output to find the line showing the failure message:

```
[2019-08-14 07:09:52,12] [error] WorkflowManagerActor Workflow dd77316f-7c18-4eb1
-aa86-e307113c1668 failed (during ExecutingWorkflowState): Job
HelloHaplotypeCaller.HaplotypeCallerGVCF:NA:1 exited with return code 2 which has
not been declared as a valid return code. See 'continueOnReturnCode' runtime
attribute for more details.
Check the content of stderr for potential additional information:
/home/username/cromwell-executions/HelloHaplotypeCaller/dd77316f-7c18-4eb1-aa86-
e307113c1668/call-HaplotypeCallerGVCF/execution/stderr.
[First 300 bytes]:Picked up _JAVA_OPTIONS: -Djava.io.tmpdir=/cromwell-
executions/HelloHaplotypeCaller/dd77316f-7c18-4eb1-aa86-e307113c1668/call-
HaplotypeCallerGVCF/tmp
.e6f08f65
USAGE: <program name> [-h]
Available Programs:
```

The log line that says Job HelloHaplotypeCaller.HaplotypeCallerGVCF:NA:1 exi
ted with return code 2 means that HaplotypeCallerGVCF was the task that failed
and, specifically, that the command it was running reported an exit code of 2. This
typically indicates that the tool you were trying to run choked on something—a syntax issue, unsatisfied input requirements, formatting errors, insufficient memory, and
so on.

The error message goes on to point out that you can find out more by looking at the
standard error (stderr) output produced by the command, and it helpfully includes
the full file path so that you easily can peek inside it. It also includes the first few lines
of the stderr log for convenience, which is sometimes enough if the tool's error output was very brief. The stderr output produced by GATK is of the more verbose
variety. So here we'll need to check out the full stderr to find out what went wrong.
Again, make sure to substitute the username and execution directory hash shown
here (dd77316f-7c18-4eb1-aa86-e307113c1668) with the ones in your output:

```
$ cat /home/username/cromwell-executions/HelloHaplotypeCaller/dd77316f-7c18-4eb1
-aa86-e307113c1668/call-HaplotypeCallerGVCF/execution/stderr
(...)
*********************************************************************
A USER ERROR has occurred: 'HaploCaller' is not a valid command.
Did you mean this?
      HaplotypeCaller
*********************************************************************
(...)
```

Ah, look at that! We wrote the name of the tool wrong; who knew? Props to GATK
for suggesting a correction, by the way; that's new in GATK4.

All command-line tools output a *return code* when they finish running as a concise way to report their status. Depending on the tool, the return code can be more or less meaningful. Conventionally, a return code of 0 indicates success, and anything else is a failure. In some cases, a nonzero code can mean the run was successful; for example, the Picard tool `ValidateSamFile` reports nonzero codes when it ran successfully but found format validation errors in the files it examined.

Other things can go wrong that we haven't covered here, such as if you have the wrong paths for input files, or if you forgot to wrap string inputs in double quotes in the inputs JSON. Again, we recommend that you experiment by making those errors on purpose, because it will help you learn to diagnose issues more quickly.

Introducing Scatter-Gather Parallelism

We're going to go over one more workflow example in this chapter to round out your first exposure to WDL and Cromwell, because we really want you to get a taste of the power of parallelization if you haven't experienced that previously. So now we're going to look at a workflow that parallelizes the operation of the `HaplotypeCaller` task we ran previously through a `scatter()` function and then merges the outputs of the parallel jobs in a subsequent step, as illustrated in Figure 8-2.

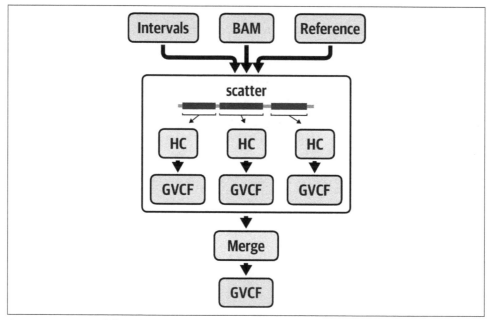

Figure 8-2. A workflow that parallelizes the execution of HaplotypeCaller.

This workflow will expose you to the magic of scatter-gather parallelism, which is a staple of genomics workflows, especially in the cloud. It will also give you an opportunity to dig into the mechanics of stringing multiple tasks together based on their inputs and outputs in a bit more detail than we covered earlier.

Exploring the WDL

Here's a full WDL that parallelizes the operation of `HaplotypeCallerGVCF` over subsets of intervals. In the `nano` editor, open it as usual:

```
$ nano $WF/scatter-hc/scatter-haplotypecaller.wdl
```

Let's walk through the main sections of the script, calling out what has changed compared to the linear implementation of this workflow:

```
version 1.0

workflow ScatterHaplotypeCallerGVCF {

  input {
        File input_bam
        File input_bam_index
        File intervals_list
    }

    String output_basename = basename(input_bam, ".bam")

    Array[String] calling_intervals = read_lines(intervals_list)

    scatter(interval in calling_intervals) {
        call HaplotypeCallerGVCF {
            input:
                input_bam = input_bam,
                input_bam_index = input_bam_index,
                intervals = interval,
                gvcf_name = output_basename + ".scatter.g.vcf"
        }
    }
    call MergeVCFs {
        input:
            vcfs = HaplotypeCallerGVCF.output_gvcf,
            merged_vcf_name = output_basename + ".merged.g.vcf"
    }

    output {
        File output_gvcf = MergeVCFs.mergedGVCF
    }
}

task HaplotypeCallerGVCF {
```

```
input {
        String docker_image
        String java_opt

        File ref_fasta
        File ref_index
        File ref_dict
        File input_bam
        File input_bam_index
        String intervals
        String gvcf_name
    }

    command {
        gatk --java-options ${java_opt} HaplotypeCaller \
            -R ${ref_fasta} \
            -I ${input_bam} \
            -O ${gvcf_name} \
            -L ${intervals} \
            -ERC GVCF
    }

    output {
        File output_gvcf = "${gvcf_name}"
    }

 runtime {
        docker: docker_image
    }
}

task MergeVCFs {

    input {
        String docker_image
        String java_opt

        Array[File] vcfs
        String merged_vcf_name
    }

    command {
        gatk --java-options ${java_opt} MergeVcfs \
            -I ${sep=' -I' vcfs} \
            -O ${merged_vcf_name}
    }

    output {
        File merged_vcf = "${merged_vcf_name}"
    }

    runtime {
```

```
        docker: docker_image
    }
}
```

The most obvious difference is that now a lot more is happening in the `workflow` block. The core of the action is this subset (collapsing the two `call` blocks for readability):

```
scatter(intervals in calling_intervals) {
        call HaplotypeCallerGVCF { ... }
}
call MergeVCFs { ... }
```

Previously, we were simply making one call to the `HaplotypeCallerGVCF` task, and that was it. Now, you see that the call to `HaplotypeCallerGVCF` is subordinated to a higher-level action, under this line, which opens a `scatter` block:

```
scatter(intervals in calling_intervals) {
```

For such a short line, it does a lot of work: this is how we parallelize the execution of `HaplotypeCaller` over subsets of intervals, which are specified in the parentheses. The `calling_intervals` variable refers to a list of intervals, and the `HaplotypeCal lerGVCF` task will be run as a separate invocation on each interval provided in that file. This might look a lot like a *for loop*, if you're familiar with that scripting construct, and indeed it is similar in the sense that its purpose is to apply the same operation to each element in a list. However, the scatter is specifically designed to allow independent execution of each operation, whereas a for loop leads to linear execution in which each operation can be run only after the previous one (if any) has run to completion.

So now that you understand what the `scatter()` instruction does, let's look into the `HaplotypeCallerGVCF` task itself, starting with how it's being called:

```
call HaplotypeCallerGVCF {
        input:
            input_bam = input_bam,
            input_bam_index = input_bam_index,
            intervals = intervals,
            gvcf_name = output_basename + ".scatter.g.vcf"
    }
```

In the previous version of the workflow, the call to the task was just that: the `call` keyword followed by the task name. In this version, the statement includes an `input` block that specifies values for some of the variables that are required by the task: the input BAM file and its index, the intervals file, and the name of the GVCF output file. Speaking of which, you'll notice that the `basename` function call that we use to generate a name for the output is now happening within this `input` block instead of happening within the task. If you compare the task definitions for `HaplotypeCallerGVCF` between this version of the workflow and the previous one, that's the only difference. Let's make a mental note of that because it will come up again in a few minutes.

Working with Intervals

GATK provides a lot of flexibility for specifying intervals; the `--intervals` argument can recognize simple strings, interval list files in several formats, and even VCF files. As a result, it is possible to scatter the execution of a tool like `HaplotypeCaller` in several ways. Here, we're showing the simplest option, which involves a single text file containing interval strings.

This works great if you have a small number of intervals, but it can become unwieldy if you are using lists with many intervals; for example, a list of exome targets, which typically contains thousands of very short intervals. If you were to use that directly here, you would end up with thousands of parallel jobs that run so quickly that the overhead of managing those jobs would be prohibitively inefficient, especially in the cloud context, where each requires spinning up a separate VM. Instead, we recommend extracting subsets of intervals into separate files, and providing a list of those separate files to `calling_intervals`. Here's the good news: you don't need to change anything in the WDL itself for this to work.

Now let's talk about what follows the `scatter` block containing the `HaplotypeCal lerGVCF` call: a call to a new task named `MergeVCFs`. To be clear, this call statement is *outside* of the `scatter` block, so it will be run only once:

```
call MergeVCFs { ... }
```

You probably already have a pretty good idea of what this task is for based on its name, but let's pretend it's not that obvious and follow the logical path for deciphering the structure of the workflow. Peeking ahead into the `MergeVCFs` task definition, we see that the command it runs is a GATK command that invokes a tool called `Mer geVcfs` (actually, a Picard tool bundled into GATK4). As input, it takes one or more VCF files (of which GVCFs are a subtype) and outputs a single merged file:

```
command {
    gatk --java-options ${java_opt} MergeVcfs \
        -I ${sep=' -I' vcfs} \
        -O ${vcf_name}
}
```

We can infer this from the command arguments and the variable names, and confirm it by looking up the `MergeVcfs` tool documentation on the GATK website.

One novel point of WDL syntax to note here: the -I ${sep=' -I' vcfs} formulation is how we deal with having a list of items of arbitrary length that we need to put into the command line with the same argument for each item. Given a list of files [FileA, FileB, FileC], the preceding code would generate the following portion of the command line: -I FileA -I FileB -I FileC.

The MergeVCFs task definition also tells us that this task expects a list of files (technically expressed as Array[File]) as its main input. Let's look at how the task is called in the workflow block:

```
call MergeVCFs {
        input:
        vcfs = HaplotypeCallerGVCF.output_gvcf,
        merged_vcf_name = output_basename + ".merged.g.vcf"
    }
```

This might feel familiar if you remember the ReadItBackToMe task in the HelloWorld Again workflow. Much as we did then, we're assigning the vcfs input variable by referencing the output of the HaplotypeCallerGVCF task. The difference is that in that case, we were passing a single-file output to a single-file input. In contrast, here we're referencing the output of a task call that resides inside a scatter block. What does that even look like?

Excellent question. By definition, each separate invocation of the task made within the scatter block generates its own separate output. The neat thing about the scatter construct is that those separate outputs are automatically collected into a list (technically an array) under the name of the output variable specified by the task. So here, although the output value of a single invocation of the HaplotypeCallerGVCF task, named HaplotypeCallerGVCF.output_gvcf, is a single GVCF file, the value of HaplotypeCallerGVCF.output_gvcf referenced in the workflow block is a list of the GVCF files generated within the scatter block. When we pass that reference to the next task call, we're effectively providing the full list of files as an array.

Order of Operations

The order calls are executed in is entirely determined by the dependencies between the inputs and outputs of task calls. Because MergeVCFs takes in the outputs of the scattered HaplotypeCallerGVCF call, it can start running only after all of the invocations of HaplotypeCallerGVCF have finished and their outputs are available. This means that we could place the MergeVCFs call statement first in the workflow block, and it would still be run last. However, we prefer to write our workflows in logical order for readability and encourage you to do the same, unless you're some kind of sadist.

Let's finish this exploration by calling out a few more details. First, you might notice that we explicitly declare the output of the MergeVCFs call as the final output of the workflow in the workflow-level output block. This is technically not required, but it is good practice. Second, the variable declarations for the BAM file, its index, and the intervals file were all pushed up to the workflow level. In the case of the BAM file and its index, this move allows us to generate the names of the output files in the input blocks of both task calls, which among other advantages gives us more flexibility if we want to put our task definitions into a common library. For something like that, we want the tasks to be as generic as possible, and leave details like file-naming conventions up to the workflow implementations. As for the intervals file, we need to have it available at the workflow level in order to implement the scatter block.

Finally, you should now be able to generate the *inputs* JSON, fill it out based on the previous exercise, and run the workflow using the same setup as previously. We included a prefilled JSON if you want to save yourself the hassle of filling in file paths; just make sure to use the *.local.inputs.json* version rather than the *.gcs.inputs.json*, which we use in Chapter 10.

Here's the Cromwell command using the prefilled local inputs JSON. Make sure to execute this in your home directory so that the relative paths in the *inputs* JSON file match the location of the data files:

```
$ java -jar $BIN/cromwell-53.1.jar \
    run $WF/scatter-hc/scatter-haplotypecaller.wdl \
    -i $WF/scatter-hc/scatter-haplotypecaller.local.inputs.json
```

When you run this command, you might notice the scatter jobs running in parallel on your VM. This is great because it means jobs will finish sooner. However, what happens if you're trying to run five hundred scatter calls across a full genome? Running these all in parallel is going to cause problems; if it doesn't outright crash, it will at least grind your VM to a halt as it swaps RAM to disk frantically. The good news is there are a couple solutions for this. First, you can control the level of parallelism allowed by the "local" backend for Cromwell, as described in the online documentation (*https://oreil.ly/8F4hp*). Alternatively, you can use a different backend that is designed to handle this kind of situation gracefully. In Chapter 10, we show you how to use the Google backend to automatically send parallel jobs to multiple VMs.

Generating a Graph Diagram for Visualization

As a coda to this exercise, let's learn to apply one more Womtool utility: the graph command. The workflows we've looked at so far have been quite simple in terms of the number of steps and overall plumbing. In the next chapter (and out in the real world), you will encounter more complex workflows, for which it might be difficult to build a mental model based on the code alone. That's where it can be incredibly helpful to be able to generate a visualization of the workflow. The Womtool graph command allows you to generate a graph file in *.dot* format and visualize it using generic graph visualization tools:

```
$ java -jar $BIN/womtool-53.1.jar \
    graph $WF/scatter-hc/scatter-haplotypecaller.wdl \
    > ~/sandbox-8/scatter-haplotypecaller.dot
```

The *.dot* file is a plain-text file, so you can view it in the terminal; for example:

```
$ cat ~/sandbox-8/scatter-haplotypecaller.dot
digraph ScatterHaplotypeCallerGVCF {
  #rankdir=LR;
  compound=true;
  # Links
  CALL_HaplotypeCallerGVCF -> CALL_MergeVCFs
  SCATTER_0_VARIABLE_interval -> CALL_HaplotypeCallerGVCF
  # Nodes
  CALL_MergeVCFs [label="call MergeVCFs"]
  subgraph cluster_0 {
    style="filled,solid";
    fillcolor=white;
    CALL_HaplotypeCallerGVCF [label="call HaplotypeCallerGVCF"]
    SCATTER_0_VARIABLE_interval [shape="hexagon" label="scatter over String as interval"]
  }
}
```

However, that's not terribly visual, so let's load it into a graph viewer. There are many available options, including the very popular open source package Graphviz (*https:// oreil.ly/FS_SR*). You can either install the package on your local machine or use it through one of its many online implementations (*https://oreil.ly/WSMml*). To do so, simply copy the contents of the *.dot* file into the text window of the visualizer app, and it will generate the graph diagram for you, as shown in Figure 8-3.

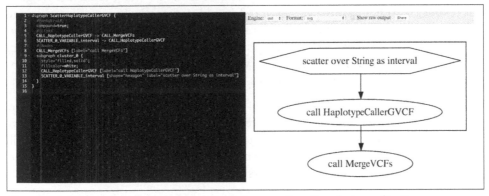

Figure 8-3. Visualizing the workflow graph in an online Graphviz application.

When we visualize a workflow graph like this, we typically look at the ovals first, which represent calls to tasks in the workflow. Here you can see that the two task calls in our workflow are indeed present, HaplotypeCallerGVCF and MergeVCFs. The direction of the arrows connecting the ovals indicates the flow of execution, so in

Figure 8-3, you can see that the output of `HaplotypeCallerGVCF` is going to be the input to `MergeVCFs`.

Interestingly, the call to `HaplotypeCallerGVCF` is displayed with a box around it, which means it is under the control of a modifying function. The modifier function is represented as a hexagon, and here you can see the hexagon is labeled "scatter over String as interval." That all makes sense because we just discussed how in this workflow the execution of the `HaplotypeCaller` task is scattered over a list of intervals.

In this case, the workflow was straightforward enough that the graph visualization didn't really tell us anything we didn't already know, but when we tackle more complex workflows in the next chapter, graph visualization is going to be a crucial part of our toolkit.

 This concludes the exercises in this chapter, so don't forget to stop your VM; otherwise, you'll be paying just to have it idly ponder the futility of existence.

Wrap-Up and Next Steps

In this chapter, we looked at how to string individual commands into a simple Hello World workflow, and we executed it using Cromwell in one-shot run mode on the single VM that we had set up in Chapter 4. We covered the basics of interpreting Cromwell's terminal output and finding output files. We iterated on the original `HelloWorld` workflow, adding variables and an additional task. Then, we moved on to examine more realistic workflows that run real GATK commands and use scatter-gather parallelism, though still at a fairly small scale. Along the way, we exercised key utilities for generating JSON templates, validating WDL syntax, testing error handling, and generating graph visualizations.

However, we only scratched the surface of WDL's capabilities as a workflow language, so now it's time to move on to some more sophisticated workflows. In Chapter 9, we examine two mystery workflows and try to reverse engineer what they do, which will give you the opportunity to hone your detective skills and also learn some useful patterns that are used in real genomics analysis workflows.

Deciphering Real Genomics Workflows

In Chapter 8, we showed you how to string commands into workflows by using the WDL language, and we had you practice running those workflows on your cloud VM using Cromwell. Throughout that chapter, we used fairly simple example workflows in order to focus on the basic syntax and rules of WDL. In this chapter, we switch gears and tackle workflows that are more complex than what you've seen so far. Rest assured, we neither expect you to instantly master all of the intricacies involved nor to memorize the various code features that you'll encounter.

Our main goal is to expose you to the logic, patterns, and strategies used in real genomics workflows and, in the process, present you with a methodology for deciphering new workflows of arbitrary complexity. To that end, we've selected two workflows from the gatk-workflows (*https://oreil.ly/D0Ofp*) collection, but we won't tell you up front what they do. Instead, for each mystery workflow, we tell you the functionality you're going to learn about and then walk you through a series of steps to learn what the workflow does and how the functionality that we're interested in is implemented. This won't turn you into a WDL developer overnight, but it will equip you with the skills to decipher other people's workflows and, with a bit of practice, learn to modify them to serve your own purposes if needed.

Mystery Workflow #1: Flexibility Through Conditionals

Our first mystery workflow gives you the opportunity to learn how to use conditional statements to control which tasks are run and under what conditions. It also shows you how to increase flexibility for controlling the workflow's parameters and outputs. As we work through this case, you'll encounter many familiar elements that you encountered in Chapter 8. However, unlike in that chapter, here we don't immediately explain all of the code; in fact, there is some code that we won't cover at all. Much of this exercise involves focusing on the more important parts and ignoring details that don't matter for what we're trying to figure out.

You're going to work in your VM with the same setup as in Chapter 8, running commands from the home directory. To begin, let's set up an environment variable in your VM shell to make the commands shorter:

```
$ export CASE1=~/book/code/workflows/mystery-1
```

Let's also create a sandbox directory for storing outputs:

```
$ mkdir ~/sandbox-9
```

And with that, you're ready to dive in.

Mapping Out the Workflow

When you encounter a new workflow, it can be tempting to dive straight into the WDL file and try to make sense of it by directly reading the code. However, unless you're very familiar with the language (or the author did an unusually great job of documenting their work), this can end up being frustrating and overwhelming. We propose a two-step approach that can help you to figure out the overall purpose and structure of a workflow. Start by generating a graph diagram of the workflow tasks. Then, identify which parts of the code correspond to each of the main components in the graph. This will provide you with a map of the overall workflow and a starting point for digging deeper into its plumbing.

Generating the graph diagram

You may recall that in Chapter 8 we discussed how the various connections between tasks in a workflow form a directed graph. The workflow execution engine uses that graph to identify the dependencies between tasks and, from there, determine in what order it should run them. The good news is that we can also take advantage of that information, as we already did in Chapter 8 when we generated the graph diagram for the HaplotypeCaller workflow.

Let's do the same thing now for our mystery workflow; run the womtool graph utility on the WDL file:

```
$ java -jar $BIN/womtool-53.1.jar graph $CASE1/haplotypecaller-gvcf-gatk4.wdl \
    > ~/sandbox-9/haplotypecaller-gvcf-gatk4.dot
```

This produces a DOT file that describes the structure of the workflow in JSON syntax. Open the DOT file with cat to see the contents, which should look like this:

```
$ cat ~/sandbox-9/haplotypecaller-gvcf-gatk4.dot
digraph HaplotypeCallerGvcf_GATK4 {
  #rankdir=LR;
  compound=true;
  # Links
  CALL_HaplotypeCaller -> CALL_MergeGVCFs
```

```
    SCATTER_1_VARIABLE_interval_file -> CALL_HaplotypeCaller
    CALL_CramToBamTask -> CALL_HaplotypeCaller
    # Nodes
    subgraph cluster_0 {
      style="filled,dashed";
      fillcolor=white;
      CALL_CramToBamTask [label="call CramToBamTask"]
      CONDITIONAL_0_EXPRESSION [shape="hexagon" label="if (is_cram)" style="dashed" ]
    }
    CALL_MergeGVCFs [label="call MergeGVCFs"]
    subgraph cluster_1 {
      style="filled,solid";
      fillcolor=white;
      CALL_HaplotypeCaller [label="call HaplotypeCaller"]
      SCATTER_1_VARIABLE_interval_file [shape="hexagon" label="scatter over File as interval_file"]
    }
}
```

This is a little difficult to directly interpret, so let's use Graphviz to make a visual representation. Go to GraphvizOnline (*https://oreil.ly/TpWlW*) and then copy and paste the contents of the DOT file into the editor panel on the left. The panel on the right should refresh and display the visual rendering of the graph, as shown in Figure 9-1.

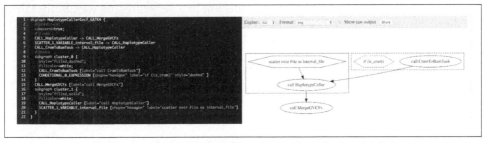

Figure 9-1. Graph description in JSON (left) and visual rendering (right).

Focus on the ovals first because they represent the calls to tasks in the workflow. We see that this workflow involves calls to three tasks: CramToBamTask, HaplotypeCaller, and MergeGVCFs. The first two calls are each in a box, which we know from Chapter 8 means there is some kind of control function that affects the call. Notice that the box around the call to CramToBamTask uses dotted lines, and that it also contains a hexagon labeled if (is_cram). Feel free to speculate and write down what you think that might do; you'll find out if you were correct when we unpack that a little further along in the chapter. The other boxed call, to HaplotypeCaller, is under the control of a scatter function, which you should recognize from the last workflow we examined in Chapter 8. Indeed, if you ignore the CramToBamTask part, this workflow looks like the same scattered HaplotypeCaller example we reviewed then, which parallelized HaplotypeCaller over a list of intervals and then combined the per-interval GVCF outputs using MergeGVCFs to produce the final GVCF.

So, what's up with that CramToBamTask call? Well, do you remember what CRAM is? As we mentioned in the genomics primer in Chapter 2, it's a compression format for storing sequencing read data in smaller files than BAM format. Some tools like `Haplo typeCaller` are not yet able to work with CRAM files with absolute reliability, so if your input is in CRAM format, you need to convert it to BAM first. Based on the name of the task, it sounds like that's exactly what `CramToBamTask` does. And we would want to run it only on actual CRAM files, so it makes sense to put it under the control of a conditional switch, `if (is_cram)`, which asks: "Is this a CRAM file?" and calls the conversion task only if the answer is yes.

To summarize, based on the workflow graph, the names of tasks and variables, and some logical inferences, we can suppose that this mystery workflow is in fact very similar to the parallelized `HaplotypeCaller` workflow that you became familiar with in Chapter 8, but with a file format conversion step added at the beginning to handle CRAM files. That sounds reasonable, right? Let's see if we can confirm this hypothesis and expand our understanding further.

Identifying the code that corresponds to the diagram components

Now we're going to begin looking at the code, but in a very directed way. Rather than scrolling through all of it sequentially, we're going to look up specific elements of code based on what we see in the graph diagram. Here's an outline of the method:

1. List all tasks referenced in the graph (in ovals):

   ```
   CramToBamTask
   HaplotypeCaller
   MergeGVCFs
   ```

2. Open the WDL file in your text editor and search for the `call` statements for each task:

   ```
   call CramToBamTask {        line 68
   call HaplotypeCaller {      line 84
   call MergeGVCFs {           line 100
   ```

3. For each call, capture the `call` statement and a few lines from the input definitions. If the task reference is shown inside a box in the graph, include the text that's also in the box, like `if (is_cram)` in Figure 9-1. Combine the code you captured into a table with screenshot slices of the graph diagram, as shown in Table 9-1.

Table 9-1. Calls to CramToBamTask, HaplotypeCaller, and MergeGVCFs

Calls in graph	Calls in workflow code
if (is_cram) ··> call CramToBamTask	```if (is cram) {` ` call CramToBamTask {` ` input:` ` input_cram = input_bam,` ` ...` ` }` `}```
scatter over File as interval_file → call HaplotypeCaller	```scatter (interval_file in` `scattered_calling_intercals` `) {` ` ...` ` call HaplotypeCaller {}` ` input:` ` input_am =` `select_first([CramToBamTask.output_bam,` `input_bam]),` ` ...` ` }` `}```
↓ call MergeGVCFs	```call MergeGVCFs {` ` input:` ` input_vcfs =` `HaplotypeCalle.output_vcf,` ` ...` `}```

This captures a lot of useful information about the key elements of code that determine how the tasks are connected and how their operation is controlled. In the next section, we dive deeper into these snippets of code as we attempt to reverse engineer how this workflow works.

You might need to tweak this approach and use your judgment regarding how many lines of input definitions or other code to include. For example, you might need to look up the task definitions to guide your choice of which inputs to include given that some are more informative than others. With practice, you will learn to reduce complex workflows to the key parts of their plumbing that matter.

Confirm What CramToBamTask Does for Extra Credit

In the main text of this section, we're assuming that the task names are a reliable reflection of the commands that they wrap. If you'd like to check for yourself, scroll down through the workflow (or use text search) to find the task definition for CramTo BamTask. You might feel like there's a lot going on in there that you don't recognize, and that's perfectly normal. For now, don't try to decipher everything; just zoom straight to the most important piece, the command block:

```
command {
    set -e
    set -o pipefail

    ${samtools_path} view -h -T ${ref_fasta} ${input_cram} |
    ${samtools_path} view -b -o ${sample_name}.bam -
    ${samtools_path} index -b ${sample_name}.bam
    mv ${sample_name}.bam.bai ${sample_name}.bai
}
```

Again, a lot is happening here, so let's break it down line by line, making a few assumptions along the way based on variable names:

```
${samtools_path} view -h -T ${ref_fasta} ${input_cram} |
```

We can reasonably infer that samtools_path refers to the popular samtools package, which provides tools for manipulating files in SAM, BAM, and CRAM format. So this line must run the view command from the samtools package on ${input_cram}, which presumably refers to an input file in CRAM format. Using samtools view is a common way to read in data before applying some transformations to it, and sure enough, there's a pipe character (|) at the end of the line, so let's see what we're going to apply in the next line:

```
${samtools_path} view -b -o ${sample_name}.bam -
```

It's a second invocation of samtools view, but this time the command specifies an output file (-o) preceded by the -b flag and with a .bam extension for the filename. This instructs samtools to write the output in BAM format. Hey, that amounts to converting the original file from CRAM to BAM! But wait, there's more:

```
${samtools_path} index -b ${sample_name}.bam
```

This runs "samtools index" on the newly created BAM file, which makes sense. GATK tools are typically able to create index files for output files on the fly, but with samtools, you need to run the indexing command separately. So we do that, and we're done. Right? No, this is bioinformatics, so of course we're not done yet. We still need to rename the index because samtools names BAM index files with the pattern ${sample_name}.bam.bai, but GATK expects the index filename to follow the pattern ${sample_name}.bai (sigh):

```
mv ${sample_name}.bam.bai ${sample_name}.bai
```

OK, so we use the common Unix mv utility to do that final renaming. *Now* we're done. And we were correct: as Table 9-1 shows, this command takes in a CRAM file and converts it to a BAM file with an appropriately named index.

By the way, are you surprised that this command block involves multiple command lines? Yes, you can do that in WDL: basically, anything you can stick in a Bash script is OK to include in a WDL command block. At runtime, the system will execute each line in turn as if you were running them manually in your terminal. In fact, the first two lines in this block, set -e and set -o pipefail, instruct the system how to handle errors when you ask it to run multiple commands: stop at the first error or try to keep running the rest? Here we're telling it to stop at the first error and consider the entire task to have failed.

Finally, if you're wondering why there's not a way to do all that with a simple one-line command, well, there actually *is* a way to do that: it's a GATK4 tool called Print Reads. However, the original author of this workflow chose to use this samtools construct instead. They might or might not have had a good reason to do so; the point is that this is the sort of thing you will encounter all the time in real genomics workflows, so we might as well show you how to make sense of it.

We're going to trust that the HaplotypeCaller and MergeGVCFs tasks are doing what their names suggest, but feel free to check them, as well—they should look very similar (though not identical) to the analogous tasks we examined in Chapter 8.

Reverse Engineering the Conditional Switch

At this point, we're reasonably confident that we understand what the workflow is doing overall. But we don't yet understand how the conditional switch works and how the various tasks are wired together in practice. Let's begin by looking at how the workflow logic is set up in the code.

How is the conditional logic set up?

In the match-up diagram in Table 9-1, we saw the box showing the call to CramToBam Task under the control of the presumed conditional switch. We identified the corresponding code by searching for call CramToBamTask and capturing the lines around it that are likely to be involved:

```
if ( is_cram ) {
    call CramToBamTask {
        input:
            input_cram = input_bam,
            ...
    }
}
```

You can read this snippet of code as "If the `is_cram` condition is verified, call `CramTo BamTask` and give it whatever was provided as `input_bam`." This immediately brings up two important questions: how are we testing for that initial condition, and how are we handling the apparent contradiction in the `input_cram` = `input_bam` input assignment?

Let's tackle the first question by scrolling up a bit in the workflow code until we find a reference to this mysterious `is_cram` variable. And here it is on line 59:

```
#is the input a cram file?
Boolean is_cram = sub(basename(input_bam), ".*\\.", "") == "cram"
```

Can you believe our luck? There's a note of documentation that confirms that the purpose of this line of code is to answer the question "Is the input a CRAM file?" All we need to do is determine how that second line is answering it. Let's look at the left side first, before the equals sign:

```
Boolean is_cram
```

This means "I'm declaring the existence of this Boolean variable named `is_cram`." *Boolean* refers to a type of variable that can have only one of two values: True or False. This variable type is named after English mathematician George Boole, who was a big fan of logic. Right there with you, George! Boolean variables are useful for expressing and testing conditions succinctly.

The equals sign is our value assignment statement, so whatever is on the other side is going to determine the value of the variable. Here's what we see:

```
sub(basename(input_bam), ".*\\.", "") == "cram"
```

Try to ignore everything that's in parentheses on the left side and just read this as `something == "cram"`. The `A == B` syntax is a compact way of asking a question in code. You can ultimately read this as saying, "is A equal to B?" or bringing it back to this specific case, "is `something` equal to B?" So the full line now reads "My Boolean variable `is_cram` is True if `something` equals `cram`, and is False if `something` does not equal `cram`."

So, what is this `something` that we're testing? You might be able to guess it by now, but let's keep decomposing the code methodically to get to the truth of the matter:

```
sub(basename(input_bam), ".*\\.", "")
```

Hopefully, you remember the `basename()` function from Chapter 8; it's how we were able to name outputs based on the name of the input files. At some point we're going to give this workflow some real inputs to run on, and the execution engine will match up the variable and the file path like this:

```
File input_bam = "gs://my-bucket/sample.bam"
```

When we just give the basename() function a file path, as in basename(input_bam), it produces the string sample.bam. If we also give it a substring, as in base name(input_bam, ".bam"), it will additionally clip that off to produce the string sam ple. This is convenient for naming an output the same thing as the input, but with a different file extension.

Here the workflow adds another layer of string manipulation by using the sub() function (also from the WDL standard library) to replace part of the string with something else. You use sub() like this:

```
sub("string to modify", "substring or pattern we want to replace", "replacement")
```

With that information, we can now interpret the full sub() command, as seen in the workflow:

1. Take the basename of our input file:

    ```
    basename(input_bam)     "sample.bam"
    ```

2. Separate the part up to the last period:

    ```
    ".*\\."             "sample."+"bam"
    ```

3. Replace that part by nothing:

    ```
    ""                  ""+"bam"
    ```

4. Output whatever is left:

    ```
    "bam"
    ```

What is left? It's the file extension! So finally, we can say for sure that that one line of code means "My Boolean variable is_cram is True if the input file extension equals cram, and is False if the input file extension does not equal cram."

 You might argue that it would be worth adding a function to the WDL standard library to grab the file extension without having to jump through these string substitution hoops. If you feel very strongly about it, let the WDL maintainers know by posting an issue in the OpenWDL repository (*https://oreil.ly/Td9bp*), and maybe they will add it! That's how a community-driven language evolves: based on feedback from the people who use it.

Now that we know where the value of the is_cram variable comes from, we understand exactly how the flow is controlled: if we provide a file with a *.cram* extension as input to the workflow, the call to CramToBamTask will run on the file provided as input. If we provide anything else, that task will be skipped and the workflow will start at the next call, HaplotypeCaller. Switch on, switch off.

What's in a Name?

Do you remember the second question we raised when we initially looked at the `Cram ToBamTask` call? Isn't it weird that the code for the input assignments says `"input_cram = input_bam"`? Indeed, it would be better to use a less specific variable name for the input file in order to reflect that the workflow can't "know" the file format until it starts running. However, this particular workflow was originally written to run directly on BAM files and then the conversion task was added at a later date by someone who decided not to mess with the existing variable name. That teaches us a lesson about the meaningfulness of task and variable names: you can't assume they're always 100% reliable.

OK, what's next? Well, before you go running off to the next feature, we have some more work to do to fully understand how the conditional switch affects the rest of the workflow. Conditionals are a really neat way to make workflows more flexible and multipurpose, but they tend to have a few side effects that you need to understand and watch out for. Let's go over two important questions that we need to ask ourselves when we're dealing with conditionals.

Does the conditional interfere with any assumptions we're making anywhere else?

If we're using a conditional to allow multiple file types to be used as input, we can't know ahead of time what the extension might be for any given run. Yet a lot of genomics workflows, including this one, use `basename()` to name outputs based on the input name. That typically involves specifying what file extension to trim off the end of the input filename. The workflow must deal with that ambiguity explicitly.

At this point, you could do a text search for "basename" to find the code we need to investigate, but in our case, what we're looking for is on the very next line of code, after the `is_cram` variable assignment:

```
String sample_basename = if is_cram then basename(input_bam, ".cram") else
basename(input_bam, ".bam")
```

This is another variable declaration, stating the variable's type (`String`) and name (`sample_basename`) on the left side of the equals sign, with a value assignment on the right. Let's look at that value assignment in more detail:

```
if is_cram then basename(input_bam, ".cram") else basename(input_bam, ".bam")
```

If you read that out loud, it's almost reasonable English, and it takes only a few additional words of interpretation to make complete sense:

"If the input is a CRAM file, run the `basename` function with the *.cram* extension; otherwise, run it with the *.bam* extension."

Problem solved! It's not the only way to achieve that result in WDL but it's probably the most readily understandable. In fact, if you look carefully at the next few lines in the workflow, you should be able to find and interpret another (unrelated) conditional variable assignment that's doing something very similar, though that one comes into play at the other end of the pipeline.

How does the next task know what to run on?

Depending on the format of the input file, `HaplotypeCaller` will be either the first or the second task called in the workflow. If it's the first, it should run on the original input file; if it's the second, it should take the output of the first task. How does the workflow code handle that?

Have another look at Table 9-1, which showed how the code maps to the graph diagram. This was the code we highlighted at the time for the call to `HaplotypeCaller`:

```
scatter (...) {
  ...
  call HaplotypeCaller {
      input:
        input_bam = select_first([CramToBamTask.output_bam, input_bam]),
        ...
  }
}
```

And there it is: in the input section of the call statement, we see that the `input_bam` is being set with yet another function, `select_first()`. The name suggests it's in charge of selecting something that comes first, and we see it's reading in an array of values, but it's not super obvious what that means, is it? If we just wanted it to use the first element in the array, why give it that array in the first place?

Yep, this is a bit of an odd one. The subtlety is that `select_first()` is designed to tolerate missing values, which is necessary when you're dealing with conditionals: depending on what path you follow, you might generate some outputs but not others. Most WDL functions will choke and fail if you try to give them an output that doesn't exist. In contrast, `select_first()` will simply ask: what else do you have that might be available?

So, in light of that information, let's look at that value assignment again in more detail. Here's a handy way to represent how the code can be broken into segments of meaning:

input_bam = select_first([	CramToBamTask.output_bam	, input_bam])
The input file for `HaplotypeCaller` should be...	...the output from `CramToBamTask`...	...but if that doesn't exist, just use the original input file.

At the moment in time when the workflow execution engine is evaluating what tasks need to be run and on what inputs/outputs, the output of the first task doesn't exist yet. However, that output might be created at a later point, so the engine can't just start running `HaplotypeCaller` willy-nilly on what might be a CRAM file. It must check the input file format to resolve the ambiguity, determine what tasks will run, and what outputs to expect. Having done that, the engine is able to evaluate the result of the `select_first()` function. If it determines that the `CramToBamTask` will run, it will hold off on running the `HaplotypeCaller` task until the *CramToBamTask.output_bam* has been generated. Otherwise, it can go ahead with the original input.

Hopefully, that makes more sense now. Though it might prompt a new question in your mind...

Can we use conditionals to manage default settings?

Yes, yes you can! If you skimmed through the workflow-level inputs, you might have noticed pairs of lines like this that look unusually complex for what should be straightforward variable declarations:

```
String? gatk_docker_override
String gatk_docker = select_first([gatk_docker_override,
                            "us.gcr.io/broad-gatk/gatk:4.1.0.0"])
```

The first line in the pair simply declares a variable named `gatk_docker_override`, which was marked as optional by adding the question mark (?) after the variable type (`String`). Based on its name, it sounds like this variable would allow us to override the choice of Docker container image used to run GATK tasks, presumably by specifying one in the JSON file we hand to Cromwell at runtime. This implies that there's a preset value somewhere, but we haven't seen a variable declaration for that yet.

We find it in the second line in the pair, which declares a variable named `gatk_docker` and assigns it a value using the `select_first()` function. Let's try breaking down that line the same way we did earlier, with a few adaptations based on context:

`String gatk_docker = select_first([`	`gatk_docker_override`	`, "us.gcr.io/broad-gatk/gatk:4.1.0.0"])`
The GATK Docker image should be...	...what we provide in the JSON file of inputs...	...but if we didn't provide one, just use `us.gcr.io/broad-gatk/gatk:4.1.0.0` by default.

Isn't that neat? This is not the only way to set and manage default values in WDL, and it's admittedly not the most readily understandable if you're not familiar with the `select_first()` function. However, it's a very convenient way for the workflow author to peg their workflow to specific versions of software, environment expectations, resource allocations, and so on, while still allowing a lot of flexibility for using alternative settings without having to modify the code. You will encounter this pattern frequently in most up-to-date GATK workflows produced by the Broad Institute.

That wraps up our exploration of Mystery Workflow #1, which turned out to be a rather garden-variety `HaplotypeCaller` workflow hopped up with a few conditional statements. It had a few twists that mostly had to do with the consequences of using conditionals; but overall, the complexity was moderate. Let's see if Mystery Workflow #2 can take us a little bit further up the scale of WDL complexity.

Mystery Workflow #2: Modularity and Code Reuse

Oh yes, now we're in the big leagues: without spoiling the story, we can tell you that this one is a major pipeline that runs in production at the Broad Institute. It does a ton of work and involves many tools. It's going to teach us a lot about how to build modular pipelines and minimize duplication of code. Let's break it down.

First, make an environment variable pointing to the code, just as you did for the first workflow:

```
$ export CASE2=~/book/code/workflows/mystery-2
```

We can use the same sandbox as the first workflow, *~/sandbox-9/*.

Mapping Out the Workflow

Let's see if we can apply the same approach as for the previous workflow: generate the graph diagram and then map the call statements in the code back to the task bubbles in the diagram.

Generating the graph diagram

You know the drill: first, let's whip out `womtool graph` to make the DOT file:

```
$ java -jar $BIN/womtool-53.1.jar graph $CASE2/WholeGenomeGermlineSingleSample.wdl \
    > ~/sandbox-9/WholeGenomeGermlineSingleSample.dot
```

Next, open the DOT file; does it look any longer than the other one to you?

```
$ cat ~/sandbox-9/WholeGenomeGermlineSingleSample.dot
digraph WholeGenomeGermlineSingleSample {
 #rankdir=LR;
 compound=true;
 # Links
 CALL_UnmappedBamToAlignedBam -> CALL_BamToCram
 CALL_UnmappedBamToAlignedBam -> CALL_CollectRawWgsMetrics
 CALL_UnmappedBamToAlignedBam -> CALL_CollectWgsMetrics
 CALL_UnmappedBamToAlignedBam -> CALL_AggregatedBamQC
 CALL_UnmappedBamToAlignedBam -> CALL_BamToGvcf
```

```
CALL_AggregatedBamQC -> CALL_BamToCram
# Nodes
CALL_AggregatedBamQC [label="call AggregatedBamQC";shape="oval";peripheries=2]
CALL_BamToGvcf [label="call BamToGvcf";shape="oval";peripheries=2]
CALL_UnmappedBamToAlignedBam [label="call
UnmappedBamToAlignedBam";shape="oval";peripheries=2]
CALL_BamToCram [label="call BamToCram";shape="oval";peripheries=2]
CALL_CollectRawWgsMetrics [label="call CollectRawWgsMetrics"]
CALL_CollectWgsMetrics [label="call CollectWgsMetrics"]
}
```

Hmm, that doesn't seem all that long. Considering we told you it involves a lot of tools, the graph file seems rather underwhelming. Let's visualize it: copy the contents of the graph file into the panel on the left side in the online Graphviz app (*https:// oreil.ly/TpWlW*), as you've done before to produce the graph diagram, as depicted in Figure 9-2.

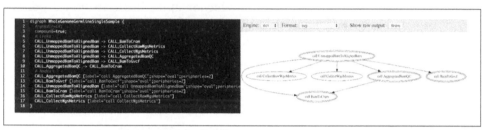

Figure 9-2. Visual rendering of the workflow graph.

What's the first thing that is visually different compared to the previous workflow diagrams we've looked at?

There are no boxes, and some of the ovals representing the calls to tasks have a double outline. We don't have any grounding yet for guessing what the double outline signifies, so let's make a note of that as something to solve. However, we do know from the two previous workflows we looked at that scatter blocks show up as boxes, so the absence of boxes in this diagram suggests that, at the very least, this workflow does not have any parallelized steps. Does that seem likely to you given how much we've been talking up the parallelization capabilities of the cloud? (Yes; we know that's called *leading the witness* (*https://oreil.ly/zkdiz*).)

Just as we did for the first workflow, let's try to form a hypothesis about what the workflow does based on the names of the task calls and their relative positions in the graph diagram.

The very first oval is labeled call UnmappedBamToAlignedBam, which suggests a transformation from the unmapped BAM file format that we referenced (ever so briefly) in Chapters 2 and 6 to an aligned BAM file—or as we'll call it for consistency, a mapped BAM file. Based on what you learned of the data preprocessing pipeline in

Chapter 6, that sounds like it could correspond to the step that applies the mapping to the raw unmapped data.

That first oval then connects to all five other ovals in the diagram, including the last one, which seems mighty strange but OK. Four of those ovals are on the same level, one step down from the first, and seem to be completely independent of one another. The two on the left have names that suggest they collect metrics related to WGS data (`call CollectWgsMetrics` and `call CollectRawWgsMetrics`), which would make sense as a quality-control step applied to evaluate the mapped BAM produced by the first call. In addition, they are both end nodes on their respective branches and have a single-line outline, so there's probably not much more to them than meets the eye.

The third oval from the left, labeled `call AggregatedBamQC`, also evidently involves quality control, but what is the "aggregated BAM" to which it is applied? If the original data was divided among several read groups as recommended in the GATK Best Practices, it would make sense to have an "aggregated BAM" after the read group data has been merged per sample. However…do you remember from Chapter 6 when the merge step happens in the Best Practices pipeline? It happens *after* the mapping, during the duplicate marking step. Unless we're totally off base in our interpretation of these task names, this would suggest that the duplicate marking and merging is happening within the same task as the mapping. That is technically feasible with a WDL workflow but generally not advisable from the point of view of performance: the mapping can be massively parallelized, but the duplicate marking cannot, so it would be a bad idea to bundle both of them for execution within the same task. Hold on to that thought.

Interestingly, the `call AggregatedBamQC` task does have a child node one more level down; it links to the oval labeled `call BamToCram`, which is also dependent on the first oval. `BamToCram` by itself sounds like the exact opposite of the `CramToBamTask` that you encountered in the previous workflow, so let's venture to guess that this one allows you to create a CRAM file for archival purposes based on the mapped BAM created by the first step. Presumably, if the data in that file is considered ready for archival, it must have already been fully preprocessed, which lends further support to the idea that that first step is doing much more than just mapping the reads to the reference. As to the contribution of the `AggregatedBamQC` task, let's suppose for now that there is something that affects the cramming step that is decided based on the results of the quality control analysis.

Finally, that just leaves the rightmost oval on the second level, an end node that is labeled `call BamToGvcf`. That sounds very familiar indeed; would you care to take a guess? Are you sick of seeing `HaplotypeCaller` everywhere yet?

So just based on the graph diagram, the names of the calls, and the way they connect to one another, we've provisionally deduced that this pipeline takes in raw unmapped sequencing data (which is probably whole genome sequence), maps it, applies some quality control, creates a CRAM archive, and generates a GVCF of variant calls that will be suitable for joint calling. That's a lot of work for such an apparently minimalistic pipeline; we should probably dig into that.

Identifying the code that corresponds to the diagram components

The book says that we must match the code with the diagram now, and you can't argue with a book, so in your text editor, go ahead and open the WDL file and follow the same steps as you did for the first workflow. List all tasks referenced in the graph (in ovals) and then, for each task, search for the call statement. For each call, capture the call statement and a few lines from the input definitions. Combine the code you captured into a table with screenshot slices of the graph diagram, as shown in Table 9-2.

Table 9-2. Workflow calls in the graph (left) and in WDL code (right)

Workflow calls in the graph	Workflow calls in WDL code
call UnmappedBamToAlignedBam	```call ToBam.UnmappedGamToAlignedBam {` ` input:` ` sample_and_unmapped_bams =` `sample_and_unmapped_bams,` ` ...` `}```
call AggregatedBamQC	```call AggregatedQC.AggregatedBamQC {` ` input:` ` base_recalibrated_bam =` `UnmappedBamToAlignedBam.output_bam,` ` ...` `}```
call BamToCram	```call ToCram.BamToCram as BamToCram {` ` input:` ` input_bam =` `UnmappedBamToAlignedBam.output_bam,` ` ...` `}```

Workflow calls in the graph	Workflow calls in WDL code
| ( call CollectWgsMetrics ) | ```
#QC the sample WGS metrics (stringent thresholds)
call QC.CollectWgsMetrics as
CollectWgsMetrics{
  input:
    input_bam =
UnmappedBamToAlignedBam.output_bam,
  ...
}
``` |
| ( call CollectRawWgsMetrics ) | ```
#QC the sample WGS metrics (common thresholds)
call QC.CollectRawWgsMetrics as
CollectWgsMetrics{
  input:
    input_bam =
UnmappedBamToAlignedBam.output_bam,
  ...
}
``` |
| ( call BamToGvcf ) | ```
call ToGvcf.VariantCalling as BamToGvcf {
  input:
    ...
    input_bam =
UnmappedBamToAlignedBam.output_bam,
    ...
}
``` |

Looking at the code for these calls, something odd pops right up. The task names look a bit weird. Check out the first one:

```
call ToBam.UnmappedBamToAlignedBam
```

Did you know periods are allowed in task names in WDL? (Spoiler: they're not.)

You can try searching for `ToBam.UnmappedBamToAlignedBam` to find the task definition; usually that can be helpful for figuring what's going on here. But here's a twist: you're not going to find it in this file. Go ahead, scroll down as much as you want; this WDL workflow does not contain any task definitions. (What?)

However, searching for either the first part of the name, `ToBam`, or the second, `Unmap pedBamToAlignedBam`, does bring up this line in a stack of other similar lines that we ignored when we originally started looking at the workflow:

```
import "tasks/UnmappedBamToAlignedBam.wdl" as ToBam
```
```

Aha. That's new; it looks like we're referencing other WDL files. In the preceding line, the import part points to another WDL file called *UnmappedBamToAlignedBam.wdl*, located in a subdirectory called *tasks*. Now things are starting to make sense: when the workflow engine reads that, it's going to import whatever code is in that file and consider it part of the overall workflow. The `as ToBam` part of this import statement assigns an alias so that we can refer to content from the imported file without having to reference the full file path. For example, this was the call we encountered earlier:

```
call ToBam.UnmappedBamToAlignedBam
```

We can interpret this as saying, "From the ToBam code, call UnmappedBamToAligned Bam." So what is UnmappedBamToAlignedBam? Is it a task? Have a quick peek inside the WDL file referenced in the import statement and then take a deep breath. It's a full workflow in its own right, called UnmappedBamToAlignedBam:

```
# WORKFLOW DEFINITION
workflow UnmappedBamToAlignedBam {
```

We're dealing with a workflow of workflows? Voltron, assemble!

This is actually a happy surprise. We suspected coming in that this was going to be a rather complex pipeline, even though the graph diagram initially gave us the wrong impression. So the discovery that it uses imports to tie together several simpler workflows into a mega-workflow is a good thing because it's going to allow us to investigate the pipeline in smaller chunks at a time. In effect, the workflow authors have already done the hard work of identifying the individual segments that can be logically and functionally separated from the others.

Now we need to see how the workflow nesting works, beyond just the import statements. Surely there's some additional wiring that we need to understand, like how are outputs passed around from one subworkflow to the next?

Unpacking the Nesting Dolls

We begin by digging down one level, into one of the subworkflows, to better understand how importing works and how the subworkflow connects to the master workflow. Which subworkflow should we pick? The VariantCalling workflow is a great option because it's likely to be the most comfortably familiar. That way we can focus on understanding the wiring rather than a plethora of new tools and operations. But before we delve any further into the code, let's map this subworkflow at a high level.

What is the structure of a subworkflow?

The good news is that subworkflows are, in fact, perfectly normal in terms of their WDL structure and syntax, so you can run all the usual womtool commands on them, and you can even run them on their own. Once more, womtool graph comes in handy:

```
$ java -jar $BIN/womtool-53.1.jar graph $CASE2/tasks/VariantCalling.wdl \
    > ~/sandbox-9/VariantCalling.dot
```

Copy the DOT file contents into the online Graphviz website to generate the graph diagram. As you can see in Figure 9-3, this workflow looks a lot like the now-classic scattered `HaplotypeCaller` workflow that we've been exploring from all angles, with a few additions. In particular, it has a few conditional switches that you can investigate in detail if you'd like to get more practice with conditionals.

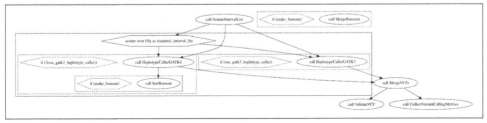

Figure 9-3. Graph diagram of the VariantCalling.wdl workflow.

Switching It Up

When you scroll through the *VariantCalling.wdl* code, you might notice that one of the conditional switches uses a Boolean variable called `use_gatk3_haplo type_caller`, which toggles the flow between two calls:

```
if (use_gatk3_haplotype_caller) {
    call Calling.HaplotypeCaller_GATK35_GVCF as HaplotypeCallerGATK3 { ... }
}

if (!use_gatk3_haplotype_caller) {
    call Calling.HaplotypeCaller_GATK4_VCF as HaplotypeCallerGATK4 { … }
}
```

This makes it possible to switch between using the GATK 3.5 and GATK4 versions of `HaplotypeCaller`. As of this writing, the Broad Institute's production pipeline still uses the GATK 3.5 version of `HaplotypeCaller` for operational reasons (mainly for continuity). However, the GATK4 version has been fully validated and is the recommended choice for new projects, so the conditional switch allows the team to maintain a single pipeline that can be used by everyone, internally and externally.

Now open up the *VariantCalling.wdl* file in your text editor and scroll through it briefly. It has a workflow definition section, `workflow VariantCalling {...}`. If you're up for it, you can run through the exercise of mapping the code for call statements to their position in the diagram; we won't do it explicitly here. We're more interested in the fact that this subworkflow has import statements of its own. Still no task definitions, though, except for one all the way at the bottom. Does that mean there's going to be yet another level of nested workflows? We've got to find out where those task definitions are hiding.

Where are the tasks defined?

Let's pick the first call statement we come across and hunt down whatever it points to. Scrolling down from the top, we come to this call:

```
call Utils.ScatterIntervalList as ScatterIntervalList
```

You should recognize the pattern from the top-level workflow: this means that we need to look for an import statement aliased to `Utils`. Sure enough, we find this toward the top of the file:

```
import "tasks/Utilities.wdl" as Utils
```

In your text editor, open the *Utilities.wdl* file, and you should see the following (some lines omitted):

```
version 1.0

## Copyright Broad Institute, 2018
##
## This WDL defines utility tasks used for processing of sequencing data.
##
...

# Generate sets of intervals for scatter-gathering over chromosomes
task CreateSequenceGroupingTSV {
  input {
    File ref_dict
...
```

Wait, it's just a bunch of tasks? No workflow definition at all? That's...well, kind of a relief. We seem to have found the deepest level of our nested structure, at least on this branch. To confirm, search for the `ScatterIntervalList` task in the *Utilities.wdl* file, and, sure enough, there is its task definition starting on line 77:

```
task ScatterIntervalList {
  input {
    File interval_list
    Int scatter_count
    Int break_bands_at_multiples_of
  }

  command <<<
...
```

This task's `command` block is pretty wild—first it runs a Picard tool via a Java command and then some bulk Python code that generates lists of intervals on the fly. We won't analyze its syntax in detail; just note that it's the sort of thing you can do within a WDL task.

The takeaway here is that this *Utilities.wdl* file functions as a library of tasks. Think for a moment about the implications: this means that you can store task definitions for commonly used tasks in a library and simply import them into individual workflows with a one-line statement, rather than rewriting or copying the code over each time. Then, when you need to fix or update a task definition, you need to do it only once rather than having to update all the copies that exist in different workflows. To be clear though, you're not obligated to use subworkflows in order to use imports for tasks. For an example of non-nested workflows that share a task library, have a look at the workflows on GitHub (*https://oreil.ly/6qVe3*).

WDL Libraries Can Also Collect Custom Struct Variables

You might have noticed that some of the inputs in the master workflow were declared with unusual variable types. The WDL variable types we've used so far have been very generic: `String`, `File`, `Boolean`, and now, `SampleAndUnmappedBams`? That seems oddly specific for a variable type, right?

```
workflow WholeGenomeGermlineSingleSample {
  input {
    SampleAndUnmappedBams sample_and_unmapped_bams
    GermlineSingleSampleReferences references
    PapiSettings papi_settings
...
```

These variable types are called *structs*, short for *construct*; they're special in the sense that they're custom-built by the workflow author. They enable us to bundle groups of variables into a single variable so that it's easier to pass them around as task inputs or outputs. This is very convenient when you have a set of inputs that always travel together, and you don't want to have to specify every one of them individually in each call's input definitions. For example, the aforementioned `SampleAndUnmappedBams` struct bundles metadata like the name of a sample, the name of the final output file, and so on with the actual sequence data files for a given sample:

```
struct SampleAndUnmappedBams {
  String base_file_name
  String final_gvcf_base_name
  Array[File] flowcell_unmapped_bams
  String sample_name
  String unmapped_bam_suffix
}
```

Just like tasks, structs can be defined within the WDL file that houses your workflow, or in a separate library that contains only structs (or a mix of structs and tasks).

Hey, look at how far we've come; this chapter is almost over. You should feel pretty good about yourself. You've worked through almost everything you need in order to understand imports and use them in your own work. But there's one more thing we want to make sure is really clear, and it has to do with connecting inputs and outputs between subworkflows.

How is the subworkflow wired up?

As you might recall, this subworkflow is imported as follows in the master workflow:

```
import "tasks/VariantCalling.wdl" as ToGvcf
```

As noted previously, this makes the code within the imported file available to the workflow interpreter and assigns the ToGvcf alias to that code so that we can refer to it elsewhere. We see this in the following call statement:

```
call ToGvcf.VariantCalling as BamToGvcf {
    input:
    ...
    input_bam = UnmappedBamToAlignedBam.output_bam,
```

Now that we know VariantCalling is a workflow defined within of the imported WDL, we can read the call like this in plain English:

> From the pile of imported code named ToGvcf, call the workflow defined as Variant-Calling, but name it BamToGvcf within the scope of this master workflow. Also, for the input BAM file it expects, give it the output from the UnmappedBamToAlignedBam workflow.

You might notice that there's an interesting difference between how the UnmappedBam ToAlignedBam workflow and the VariantCalling workflow are treated: the latter is assigned an alias, whereas the former is not. Instead, UnmappedBamToAlignedBam retains the original name given in the WDL where it is defined. Based on the naming patterns we've observed so far, our guess is that the authors of the workflow wanted to identify the high-level segments by the file formats of the main inputs and outputs at each stage. This is a very common pattern in genomics; it's not the most robust strategy, but it certainly is concise and does not require too much domain understanding to decipher beyond the file formats themselves.

Anyway, the practical consequence is that we point to the outputs of the subworkflow by either its original name or by its alias if it has one. To be clear, however, we can point only to outputs that were declared as workflow-level outputs in the subworkflow. If any tasks in the subworkflow were not declared at the workflow level, but only at the task level, they will be invisible to the master workflow and to any of its other subworkflows.

Aliases are a convenient feature of WDL that are useful beyond dealing with imports; they can also be used to call the same task in different ways within the same workflow.

As we've already observed, the import wiring system also allows us to call individual tasks from an imported WDL by referring to the alias given to the WDL file, as in this call statement to one of the tasks in the *Utility.wdl* library:

```
call Utils.ScatterIntervalList as ScatterIntervalList
```

To reprise our model for loose translation to plain English, this reads as follows:

> From the pile of imported code named Utils, call the task defined as ScatterInterval-List, but name it ScatterIntervalList instead of Utils.ScatterIntervalList within the scope of this master workflow.

The twist here is that if we don't provide an alias, we must refer to the task as `Utils.ScatterIntervalList` everywhere. When we define this alias, we can just refer to the task directly as `ScatterIntervalList`. This is a subtle but important difference compared to the way workflow imports work. Note that we could also use a different name from the original if we wanted, like `ScatterList` for short, or `Samantha` if we're actively trying to mess with anyone who might read our code (but please don't do that).

Finally, remember that some of the ovals in the top-level workflow have a double outline, including the one labeled `UnmappedBamToAlignedBam`, which happens to be a workflow—do you think that could possibly be related? Let's check out the other calls. The `AggregatedBamQC` oval had a double outline, so look up the import statement to identify the corresponding WDL file (*tasks/AggregatedBamQC.wdl*) and peek inside. Sure enough, it's another workflow. Now look up the import statement for one of the ovals that had a single outline, for example `CollectWgsMetrics`, and peek inside the file (*tasks/Qc.wdl*). Ooh, no workflow declaration, just tasks. It seems we've identified what the double outline means: it denotes that the call points to a nested workflow, or subworkflow. Now that's really useful.

It would be great to also be able to recognize in the diagram whether an oval with a single outline is a true task or whether it's pointing to a library of tasks. Something to request from the Cromwell development team.

To recap, we elucidated the structure of this complex pipeline, which implements the full GATK Best Practices for germline short variant discovery in whole genome sequencing data. It combines data preprocesssing (read alignment, duplicate marking, and base recalibration) with the scattered `HaplotypeCaller` we've been working with previously, and adds on a layer of quality control.

As it happens, the original form of this pipeline was a massive, single-WDL implementation, but it later was split into multiple WDL files to be more readable and manageable. The current pipeline utilizes import statements with several levels of nesting of both workflows and tasks. The result is a high level of code modularity and reusability that make it a good model for developing new pipelines, especially complex ones.

If you're looking for more practice analyzing the use of imports with nested workflows and task libraries, try to map out the complete structure of this monster pipeline, including the preprocessing and QC workflows. When you're done with that, take a look at the exome version (*https://oreil.ly/OMrc9*) of this pipeline, which follows the same model and, in fact, reuses some of the same code! See if you can combine them into a single codebase that maximizes code reuse.

 With that, it's once again time to stop your VM to avoid paying Google Cloud for the privilege of having a reserved machine idling away for nothing.

Wrap-Up and Next Steps

In this chapter, we explored real genomics workflows that apply advanced WDL features to do some seriously interesting things. We began with an exploration of how you can use conditionals to control the flow of a WDL workflow, branching the analysis based on an input file type. We then looked at a rather complex production workflow from the Broad Institute that uses subworkflows, structures, and task libraries to break the workflow into more manageable (and reusable) pieces. Along the way, we focused on the skills and techniques you need to take a complex workflow, pick it apart, and understand it component by component. This will help not only in understanding the complex production Best Practices workflows that we've just started to examine, but it will serve you well when building your own workflows. In this chapter, we didn't actually run the workflows because our focus was on developing your skills as a workflow detective. In Chapter 10, we dive into the execution of complex workflows but, this time, we show you how to take advantage of the full power of the cloud instead of just a single VM.

Running Single Workflows at Scale with Pipelines API

In Chapter 8, we started running workflows for the first time, working on a custom virtual machine in GCP. However, that single-machine setup didn't allow us to take advantage of the biggest strength of the cloud: the availability of seemingly endless numbers of machines on demand! So in this chapter, we use a service offered by GCP called *Genomics Pipelines API* (PAPI), which functions as a sort of job scheduler for GCP Compute Engine instances, to do exactly that.

First, we try simply changing the Cromwell configuration on our VM to submit job execution to PAPI instead of the local machine. Then, we try out a tool called WDL_Runner that wraps Cromwell and manages submissions to PAPI, which makes it easier to "launch and forget" WDL executions. Both of these options, which we explore in the first half of this chapter, will open the door for us to run full-scale GATK pipelines that we could not have run on our single-VM setup in Chapter 9. Along the way, we also discuss important considerations such as runtime, cost, portability, and overall efficiency of running workflows in the cloud.

Introducing the GCP Genomics Pipelines API Service

The Genomics Pipelines API is a service operated by GCP that makes it easy to dispatch jobs for execution on the GCP Compute Engine without having to actually manage VMs directly. Despite its name, the Genomics Pipelines API is not at all specific to genomics, so it can be used for a lot of different workloads and use cases. In general, we simply refer to it as the Pipelines API, or PAPI. It is possible to use PAPI to execute specific analysis commands directly as described in the Google Cloud documentation (*https://oreil.ly/l3OBw*), but in this chapter our goal is to use PAPI in the context of workflow execution with Cromwell, which is illustrated in Figure 10-1.

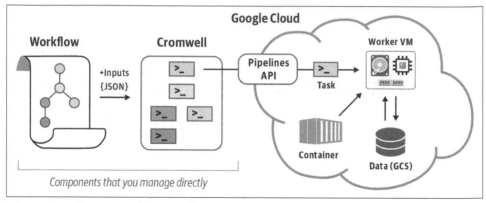

Figure 10-1. Overview of Cromwell + PAPI operation.

Just as in Chapter 8, we provide a WDL that describes our workflow to the Cromwell engine, which then interprets the workflow and generates the individual jobs that need to be executed. What's new is that instead of having Cromwell hand the jobs over to the local executor of the machine it is itself running on, we're going to point it to PAPI, as shown in Figure 10-1.

For every job that Cromwell sends to PAPI, the service will create a VM on Google Compute Engine with the specified runtime attributes (CPU, RAM, and storage), set up the Docker container specified in the WDL, copy input files to the VM's local disk, run the command(s), copy any outputs and logs to their final location (typically a GCS bucket), and finally, delete the VM and free up any associated compute resources. This makes it phenomenally easy to rapidly marshal a fleet of custom VMs, execute workflow tasks, and then walk away with your results without having to worry about managing compute resources, because that's all handled for you.

As we wrap up this book, Google Cloud is rolling out an updated version of this service under a new name, "Life Sciences API." Once we've had a chance to try it out, we'll write a blog post about the new service and how to adapt the book exercises to use it. In the meantime, we expect that the Genomics Pipelines API will remain functional for the foreseeable future.

Enabling Genomics API and Related APIs in Your Google Cloud Project

To run the exercises in this chapter, you need to have three APIs enabled in your Google Cloud Project: Genomics API (*https://oreil.ly/YYEiu*), Cloud Storage JSON API (*https://oreil.ly/K9jxa*), and Compute Engine API (*https://oreil.ly/Byzox*). You can use the direct links that we've included, or you can go to the APIs & Services (*https://oreil.ly/tDGTi*) section of the Google Cloud console and click the "+ Enable APIs and Services" button. Clicking it will take you to the API Library (*https://oreil.ly/73Fg6*), where you can search for each API by name. If you find yourself confused by APIs that have similar names in the search results, simply check the logos and descriptions shown in Figure 10-2.

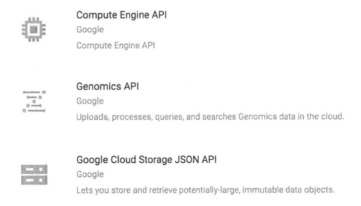

Figure 10-2. Logos and descriptions for the three required APIs: Genomics API, Cloud Storage JSON API, and Compute Engine API.

On each API's page, you will see a blue button that reads either Enable or Manage. The latter means that the API is already enabled in your project and you don't need to do anything. If you see the Enable button, click it to enable the API. You might see a message that says "To use this API, you may need credentials. Click 'Create credentials' to get started." but you can ignore it—GCP and Cromwell will handle authentication without requiring you to supply additional credentials.

 To use the Compute Engine API, you must have a means of payment set up in the Billing section even if you're using the free credits. If you followed the instructions in Chapter 4, you should be all set; but if you skipped straight to here, you'll need to go back and follow the billing setup in that chapter.

Directly Dispatching Cromwell Jobs to PAPI

In Chapter 8, we were running Cromwell without a configuration file, so by default it sent all jobs to the machine's local executor. Now we're going to provide a configuration file that points Cromwell to PAPI for execution.

Fun fact: you can configure Cromwell to run jobs through PAPI from anywhere, whether you're running Cromwell on your laptop, on a local server, or even on a VM on a different cloud platform. Here, we show you how to do it from your VM using GCP given that you already have it set up appropriately (you have Cromwell installed and you're authenticated through GCP), but the basic procedure and requirements would be the same anywhere else.

Configuring Cromwell to Communicate with PAPI

The Cromwell documentation has a section on using GCP (*https://oreil.ly/MnckS*) that includes a template configuration file for running on PAPI, which we've copied to the book bundle for your convenience. You can find it at *~/book/code/config/google.conf* on your VM if you followed the instructions provided in Chapter 4. If you open up the *google.conf* file in a text editor, this is what it looks like:

```
$ cat ~/book/code/config/google.conf
```

```
# This is an example configuration file that directs Cromwell to execute
# workflow tasks via the Google Pipelines API backend and allows it to retrieve
# input files from GCS buckets. It is intended only as a relatively simple example
# and leaves out many options that are useful or important for production-scale
# work. See https://cromwell.readthedocs.io/en/stable/backends/Google/ for more
# complete documentation.

engine {
  filesystems {
    gcs {
      auth = "application-default"
      project = "<google-billing-project-id>"
    }
  }
}

backend {
  default = PAPIv2

  providers {
    PAPIv2 {
      actor-factory = "cromwell.backend.google.pipelines.v2alpha1.PipelinesApi
      LifecycleActorFactory"
      config {
        # Google project
        project = "<google-project-id>"

        # Base bucket for workflow executions
        root = "gs://<google-bucket-name>/cromwell-execution"

        # Polling for completion backs-off gradually for slower-running jobs.
        # This is the maximum polling interval (in seconds):
        maximum-polling-interval = 600

        # Optional Dockerhub Credentials. Can be used to access private docker images.
        dockerhub {
          # account = ""
          # token = ""
        }

        # Number of workers to assign to PAPI requests
        request-workers = 3

        genomics {
          # A reference to an auth defined in the `google` stanza at the top. This auth is used
```

```
      # to create
      # Pipelines and manipulate auth JSONs.
      auth = "application-default"

      # Endpoint for APIs, no reason to change this unless directed by Google.
      endpoint-url = "https://genomics.googleapis.com/"

      # Pipelines v2 only: specify the number of times localization and delocalization
      # operations should be attempted
      # There is no logic to determine if the error was transient or not, everything
      # is retried upon failure
      # Defaults to 3
      localization-attempts = 3

    }

    filesystems {
      gcs {
        auth = "application-default"
        project = "<google-billing-project-id>"
        }
      }
    }

    default-runtime-attributes {
      cpu: 1
      failOnStderr: false
      continueOnReturnCode: 0
      memory: "2048 MB"
      bootDiskSizeGb: 10
      # Allowed to be a String, or a list of Strings
      disks: "local-disk 10 SSD"
      noAddress: false
      preemptible: 0
      zones: ["us-east4-a", "us-east4-b"]
    }
   }
  }
 }
}
```

We know that's a lot to take in if you're not used to dealing with this sort of thing; the good news is that you can ignore almost all of it, and we're going to walk you through the bits that you do need to care about. As a heads-up before we begin, be aware that you're going to need to edit the file and save the modified version as *my-google.conf* in your sandbox directory for this chapter, which should be *~/sandbox-10*. Go ahead and create that directory now. We recommend using the same procedure as you used in earlier chapters to edit text files:

```
$ mkdir ~/sandbox-10
$ cp ~/book/code/config/google.conf ~/sandbox-10/my-google.conf
```

Now set environment variables to point to your sandbox and related directories:

```
$ export CONF=~/sandbox-10
$ export BIN=~/book/bin
$ export WF=~/book/code/workflows
```

In Chapter 4, we also defined a BUCKET environment variable. If you are working in a new terminal, make sure that you also redefine this variable because we use it a little later in this chapter. Replace *my-bucket* in the following command with the value you used in Chapter 4:

```
$ export BUCKET="gs://my-bucket"
```

Now that our environment is set up, we can get to work. First, let's identify what makes this configuration file point to GCP and PAPI as opposed to another backend. Go ahead and open your copy of the configuration file now:

```
$ nano ~/sandbox-10/my-google.conf
```

You can see there are many references to Google throughout the file, but the key setting is happening here:

```
backend {
  default = PAPIv2

  providers {
    PAPIv2 {
```

Here, PAPIv2 refers to the current name and version of the PAPI.

Scrolling deeper into that section of the file, you'll find two settings, the project ID and the output bucket, which are very important because you must modify them in order to get this configuration file to work for you:

```
config {
        // Google project
        project = "<google-project-id>"

        // Base bucket for workflow executions
        root = "gs://<google-bucket-name>/cromwell-execution"
```

You need to replace *google-project-id* to specify the Google Project you're working with, and replace *google-bucket-name* to specify the location in GCS that you want Cromwell to use for storing execution logs and outputs. For example, for the test project we've been using so far, it looks like this when it's filled in:

```
config {
        // Google project
        project = "ferrous-layout-260200"

        // Base bucket for workflow executions
        root = "gs://my-bucket/cromwell-execution"
```

Finally, you also need to provide the project ID to charge when accessing data in GCS buckets that are set to the requester-pays mode, as discussed in Chapter 4. It appears in two places in this configuration file: once toward the beginning and once toward the end:

```
gcs {
    auth = "application-default"
    project = "<google-billing-project-id>"
}
```

Specifically, you need to replace *google-billing-project-id* to specify the billing project to use for that purpose. This setting allows you to use a different billing project from the one used for compute, for example if you're using separate billing accounts to cover different kinds of cost. Here, we just use the same project as we just defined earlier:

```
gcs {
    auth = "application-default"
    project = "ferrous-layout-260200"
}
```

Make sure to edit both occurrences of this setting in the file.

When you're done editing the file, be sure to save it as *my-google.conf* in your sandbox directory for this chapter, *~/sandbox-10*. You can give it a different name and/or put it in a different location if you prefer, but then you'll need to edit the name and path accordingly in the command line in the next section.

Running Scattered HaplotypeCaller via PAPI

You have one more step to do before you can actually launch the Cromwell `run` command to test this configuration: you need to generate a file of credentials that Cromwell will give to PAPI to use for authentication. To do so, run the following `gcloud` command:

```
$ gcloud auth application-default login
```

It's similar to the `gcloud init` command you ran in Chapter 4 to set up your VM, up to and including the little lecture about security protocols. Follow the prompts to log in via your browser and copy the access code to finish the process. The credentials file, which the system refers to as Application Default Credentials (ADC), will be created in a standard location that the `gcloud` utilities can access. You don't need to do anything more for it to work.

We provide a JSON file of inputs that is prefilled with the paths to test files (see *$WF/ scatter-hc/scatter-haplotypecaller.gcs.inputs.json*). You can check what it contains; you'll see the files are the same as you've been using locally, but this time we're pointing Cromwell to their locations in GCS (in the book bucket):

```
$ cat $WF/scatter-hc/scatter-haplotypecaller.gcs.inputs.json
```

With that in hand, it's time to run the following Cromwell command:

```
$ java -Dconfig.file=$CONF/my-google.conf -jar $BIN/cromwell-53.1.jar \
    run $WF/scatter-hc/scatter-haplotypecaller.wdl \
    -i $WF/scatter-hc/scatter-haplotypecaller.gcs.inputs.json
```

This command calls the same workflow that we last ran in Chapter 8, *scatter-haplotypecaller.wdl*, but this time we're adding the -Dconfig.file argument to specify the configuration file, which will cause Cromwell to dispatch the work to PAPI instead of the local machine. PAPI in turn will orchestrate the creation of new VMs on Compute Engine to execute the work. It will also take care of pulling in any input files and saving the logs and any outputs to the storage bucket you specified in your configuration file. Finally, it will delete each VM after its work is done.

Local Paths Versus Cloud Storage Paths

Notice that in this exercise, we are using a different *inputs* JSON file compared to the command we ran in Chapter 8. This one, *scatter-haplotypecaller.gcs.inputs.json*, points to the locations of the original files that we provide in the gs://genomics-in-the-cloud bucket, instead of the local copies you put on the VM in Chapter 4. The subtlety here is that in Chapter 8, Cromwell was executing the commands locally on your VM, so it was able to use the local copies of the files. Now that it's outsourcing the work to other VM instances, you must provide paths to files in GCS because the local storage of the VM you're working with is not accessible to those other VMs. Figure 10-3 illustrates these relationships.

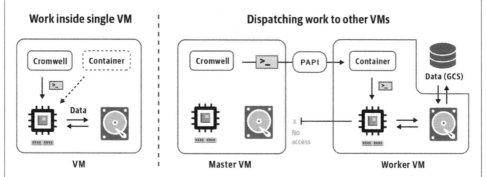

Figure 10-3. Side-by-side comparison of local versus PAPI execution.

If you want to experiment a little, you can try removing the PAPI configuration and running the workflow locally on your VM with the *inputs* JSON file that has the paths to the bucket. The workflow will fail with an error that should look like this (right before the stack trace begins):

```
Could not build the path
"gs://genomics-in-the-cloud/v1
/data/germline/intervals/snippet-intervals-min.list". It
may refer to a filesystem not supported by this instance of
Cromwell. Supported filesystems are: HTTP, LinuxFileSystem.
```

This shows that by default, Cromwell does not recognize GCS as a source of input files, even if it's running on a VM instance in GCP. That's why the configuration file we used earlier explicitly lists GCS as a supported filesystem:

```
filesystems {
    gcs {
```

This filesystem declaration shows up in two places: within both the top-level engine and backend blocks, respectively. The one under engine specifies that the Cromwell engine itself can pull files from GCS when interpreting the WDL script. This is necessary in our example because the engine must retrieve the intervals list from GCS in order to process the scatter instruction. In contrast, the one under backend enables the calls executed through PAPI to use files in GCS.

In principle, you could create a hybrid configuration that points to the local system for execution but allows the use of GCS as a supported filesystem. Be aware, however, that this will cause Cromwell to localize copies of those files to the instance's local storage. If you were to run workflows with such a configuration from outside GCP, you (or the owner of the bucket) would be charged egress fees. This could be quite costly if you were to do that on large-scale data.

Monitoring Workflow Execution on Google Compute Engine

This all sounds great, but how can you know what is actually happening when you run that command? As usual Cromwell will output a lot of information to the terminal that is rather difficult to parse, so let's walk through a few approaches that you can take to identify what is running where.

First, you should see in the early part of the log output that Cromwell has correctly identified that it needs to use PAPI. It's not super obvious, but you'll see a few lines like these that mention PAPI:

```
[2019-12-14 18:37:49,48] [info] PAPI request worker batch interval is
33333 milliseconds
```

Scrolling a bit farther down, you'll find a line that references Call-to-Backend assignments, which lists the task calls from the workflow:

```
[2019-12-14 18:37:52,56] [info] MaterializeWorkflowDescriptorActor [68271de1]:
Call-to-Backend assignments: ScatterHaplotypeCallerGVCF.HaplotypeCallerG
VCF -> PAPIv2, ScatterHaplotypeCallerGVCF.MergeVCFs -> PAPIv2
```

The -> PAPIv2 bit means that each of these tasks was dispatched to PAPI (version 2) for execution. After that, past the usual detailed listing of the commands that Cromwell generated from the WDL and *inputs* JSON file, you'll see many references to PipelinesApiAsyncBackendJobExecutionActor, which is another component of the PAPI system that is handling your workflow execution.

Of course, that tells you only that Cromwell is dispatching work to PAPI, and that PAPI is doing something in response, but how do you know what it's doing? The most straightforward way is to go to the Compute Engine console (*https://oreil.ly/-mFEu*), which lists VM instances running under your project. At the very least, you should see the VM that you have been using to work through the book exercises, whose name you should recognize. And if you catch them while they're running, there can be up to four VM instances with obviously machine-generated names (*google-pipelines-worker-xxxxx…*), as shown in Figure 10-4. Those are the VMs created by PAPI in response to your Cromwell command.

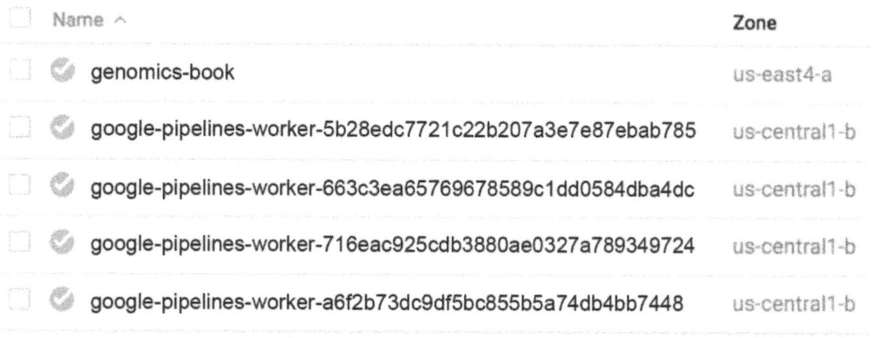

	Name ⌃	Zone
☐ ✅	genomics-book	us-east4-a
☐ ✅	google-pipelines-worker-5b28edc7721c22b207a3e7e87ebab785	us-central1-b
☐ ✅	google-pipelines-worker-663c3ea65769678589c1dd0584dba4dc	us-central1-b
☐ ✅	google-pipelines-worker-716eac925cdb3880ae0327a789349724	us-central1-b
☐ ✅	google-pipelines-worker-a6f2b73dc9df5bc855b5a74db4bb7448	us-central1-b

Figure 10-4. List of active VM instances

Why has PAPI created multiple VMs, you might ask? As a reminder, this workflow divides the work of calling variants into four separate jobs that will run on different genomic intervals. When Cromwell launched the first four jobs in parallel, PAPI created a separate VM for each of them. The only job Cromwell will have queued up for later is the merging task that collects the outputs of the four parallel jobs into a single task. That's because the merging task must wait for all the others to be complete in order for its input dependencies to be satisfied. When that is the case, PAPI will once more do the honors to get the work done and corral the final output to your bucket.

Not to put too fine a point on it, but that right there is one of the huge advantages of using the cloud for this kind of analysis. As long as you can chop the work into independent segments, you can parallelize the execution to a much larger degree than you can typically afford to do on a local server or cluster with a limited number of nodes available at any given time. Even though the cloud is not *actually* infinite (sorry, the Tooth Fairy is also not real), it does have a rather impressive capacity for accommodating ridiculously large numbers of job submissions. The Broad Institute, for example, regularly runs workflows on Cromwell across thousands of nodes simultaneously.

The zone of the PAPI-generated VMs may be different from the zone you set for the VM you created yourself, as it was in our case here. That's because we didn't explicitly specify a zone in the configuration file, so PAPI used a default value. There are several ways to control the zone where the work will be done, including at the workflow level or even at the task level within a workflow, as described in the Cromwell documentation (*https://oreil.ly/2YElr*).

You can also see an overview of Compute Engine activity in the Home Dashboard (*https://oreil.ly/ChMKK*) of your project, as shown in Figure 10-5. The advantage of the dashboard, even though it gives you only an aggregate view, is that it allows you to see past activity, not just whatever is running at the moment you're looking.

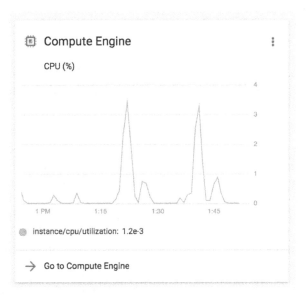

Figure 10-5. Overview of Compute Engine activity.

Finally, you can check the execution directory in the bucket that you specified in the config. You can do so either by navigating to the bucket in the Cloud Storage console (*https://oreil.ly/1iQmv*), or by using `gsutil`, as we've previously described. Take a few minutes to explore that directory and see how the outputs are structured. It should look the same as in Chapter 8 when you ran this same workflow on the VM's local system. However, keep in mind that the contents of the execution directory in the bucket don't represent what is happening in real time on the VMs. A synchronization process updates the execution logs and copies any outputs as they become available, operating at intervals that start out very short and gradually increase to avoid over-burdening the system in the case of long-running jobs. The maximum length of these intervals is customizable; you can see the relevant code in the configuration file we used earlier:

```
// Polling for completion backs-off gradually for slower-running jobs.
// This is the maximum polling interval (in seconds): maximum-polling-interval = 600
```

The interval time is measured in seconds, so you can see that, by default, the maximum time between updates of the bucket is 10 minutes.

The workflow can take up to 10 minutes to complete. At that point, you'll see the usual `finished with status 'Succeeded'` message, which is followed by the list of final outputs in JSON format:

```
[INFO] … SingleWorkflowRunnerActor workflow finished with status 'Succeeded'.
 "outputs": {
  "ScatterHaplotypeCallerGVCF.output_gvcf": "gs://genomics-book-test-99/cromwell-
  execution/ScatterHaplotypeCallerGVCF/68271de1-4220-4818-bfaa-5694551cbe81/call-
  MergeVCFs/mother.merged.g.vcf"
  },
```

You should see that this time, the final output is located in the bucket that you specified in the configuration file. The path to the output has the same structure as we described for local execution.

Now, take a moment to think about what you just achieved: using a mostly prebaked configuration file and just one command line, you kicked off a process that involved marshaling sophisticated computational resources to parallelize the execution of a real (if simple) genomics workflow. You could use that exact same process to run a full-scale workflow, like the whole genome analysis pipeline we dissected in Chapter 9, or you could hold on to your hat while we roll out a few alternative options. Specifically, later in this chapter we show you two approaches that involve "wrapping" Cromwell in additional layers of tooling that increase both the ease of use and scalability of this system.

However, before we plunge into the verdant ecosystem of Cromwell-enabling add-ons, we're going to take a bit of a side track to discuss the trade-offs and opportunities involved in running workflows on the cloud.

Understanding and Optimizing Workflow Efficiency

Did you pay attention to how much time it took to run the *scatter-haplotypecaller.wdl* workflow through PAPI? About 10 minutes, right? Do you remember how long it took to run just on your VM in Chapter 8? More like two minutes? So let's get this straight: running the same workflow with parallelization on multiple machines took five times longer than just running the same jobs on a single machine. That sounds… kind of terrible?

The good news is that it's mostly an artifact of the very small scale of the jobs we're running. We've been using a set of intervals that cover only a tiny region of the genome, and `HaplotypeCaller` itself takes a trivially small amount of time to run on such short intervals. When you run the workflow locally on your VM, there's not actually much work to do: the GATK container image and the files are already there, so all that really needs to happen is for Cromwell to read the WDL and launch the GATK commands, which, as we've mentioned, run quickly. In contrast, when you tell Cromwell to dispatch work to PAPI, you set into motion this massive machinery that involves creating VMs, retrieving container images, localizing files from GCS, and so on. That's all overhead that shows up in the form of a longer runtime. So for short tasks, the actual amount of overall runtime spent on the "real work" is dwarfed by the time spent on behind-the-scenes setup. However, that setup time is roughly constant, so for longer-running tasks (e.g., if you ran this workflow on much larger genomic intervals), the setup time ends up being just a drop in the Google bucket.

With that example in your pocket, let's take a stroll through some of the considerations you'll need to keep in mind, whether you're planning to develop your own workflows or simply using someone else's on the cloud.

Granularity of Operations

The `HaplotypeCaller` workflow that we ran here is intended to run on much longer intervals that cover huge amounts of data, so it makes perfect sense to execute it through PAPI when processing full-scale datasets. But what if part of your workflow involves short-running operations like simple file format conversion and indexing? Well, that might be an opportunity to combine several operations into a single task in the workflow. We already saw an example of this in the second workflow we examined in Chapter 9, in the `CramToBamTask` command block:

```
command {
    set -e
    set -o pipefail

    ${samtools_path} view -h -T ${ref_fasta} ${input_cram} |
    ${samtools_path} view -b -o ${sample_name}.bam -
    ${samtools_path} index -b ${sample_name}.bam
    mv ${sample_name}.bam.bai ${sample_name}.bai
}
```

The workflow authors could have separated subsets of this `command` block into separate WDL tasks to maximize modularity. For example, having a standalone BAM indexing task might have been useful and reusable elsewhere. However, they correctly identified that doing so would increase the overhead when run through PAPI, so they traded off some potential modularity in favor of higher efficiency.

Balance of Time Versus Money

It's important to understand that these kinds of trade-offs will make a difference not just in the amount of time it takes to run your pipeline, but also its cost. This is probably not something you're used to thinking about if you've worked mostly on local systems that have computational resources already paid for and your major limitation is how much quota you have been allotted. On the cloud, however, it's almost entirely pay-as-you-go, so it's worth thinking things through if you're on a budget.

For example, here's another trade-off to think about when moving workflows to the cloud: how widely should you parallelize the execution of your variant-calling workflow? You could simply parallelize it by chromosome, but those still create huge amounts of data, especially for whole genome samples, so it will take a very long time to process each, and you won't really be taking advantage of the amazing capacity for parallelism of the cloud. In addition, in humans at least, the various chromosomes have enormously different sizes; for example, chromosome 1 is about five times longer than chromosome 22, so the latter will finish much faster, and its results will sit around waiting for the rest of the chromosomes to be done.

A more efficient approach is to chop the chromosomes themselves into subsets of intervals, preferably in areas of uncertainty where the reference sequence has stretches of N bases, meaning the content of those regions is unknown, so it's OK to interrupt processing there. Then, you can balance the sizes of the intervals in such a way that most of them will take about the same amount of time to process. But that still leaves you a lot of leeway for deciding on the average length of the intervals when you design that list. The more you chop up the sequence, the sooner you can expect to have your results in hand, because you can (within reason) run all those intervals in parallel. However, the shorter the runtime of each individual interval, the more you will feel the pain—and more to the point, the cost—of the overhead involved in dispatching the work to separate VMs.

Not convinced? Suppose that you define three hundred intervals that take three hours each to analyze (these are made-up numbers, but should be in the ballpark of realistic). For each interval, you'll pay up to 10 minutes of VM overhead time during the setup phase before the actual analysis begins (oh yeah, you get charged for that time). That's 300 × 10 minutes; in other words, 50 hours' worth of VM cost spent on overhead. Assuming that you're using basic machines that cost about $0.03/hour, that amounts to $1.50 which is admittedly not the end of the world. Now let's say you turn up the dial by a factor of 10 on the granularity of your intervals, producing 3,000 shorter intervals that take 18 minutes each to analyze. You'll get your results considerably faster, but you'll pay 3,000 × 10 minutes, or 500 hours' worth of VM cost on overhead. Now you're spending $15 per sample on overhead. Across a large number of samples, that might make a noticeable dent in your budget. But we're not judging; the question is whether the speed is worth the money to *you*.

The takeaway here is that the cloud gives you a lot of freedom to find your happy point.

When the Broad Institute originally moved its entire genome analysis pipeline from its on-premises cluster to GCP, it cost about $45 to run per whole genome sample (at 30X coverage). Through a combination of workflow optimizations, and in collaboration with GCP engineers, the institute's team was able to squeeze that down to $5 per whole genome sample. These phenomenal savings translated into lower production costs, which at the end of the day means more science per research dollar.

Admittedly, the cloud pipeline does take about 23 hours to run to completion on a single sample, which is quite a bit longer than the leading accelerated solutions that are now available to run GATK faster on specialized hardware, such as Illumina's DRAGEN solution. For context, though, it typically takes about two days to prepare and sequence a sample on the wetlab side of the sequencing process, so unless you're in the business of providing urgent diagnostic services, the additional day of analysis is usually not a source of concern.

Here's the thing: *that processing time does not monopolize computational resources that you could be using for something else.* If you need more VMs, you just ask PAPI to summon them. Because there is practically no limit to the number of workflows that the operations team can launch concurrently, there is little risk of having a backlog build up when an order for a large cohort comes through—unless it's on the scale of gnomAD (*https://oreil.ly/vHPKf*), which contains more than 100,000 exome and 80,000 whole genome samples. Then, it takes a bit of advance planning and a courtesy email to your friendly GCP account manager so that they can tell their operations team to gird their loins. (We imagine the email: "Brace yourselves; gnomAD is coming.")

 The cost of the Broad's cloud-based whole genome analysis pipeline unfortunately increased to around $8 per sample following the discovery of two major computer security vulnerabilities, Spectre (*https://oreil.ly/fbaUw*) and Meltdown (*https://oreil.ly/xwa5J*), that affect computer processors worldwide. The security patches that are required to make the machines safe to use add overhead that causes longer runtimes, which will apparently remain unavoidable until cloud providers switch out the affected hardware. Over time, though, the cost of storage and compute on the cloud continues to fall, so we expect the $8 price to decrease over time.

Suggested Cost-Saving Optimizations

If you're curious to explore the workflow implementation that the Broad Institute team developed to achieve those substantial cost savings, look no further than back to Chapter 9. The optimized pipeline we've been talking about is none other than the second workflow we dissected in that chapter (the one with the subworkflows and task libraries).

Without going too far into the details, here's a summary of the three most effective strategies for optimizing WDL workflows to run cheaply on Google Cloud. You can see these in action in the *GermlineVariantDiscovery.wdl* library of tasks that we looked at in Chapter 9.

Dynamic sizing for resource allocation

The more storage you request for a VM for a given task, the more it will cost you. You can keep that cost down by requesting only the bare minimum amount of disk storage that will fit the task, but how do you deal with variability in file input sizes without having to check them manually for every sample you need to process? Good news: there are WDL functions that allow you to evaluate the size of input files going into a task at runtime (but before the VM is requested). Then, you can apply some arithmetic (based on reasonable assumptions about what the task will produce) to calculate how much disk should be allocated. For example, the following code measures the total size of the reference genome files and then calculates the desired disk size based on the amount of space needed to account for a fraction of the input BAM file (more on that in a minute) plus the reference files, plus some padding to account for the output:

```
Float ref_size = size(ref_fasta, "GB") +
size(ref_fasta_index, "GB") + size(ref_dict, "GB")
Int disk_size = ceil(((size(input_bam, "GB") + 30) / hc_scatter) + ref_size) + 20
```

See this blog post (*https://oreil.ly/P9Vxg*) for a more detailed discussion of this approach.

File streaming to GATK4 tools

This is another great way to reduce the amount of disk space you need to request for your task VMs. Normally, Cromwell and PAPI localize all input files to the VM as a prerequisite for task execution. However, GATK (starting in version 4.0) is capable of streaming data directly from GCS, so if you're running a GATK tool on a genomic interval, you can instruct Cromwell not to localize the file and just let GATK handle it, which it will do by retrieving only the subset of data specified in the interval. This is especially useful, for example, if you're running `HaplotypeCaller` on a 300 Gb BAM file but you're parallelizing its operation across many intervals to run on segments that are orders of magnitude smaller. In the previous code example, that's why the size of the input BAM file is divided by the width of the scatter (i.e., the number of intervals). To indicate in your WDL that an input file can be streamed, simply add the following to your task definition, for the relevant input variable:

```
parameter_meta {
    input_bam: {
      localization_optional: true
    }
  }
```

Note that this currently works only for files in GCS and is not available to Picard tools bundled in GATK.

Preemptible VM instances

This involves using a category of VMs, called preemptible (*https://oreil.ly/lep8v*), that are much cheaper to use than the normal pricing (currently 20% of list price). These discounted VMs come with all the same specifications as normal VMs, but here's the catch: Google can take them away from you at any time. The idea is that it's a pool of spare VMs that you can use normally as long as there's no shortage of resources. If somebody out there suddenly requests a large number of machines of the same type as you're using and there aren't enough available, some or all of the VMs that you are using will be reallocated to them, and your job will be aborted.

To head off the two most common questions, no, there is no (built-in) way to save whatever progress had been made before the interruption; and yes, you will get charged for the time you used them. On the bright side, PAPI will inform Cromwell that your jobs were preempted, and Cromwell will try to restart them automatically on new VMs. Specifically, Cromwell will try to use preemptible VMs again, up to a customizable number of attempts (three attempts by default in most Broad Institute workflows); then, it will fall back to using regular (full-price) VMs. To control the number of preemptible attempts that you're willing to allow for a given task, simply set the `preemptible:` property in the `runtime` block. To disable the use of preemptibles, set it to 0 or remove that line altogether. In the following example, the preemptible count is set using a `preemptible_tries` variable so that it can be easily customized on demand, as are the Docker container and the disk size:

```
runtime {
    docker: gatk_docker
    preemptible: preemptible_tries
    memory: "6.5 GiB"
    cpu: "2"
    disks: "local-disk " + disk_size + " HDD"
  }
```

The bottom line? Using preemptible VMs is generally worthwhile for short-running jobs because the chance of them getting preempted before they finish is typically very small, but the longer the job, the less favorable the odds become.

Platform-Specific Optimization Versus Portability

All of these optimizations are very specific to running on the cloud—and for some of them, not just any cloud—so now we need to talk about their impact on portability. As we discussed when we introduced workflow systems in Chapter 8, one of the key goals for the development of Cromwell and WDL was portability: the idea that you could take the same workflow and run it anywhere to get the same results, without having to deal with a whole lot of software dependencies or hardware requirements. The benefits of computational portability range from making it easier to collaborate with other teams, to enabling any researcher across the globe to reproduce and build on your work after you've published it.

Unfortunately, the process of optimizing a workflow to run efficiently on a given platform can make it less portable. For example, earlier versions of WDL did not include the `localization_optional: true` idiom, so to use the GATK file-streaming capability, workflow authors had to trick Cromwell into not localizing files by declaring those input variables as `String` types instead of `File` types. The problem? The resulting workflows could not be run on local systems without changing the relevant variable declarations. Instant portability fail.

The introduction of `localization_optional: true` was a major step forward because it functions as a "hint" to the workflow management system. Basically, it means, "Hey, you don't have to localize this file if you're in a situation that supports streaming, but if you're not, please do localize it." As a result, you can run the same workflow in different places, enjoy the optimizations where you can, and rest assured that the workflow will still work everywhere else. Portability win!

The other optimizations we showed you have always had this harmless "hint" status, in the sense that their presence does not affect portability. Aside from the `docker` property, which we discussed in Chapter 8, Cromwell will happily ignore any properties specified in the `runtime` block that are not applicable to the running environment, so you can safely have `preemptible`, `disk`, and so on specified even if you are just running workflows on your laptop.

Where things might get tricky is that other cloud providers such as Microsoft Azure and AWS have their own equivalents of features like preemptible VM instances (on AWS, the closest equivalents are called Spot Instances), but they don't behave exactly the same way. So, inevitably the question will arise as to whether Cromwell should use the `preemptible` property to control Spot Instances when running on AWS, or provide a separate property for that purpose. If it's the latter, where does it stop? Should every runtime property exist in as many flavors as there are cloud platforms?

Oh, no, *we* don't have an answer to that. We'll be over here heating up some popcorn and watching the developers battle it out. The best way to keep track of what a given Cromwell backend currently can and can't do is to take a look at the online backend documentation (*https://oreil.ly/7uTh6*). For now, let's get back to talking about the most convenient and scalable ways to run Cromwell on GCP.

Wrapping Cromwell and PAPI Execution with WDL Runner

It's worth repeating that you can make Cromwell submit workflows to PAPI from just about anywhere, as long as you can run Cromwell and connect to the internet. As mentioned earlier, we're making you use a GCP VM because it minimizes the amount of setup required to get to the good parts (running big workflows!), and it reduces the risk of things not working immediately for you. But you could absolutely do the same thing from your laptop or, if you're feeling a bit cheeky, from a VM on AWS or Azure. The downside of the laptop option, however, is that you would need to make sure that your laptop stays on (powered and connected) during the entire time that your workflow is running. For a short-running workflow that's probably fine; but for full-scale work like a whole genome analysis, you really want to be able to fire off a workflow and then pack up your machine for the night without interrupting the execution.

Enter the *WDL Runner*. This open source toolkit acts as a lightweight wrapper for Cromwell and PAPI. With a single command line from you, WDL Runner creates a VM in GCP, sets up a container with Cromwell configured to use PAPI, and submits your workflow to be executed via PAPI, as shown in Figure 10-6. Does any of that sound familiar? Aside from a few details, this is pretty much what you did in earlier chapters: create a VM instance, put Cromwell on it, configure it to communicate with PAPI, and run a Cromwell command to launch a workflow.

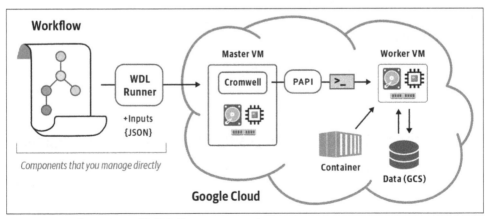

Figure 10-6. Overview of WDL Runner operation.

So at a basic level, you can view WDL Runner as a way to outsource all of that work and achieve the same result without having to manually manage a VM. But there's more: when the workflow completes, WDL Runner will transfer the execution logs to a location of your choosing and delete the VM, so it won't keep costing you money if you don't immediately turn it off. The cherry on the cake is that WDL Runner will also copy the final outputs to a location of your choosing, so you don't need to go dig through Cromwell's (annoyingly deep) execution directories to find the result you care about. Oh, and here's another cherry: WDL Runner also comes with monitoring tools that allow you to get a status summary for your workflow.

Needless to say, you can run WDL Runner from anywhere, but we're going to have you run it from your VM because it's right there and it has everything you need on it. Feel free to try it out from your laptop, as well, of course.

Setting Up WDL Runner

Good news: there's nothing to set up! We've included a copy of the WDL Runner code, which is available in GitHub (*https://oreil.ly/-upHl*), in the book bundle that you already copied to your VM. WDL Runner's main requirements are Python 2.7, Java 8, and gcloud, which are all available on your VM already.

To make the paths easier to manage, we are once again going to use environment variables. Two of them are variables that you should already have defined earlier in the chapter, $WF pointing to the *workflows* directory and $BUCKET pointing to your Google bucket. In addition, you are going to create a $WR_CONF variable pointing to the *config* directory and a $WR_PIPE variable pointing to the location of the *WDL_Runner* pipeline configuration file, *wdl_pipeline.yaml*, as follows:

```
$ export WR_CONF=~/book/code/config
$ export WR_PIPE=~/book/wdl_runner/wdl_runner
```

That's not a mistake, by the way; there are two nested directories that are both called *wdl_runner*; the lower-level one refers to code that is specific to *wdl_runner* whereas the other one refers to the overall package. If you poke around in the latter, you'll see that beside the other directory called *wdl_runner*, there is as a directory called *monitoring_tools*. We'll give you three guesses as to what's in there, and the first two don't count.

Running the Scattered HaplotypeCaller Workflow with WDL Runner

This is going to be reasonably straightforward. As a heads-up, you'll notice that the command line for launching a workflow with WDL Runner is a bit longer than the one for launching it directly with Cromwell itself. That's because WDL Runner gives you a few more options for where to store outputs and so on without having to mess around with a configuration file. Here it is:

```
$ gcloud alpha genomics pipelines run \
      --pipeline-file $WR_PIPE/wdl_pipeline.yaml \
      --regions us-east4 \
      --inputs-from-file WDL=$WF/scatter-hc/scatter-haplotypecaller.wdl,\
  WORKFLOW_INPUTS=$WF/scatter-hc/scatter-haplotypecaller.gcs.inputs.json,\
```

```
WORKFLOW_OPTIONS=$WR_CONF/empty.options.json \
    --env-vars WORKSPACE=$BUCKET/wdl_runner/test/work,\
OUTPUTS=$BUCKET/wdl_runner/test/output \
    --logging $BUCKET/wdl_runner/test/logging
```

But don't run it yet! Let's walk through it briefly so that you know what to customize versus what to leave alone.

The first line tells you that this command is primarily calling `gcloud`, which puts an interesting twist on the situation. Previously, we were running Cromwell and instructing it to communicate with Google Cloud things. Now, we're calling a Google Cloud thing to have it run Cromwell and make it communicate with Google Cloud things. You might not be entirely surprised to learn that WDL Runner was originally written by a team at GCP.

The `--pipeline-file` argument takes the path to a sort of configuration file that comes prebaked with the WDL Runner code; you would need to modify the path if you wanted to run WDL Runner from a different working directory.

The `--regions` argument, unsurprisingly, controls the region where you want the VMs to reside. Funny thing: the WDL Runner seems to require this, but it conflicts with the default `gcloud` settings that were set on your VM when you originally authenticated yourself with `gcloud`. Specifically, the command fails and returns this error message:

```
ERROR: (gcloud.alpha.genomics.pipelines.run) INVALID_ARGUMENT: Error: validating
pipeline: zones and regions cannot be specified together
```

To work around this issue, you need to unset the default zone by running the following `gcloud` command:

```
$ gcloud config set compute/zone ""
```

The `--inputs-from-file` argument lists the workflow WDL file, *inputs* JSON file, and options file. The options file allows you to specify runtime configuration settings that will apply to the entire workflow. We don't need to specify anything, but because WDL Runner requires that we provide one, we included a stub file that doesn't actually contain any settings. As an aside, make sure there is absolutely no whitespace between the key:value pairs that specify the files (even around the commas), or the command will fail with a weird error about unrecognized arguments. Yes, we try to experience all the failures so you don't have to.

The `--env-vars` argument lists your desired locations for storing the execution directory and the final outputs; these can be anything you want, as long as they are valid GCS paths starting with *gs://*. Here, we're using the $BUCKET environment variable that we originally set up in Chapter 4, and we're using a directory structure that we find convenient for managing outputs and logs. Feel free to chart your own paths, as it were. The `--logging` argument does the same thing as the previous argument but for the execution logs.

When you're happy with the look of your command, run it and then sit back and enjoy the sight of software at work. (Yes, saying this does feel a bit like tempting fate, doesn't it.) If everything is copacetic, you should get a simple one-line response that looks like this:

```
Running [projects/ferrous-layout-260200/operations/8124090577921753814].
```

The string of numbers (and sometimes letters) that follows `operations/` is a unique identifier, the operations ID, that the service will use to track the status of your submission. Be sure to store that ID somewhere because it will be essential for the monitoring step, which is coming up next.

Monitoring WDL Runner Execution

And here we are, ready to monitor the status of our workflow submission. Remember the not-so-mysterious directory called *monitoring_tools*? Yep, we're going to use that now. Specifically, we're going to use a shell script called *monitor_wdl_pipeline.sh* that is located in that directory. The script itself is fairly typical sysadmin fare, meaning that it's really painful to decipher if you're not familiar with that type of code. Yet, in a stroke of poetic justice, it's quite easy to use: you simply give the script the operations ID that you recorded in the previous step, and it will let you know how the workflow execution is going.

Before you jump in, however, we recommend that you open a new SSH window to your VM so that you can leave the script running. It's designed to continue to run, polling the `gcloud` service at regular intervals, until it receives word that the workflow execution is finished (whether successfully or not). As a reminder, you can open any number of SSH windows to your VM from the Compute Engine (*https://oreil.ly/ NMIIX*) console.

When you're in the new SSH window, move into the first *wdl_runner* directory and run the following command, substituting the operations ID you saved from the previous step:

```
$ cd ~/book/wdl_runner
$ bash monitoring_tools/monitor_wdl_pipeline.sh 7973899330424684165
Logging:
Workspace:
Outputs:
2019-12-15 09:21:19: operation not complete
No operations logs found.
There are 1 output files
Sleeping 60 seconds
```

This is the early stage output that you would see if you start the monitoring script immediately after launching the WDL Runner command. As time progresses, the script will produce new updates, which will become increasingly more detailed.

If you're not satisfied by the content of the monitoring script's output, you can always go to the Compute Engine console as we did earlier to view the list of VMs that are working on your workflow, as shown in Figure 10-7. This time, you'll see up to five new VMs with machine-generated names (in addition to the VM you created manually).

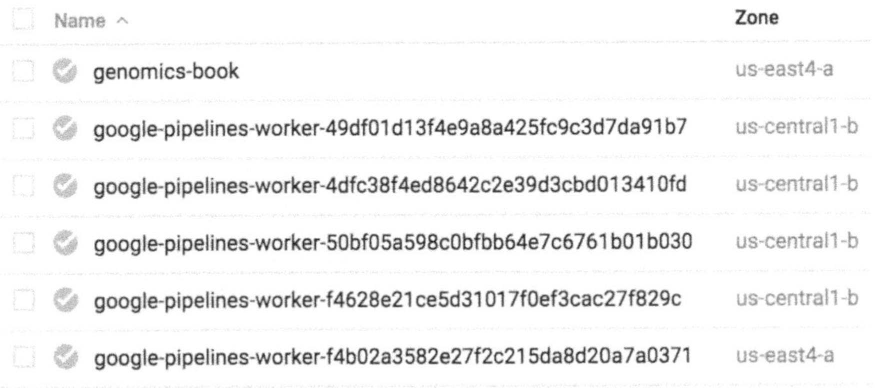

Name ∧	Zone
genomics-book	us-east4-a
google-pipelines-worker-49df01d13f4e9a8a425fc9c3d7da91b7	us-central1-b
google-pipelines-worker-4dfc38f4ed8642c2e39d3cbd013410fd	us-central1-b
google-pipelines-worker-50bf05a598c0bfbb64e7c6761b01b030	us-central1-b
google-pipelines-worker-f4628e21ce5d31017f0ef3cac27f829c	us-central1-b
google-pipelines-worker-f4b02a3582e27f2c215da8d20a7a0371	us-east4-a

Figure 10-7. List of active VM instances (WDL Runner submission).

But hang on, why would there be five VMs this time instead of four? If you're confused, have a quick look back at Figure 10-6 to refresh your memory of how this system works. This should remind you that the fifth wheel in this case is quite useful: it's the VM instance with Cromwell that is controlling the work.

Getting back to the monitoring script, you should eventually see an update that starts with operation complete:

```
2019-12-15 09:42:47: operation complete
```

There's a huge amount of output below that, though, so how do you know whether it was successful? It's the next few lines that you need to look out for. Either you see something like this and you breathe a happy sigh

```
Completed operation status information
  done: true
  metadata:
    events:
    - description: Worker released
```

or you see see the dreaded error: line immediately following done: true, and you take a deep breath before diving into the error message details:

```
Completed operation status information
  done: true
  error:
    code: 9
    message: 'Execution failed: while running "WDL_Runner": unexpected exit status 1
was not ignored'
```

We're not going to sugarcoat this: troubleshooting Cromwell and PAPI errors can be rough. If something went wrong with your workflow submission, you can find information about the error both in the output from the *monitor_wdl_pipeline.sh* script if you're using WDL Runner, and in the files saved to the bucket, which we take a look at after this.

Finally, let's take a look at the bucket to see the output produced by WDL Runner, shown in Figure 10-8.

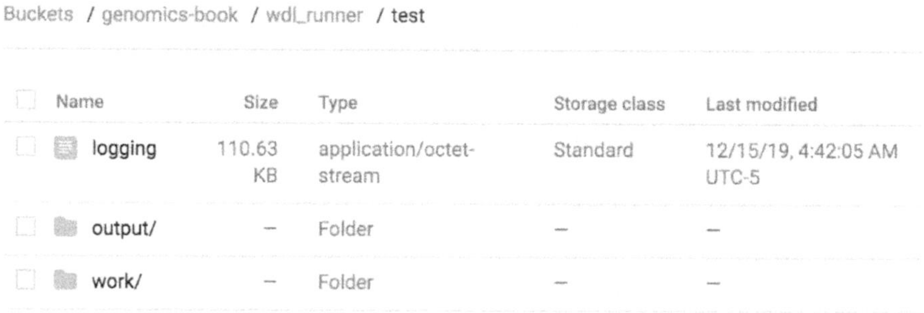

Buckets / genomics-book / wdl_runner / test

	Name	Size	Type	Storage class	Last modified
	logging	110.63 KB	application/octet-stream	Standard	12/15/19, 4:42:05 AM UTC-5
	output/	—	Folder	—	—
	work/	—	Folder	—	—

Figure 10-8. Output from the WDL Runner submission.

The *logging* output is a text file that contains the Cromwell log output. If you peek inside it, you'll see that it looks different from the log output you saw earlier when you directly ran Cromwell. That's because this instance of Cromwell was running in server mode! This is exciting: you ran a Cromwell server and didn't even know you were doing it. Not that you were really taking advantage of the server features—but it's a start. The downside is that the Cromwell server logs are extra unreadable for standard-issue humans, so we're not even going to try to read them here.

The *work* directory is simply a copy of the Cromwell execution directory, so it has the same rabbit-hole-like structure and contents as you've come to expect by now.

Finally, the *output* directory is one of the features we really love about the WDL Runner: this directory contains a copy of whatever file(s) were identified in the WDL as being the final output(s) of the workflow. This means that if you're happy with the result, you can just delete the working directory outright without needing to trawl through it to save outputs you care about. When you have workflows that produce a lot of large intermediate outputs, this can save you a lot of time—or money from not paying storage costs for those intermediates.

At the end of the day, we find WDL Runner really convenient for testing workflows quickly with little overhead while still being able to "fire and forget"; that is, launch the workflows, close your laptop, and walk away (as long as you record those operations IDs!). On the downside, it hasn't been updated in a while and is out of step with WDL features; for example, it neither supports subworkflows nor task imports because it lacks a way to package more than one WDL file. In addition, you need to be very disciplined with keeping track of those operations IDs if you find yourself launching a lot of workflows.

Here we showed you how to use WDL Runner as an example of how a project can wrap Cromwell, not as a fully-baked solution, and in the next chapter we're going to move on to a system that we like better. However, if you like the concept of wdl_run ner, don't hesitate to make your voice heard on the project GitHub page (*https:// oreil.ly/-upHl*) to motivate further development.

 What time is it? Time to stop your VM again, and this time it's especially important to do so because we won't use it in any of the book's remaining exercises. Feel free to delete your VM if you don't anticipate coming back to previous tutorials.

Wrap-Up and Next Steps

In this chapter, we showed you how to take advantage of the cloud's scaling power by running your workflows through the GCP Genomics PAPI service, first directly from the Cromwell command line and then from a light wrapper called WDL Runner. However, although these two approaches gave you the ability to run individual workflows of arbitrary complexity in a scalable way in GCP, the overhead involved in spinning up and shutting down a new instance of Cromwell every time you want to run a new workflow is highly inefficient and will scale poorly if you plan to run a lot of workflows.

For truly scalable work, we want to use a proper Cromwell server that can receive and execute any number of workflows on demand, and can take advantage of sophisticated features like call caching, which allows Cromwell to resume failed or interrupted workflows from the point of failure. Call caching requires a connection to a database and is not supported by WDL Runner. Yet, setting up and administering a server and database—not to mention operating them securely in a cloud environment—involves more complexity and tech support burden than most of us are equipped to handle.

That is why, in the next chapter, we're going to show you an alternative path to enjoying the power of a Cromwell server without shouldering the maintenance burden.

Specifically, we're going to introduce you to Terra, a scalable platform for biomedical research built and operated on top of GCP by the Broad Institute in collaboration with Verily. Terra includes a Cromwell server that you'll use to run GATK Best Practices workflows at full scale for the first time.

Running Many Workflows Conveniently in Terra

In Chapter 10, we gave you a tantalizing first taste of the power of the Cromwell plus Pipelines API combination. You learned how to dispatch individual workflows to PAPI, both directly through Cromwell and indirectly through the WDL Runner wrapper. Both approaches enabled you to rapidly marshal arbitrary amounts of cloud compute resources without needing to administer them directly, which is probably the most important lesson you can take from this book. However, as we've discussed, both approaches suffer from limitations that would prevent you from achieving the truly great scalability that the cloud has to offer.

In this chapter, we show you how to use a fully featured Cromwell server within *Terra*, a cloud-based platform operated by the Broad Institute. We begin by introducing you to the platform and walking you through the basics of running workflows in Terra. Along the way, you'll have the opportunity to experiment with the call caching feature that allows the Cromwell server to resume failed or interrupted workflows from the point of failure. With that experience in hand, you'll graduate to finally running a full-scale GATK Best Practices pipeline on a whole genome dataset.

Getting Started with Terra

You are just a few short hops away from experiencing the delights of a fully loaded Cromwell server thanks to Terra (*https://terra.bio*), a scalable platform for biomedical research operated by the Broad Institute in collaboration with Verily. Terra is designed to provide researchers with a user-friendly yet flexible environment for doing secure and scalable analysis in GCP. Under the hood, Terra includes a persistent Cromwell server configured to dispatch workflows to PAPI and equipped with a dedicated call caching database. On the surface, it provides a point-and-click interface for running and monitoring workflows as well as API access for those who prefer to interact with the system programmatically. Terra also provides rich functionality for accessing and managing data, performing interactive analysis through Jupyter Notebook, and collaborating securely through robust permissions controls. Figure 11-1 summarizes Terra's current primary capabilities.

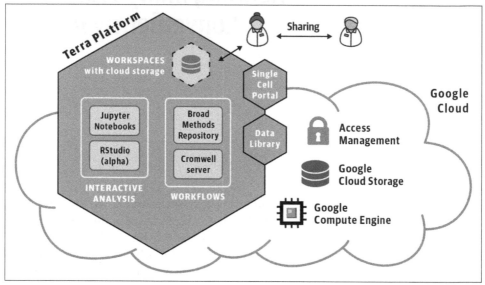

Figure 11-1. Overview of the Terra platform.

In this section, we focus on the workflow configuration, execution, and monitoring functionality through the web interface, but later in the chapter, we also dig into some interactive analysis in Jupyter.

 As of this writing, the Terra platform is under active development, so you likely will encounter differences between the screenshots here and the live web interface, as well as new features and behaviors. We'll provide updated instructions and guidance regarding any major changes in the blog for this book (*https://oreil.ly/ genomics-blog*).

Creating an Account

You can register for a Terra account for free (*https://app.terra.bio*). In the upper-left corner of your browser window, click the three-line symbol to expand the side menu and bring up the Sign in with Google button, as shown in Figure 11-2.

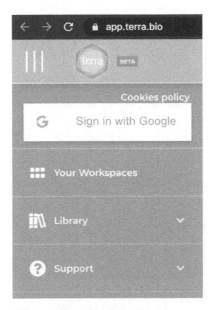

Figure 11-2. Expanded side menu showing sign-in button.

 Menu symbols and navigation patterns in Terra are often similar to those in the GCP console. Whenever you see the same symbols, you can safely assume they have the same purpose.

Sign in with the same account that you have been using so far—the one you used to set up your GCP account, and submit the registration form shown in Figure 11-3. The contact email can be different from the email you used to sign in; for example, if you are using a personal account to do the work but prefer to get email notifications sent to your work account. Email notifications you might receive from Terra mainly consist of notices about new feature releases and service status (such as planned maintenance or incident alerts). They are relatively infrequent, and you can opt out if you do not want to receive these notifications. The contact email you specify here is also used by the Terra helpdesk ticketing system, which will email you if you ask a question, report a bug, or suggest a feature.

Figure 11-3. The New User Registration form.

You'll need to accept the Terra Terms of Service. We strongly suggest that you follow the recommendation to secure your Google-linked account with two-factor authentication. Keep in mind that Terra is designed primarily to enable analysis of human patient data and includes repositories that host data funded by US government agencies with access restrictions, so security is a serious matter on the platform.

Creating a Billing Project

After you have completed the registration process, you should find yourself on the Terra portal landing page. Now you need to set up a billing project in Terra and connect it to a GCP billing account. Terra is built on top of GCP, and when you do work in Terra that incurs costs, GCP directly bills your account. The Broad Institute will not charge you any surcharges for the use of the platform.

If your institution or company has already set up billing through Terra, you may have access to GCP billing accounts or even Terra billing projects that are ready to go. An easy way check for this is to go to the Billing page (*https://oreil.ly/WYZyl*) (accessible from the side menu where you signed in) and see if there are any billing projects listed there. If not, click the blue plus symbol next to "Billing Projects" to open the billing project creation dialog, and click on the pull-down menu labeled "Select billing account" to see whether you already have access to a billing account. You can also check with your supervisor or friendly neighborhood administrator, who can grant you access if applicable. If they provide you with custom instructions for setting up your account, follow those and skip to the end of this section.

Here we assume that you do not already have access to preconfigured billing, and we show you how to connect the GCP billing account that you have been using so far. The procedure detailed in the sidebar below can be applied to any existing GCP billing account for which you have owner-level permissions.

Connecting an Existing GCP Billing Account to Use in Terra

To connect an existing billing account to your Terra account, you'll need to add the *terra-billing@terra.bio* service account as a user on your billing account. To do so, go to the GCP console Billing page (*https://oreil.ly/WEosg*) and navigate to the "Account management" tab. If you don't see a Permissions panel on the righthand side, click the "Show info panel" button located in the upper-right corner of the page. Then, under Permissions, click the "Add members" button shown in Figure 11-4.

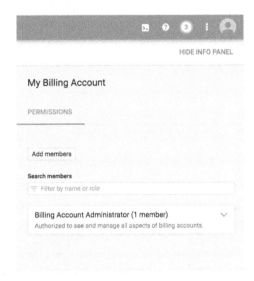

Figure 11-4. The GCP console Billing account permissions panel.

Add the *terra-billing@terra.bio* service account as a "Billing account user," as illustrated in Figure 11-5, and then click SAVE. This allows Terra to provide your billing account details to GCP in order to cover the cost of the resources you use when you do work or store data through Terra.

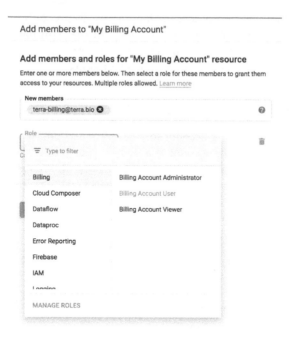

Figure 11-5. Adding the Terra billing user account as a user on a GCP billing account.

You should now be able to link this GCP billing account to a billing project in Terra.

Back on the Terra Billing page (*https://oreil.ly/WYZyl*), click the blue plus symbol next to "Billing Projects" to open the billing project creation dialog as shown in Figure 11-6. Give your new billing project a unique name and use the pull-down menu to select your GCP billing account, which should show up automatically if you successfully completed the linking procedure described above. You may need to refresh the Terra Billing page once.

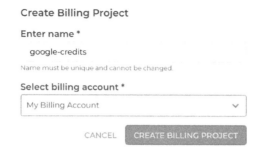

Figure 11-6. Using an existing billing account to create a billing project in Terra.

Press the button labeled "Create Billing Project" and check that your new project is now listed in the left-hand menu on the Billing page. You may see a small rotating symbol next to the project name, indicating that the project creation is in progress. Your billing project will typically be ready to use within a minute or two. You can click on the project name to view details and manage permissions, for example allowing other users to use this billing project to do work in Terra.

That completes the setup of your billing project.

After you have your billing project set up, all you need to go run some workflows is a workspace. In Chapter 13, we show you how to create your own workspaces from scratch, but for now, you're simply going to clone a workspace that we set up for you with all the essentials.

Cloning the Preconfigured Workspace

You can find the Genomics-in-the-Cloud-v1 workspace in the Library Showcase (*https://oreil.ly/RdqSW*) (which is accessible from the expandable menu on the left) or go straight to this link (*https://oreil.ly/n7oOr*). The landing page, or Dashboard, provides information about its purpose and contents. Take a minute to read the summary description if you'd like, and then we'll get to work.

This workspace is read-only, so the first thing you need to do is to create a clone that will belong to you and that you can therefore work with. To do so, in the upper-right corner of your browser, click the round symbol with three dots to expand the action menu. Select Clone, as shown in Figure 11-7, to bring up the workspace cloning form.

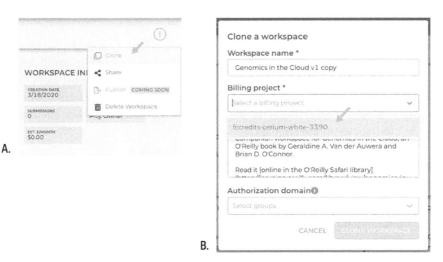

Figure 11-7. Cloning the preconfigured workspace. A) List of available actions; B) cloning form.

Select your billing project; that will determine the account GCP will bill when you do work that incurs costs. You can change the workspace name or leave it as is; workspace names must be unique within a billing project but do not need to be unique across all of Terra. Ignore the Authorization domain bit for now; that's an optional whitelisting option that you don't need at this time.

When you click the Clone Workspace button, you're automatically taken to your new workspace. Take a good look around; everything that the light touches belongs to you. There's a lot there, huh? Actually, you know what, don't spend too much time looking around. Let's keep a tight focus on our goal, which is to get you running workflows through Terra's Cromwell server.

Running Workflows with the Cromwell Server in Terra

Alright, it's showtime. Head on over to the Workflows section of the workspace, where you should see a list of available workflow configurations, as shown in Figure 11-8.

Figure 11-8. List of available workflow configurations.

These are two configurations for the same workflow, which you might recognize from the first part of their names, *scatter-hc*. That's right, it's your favorite parallelized *HaplotypeCaller* workflow. In case you're wondering, the key difference between these two configurations is that one is designed to run on a single input sample, whereas the other can run on any arbitrary number of input samples. That sounds exciting, right? Right. But let's focus on the simple case first.

Running a Workflow on a Single Sample

On the page that lists the workflow configurations, click the one called *scatter-hc.filepaths* to open the configuration page. Figure 11-9 shows the summary information, including a one-line synopsis and short description.

Figure 11-9. Viewing the workflow information summary.

In addition, the summary links to the Source of the workflow. If you follow that link, it will take you to the Broad Institute's Methods Repository (*https://oreil.ly/xBXrU*), an internal repository of workflows. Terra can also import workflows from Dockstore (*https://dockstore.org*) (as you'll see in Chapter 13), but for this exercise, we chose to use the internal repository, which is a little easier to work with on first approach.

Back on the *scatter-hc.filepaths* configuration page, have a look at the SCRIPT tab, which displays the actual workflow code. Sure enough, it's the same workflow that we've been using for a while.

As shown in Figure 11-10, the code display uses *syntax highlighting*: it colors parts of the code based on the syntax of the WDL language. It's not possible to edit the code in this window, but the aforementioned Methods Repository includes a code editor (also with syntax highlighting), which can be quite convenient for making small tweaks without too much hassle.

```
     SCRIPT       • •     INPUTS      • •     OUTPUTS     • •    RUN ANALYSIS

 1   ## This workflow runs the HaplotypeCaller tool from GATK4 in GVCF mode
 2   ## on a single sample in BAM format. The execution of the HaplotypeCaller
 3   ## tool is parallelized using an intervals list file. The per-interval
 4   ## output GVCF files are then merged to produce a single GVCF file for
 5   ## the sample, which can then be used by the joint-discovery workflow
 6   ## according to the GATK Best Practices for germline short variant
 7   ## discovery.
 8
 9   version 1.0
10
11   workflow ScatterHaplotypeCallerGVCF {
12
13       input {
14           File input_bam
15           File input_bam_index
16           File intervals_list
17       }
18
19       String output_basename = basename(input_bam, ".bam")
20
21       Array[String] calling_intervals = read_lines(intervals_list)
22
23       scatter(interval in calling_intervals) {
24           call HaplotypeCallerGVCF {
```

Figure 11-10. Viewing the workflow script.

So now you know for sure which workflow you're going to be running, but where do you plug in the inputs? When you ran this workflow from the command line, you handed Cromwell a JSON file of inputs along with the WDL file. You can see the functional equivalent here on the Inputs tab, which is shown in part in Figure 11-11. For each input variable, you can see the Task name in the leftmost column and then the name of the Variable as it is defined in the workflow script as well as its Type. Finally, the rightmost column contains the Attribute, or value, that we are giving to each variable.

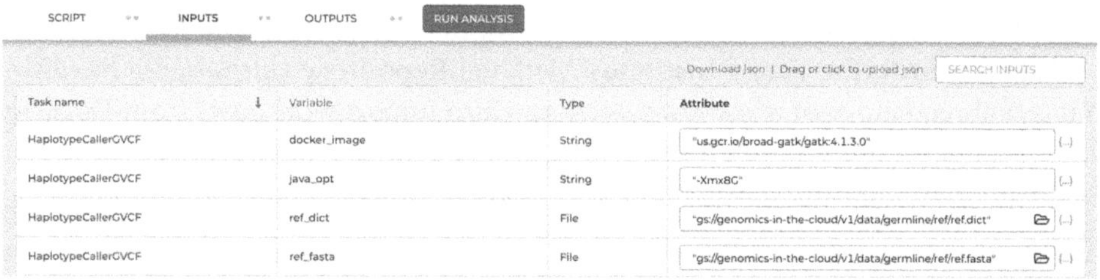

Figure 11-11. Viewing the workflow inputs.

We've prefilled the configuration form for you, so you don't need to edit anything, but please do take a moment to look at how the input values are specified. Especially for the File variables: we've provided the full paths to the locations of the files in GCS. This is truly an exact transcription of the contents of the JSON file of inputs. In fact, as you'll learn in more detail in Chapter 12, all we had to do to set this up was to upload the JSON file to populate the contents of the form.

So are you ready to click that big blue Run Analysis button? Go for it; you've earned it. A small window will pop up asking for confirmation, as shown in Figure 11-12. Press the blue Launch button and sit back while Terra processes your workflow submission.

Confirm launch

This analysis will be run by Cromwell 49.

This will launch **1** analysis.

CANCEL   LAUNCH

Figure 11-12. The workflow launch dialog.

Under the hood, the system sends the built-in Cromwell server a packet of information containing the workflow code and inputs. As usual, Cromwell parses the workflow and starts dispatching individual jobs to PAPI for execution on the Google Compute Engine (GCE), as illustrated in Figure 11-13.

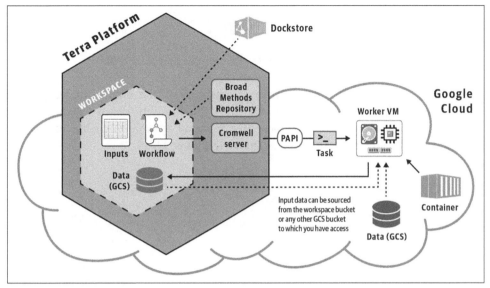

Figure 11-13. Overview of workflow submission in Terra.

Meanwhile, Terra will take you to the Job History section of the workspace, where you can monitor execution and look up the status of any past submissions—after you've had a chance to run some, that is. Speaking of which, there will be more to look at here when the workflow you just submitted is further along, so let's move on and plan to circle back to the Job History later in the chapter.

At this point, assuming everything went fine, you've essentially just replicated your earlier achievements of running the scattered *HaplotypeCaller* workflow, through PAPI, on a single sample. That's nice, but wasn't the point that we wanted to be able to run workflows on multiple samples at the same time? Why, yes; yes, it was. That's where the second configuration comes in.

Running a Workflow on Multiple Samples in a Data Table

Let's go take a look at that other configuration—the one called *scatter-hc.data-table*. As a reminder, you need to navigate to the Workflows pane and then click the configuration. You'll see mostly the same thing as with the previous one (and the Source link is exactly the same), but if you look closely, there is one important difference. As shown in Figure 11-14, this configuration is set up to "Run workflow(s) with inputs defined by data table," whereas the *.filepaths* configuration was set to "Run workflow with inputs defined by file paths" and specified the *book_sample* table as the source of data, as shown in Figure 11-9.

Figure 11-14. The second workflow is set to run on rows in a data table.

Now take a closer look at how the inputs are defined on the Inputs tab. Do you see anything unfamiliar? For most of the variables, the straightforward values (like the Docker address and the file paths) have been replaced by what looks like more variables: either `workspace.*` or `this.*`, as shown in Figure 11-15.

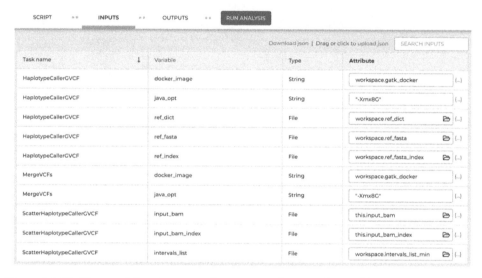

Figure 11-15. The workflow input configuration references data tables.

And that's exactly what those are, references to values that are defined somewhere else. Specifically, they point to values that are stored in metadata tables in the Data section of the workspace. Let's head over there now and see if we can shed some light on how this all works.

In the Data section, you'll find a menu of data resources that should look something like Figure 11-16. Within that menu, you should see a table called *book_sample* and another one called *Workspace Data*. We're going to start with the *Workspace Data* table, on the assumption that it's the least complicated to understand.

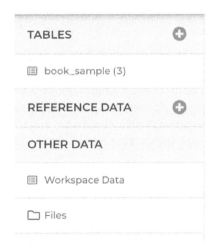

TABLES ⊕

⊞ book_sample (3)

REFERENCE DATA ⊕

OTHER DATA

⊞ Workspace Data

☐ Files

Figure 11-16. Viewing the menu of data tables on the DATA tab.

Click Workspace Data to view the contents of that table, which are shown in
Figure 11-17. As you can see, this is a fairly simple list of key:value pairs; for example,
the key *gatk_docker* is associated with the value `us.gcr.io/broad-gatk/gatk:`
`4.1.3.0`. Meanwhile, behind the scenes, this *Workspace Data* table is called *workspace*. As a result, we can use the expression `workspace.gatk_docker` in the workflow
inputs form, as shown in Figure 11-15, to refer to the value `us.gcr.io/broad-gatk/`
`gatk:4.1.3.0`, which is stored under the *gatk_docker* key in the *workspace* table.

Key	Value
gatk_docker	us.gcr.io/broad-gatk/gatk:4.1.3.0
intervals_list_full	snippet-intervals-full.list
intervals_list_min	snippet-intervals-min.list
ref_dict	ref.dict
ref_fasta	ref.fasta
ref_fasta_index	ref.fasta.fai

Figure 11-17. The Workspace Data table.

So the *workspace* keys are essentially serving as a kind of global variable that you can
use in multiple configurations within the same workspace. This is very convenient if
you want to update the version of the GATK Docker image in all of your

configurations with minimum hassle: just update the key's value in the *Workspace Data* table and you're all set.

The same benefit applies to the resource files that are typically used in multiple work-flows across a project, such as the reference genome files or the interval lists. Considering how awkward it can be to wrangle those long *gs://* file paths, it's a real blessing to be able to define them just once and then simply refer to them with short pointers everywhere else. The filenames shown in blue and underlined are full file paths to locations in GCS, even though the system shows only the filename.

 The keys defined in the *Workspace Data* table do not need to match the names of the variables from the WDL. We tend to standardize variable names and keys across our workflows in order to make it more obvious which ones go together, but it's not a requirement of the system. If you wanted, you could set up a key called *my_docker* and provide it as *workspace.my_docker* to the *gatk_docker* variable in the workflow configuration.

Now let's have a look at the other table, *book_sample*. As shown in Figure 11-18, this one is a more proper table, with multiple rows and columns. Each row is a different sample, identified by a unique (but for once, very readable) key in the *book_sample_id* column. Each sample has a file path under the *input_bam* column and another one under the *input_bam_index* column. If you peek at the paths, you might recognize these as the sample files of the family trio that we used in the joint calling exercise all the way back in Chapter 6 (Germline short variant analysis), when the world was new and you were still running individual command lines in your VM.

☐ ▾	book_sample_id ↓	input_bam	input_bam_index
☐	father	father.bam	father.bai
☐	mother	mother.bam	mother.bai
☐	son	son.bam	son.bai

Figure 11-18. The book_sample table.

So how do we plug the contents of this table into the inputs form? Well, you just learned about the workspace.*something* syntax that we encountered in Figure 11-15; now it's time to extend that lesson to elucidate the this.*something* syntax. In a nut-shell, it's the same idea, except now the pointer is this instead of workspace. and it's going to point to any row of data taken from the *book_sample* table and submitted to the workflow system for processing, instead of pointing to the *Workspace Data* table.

Too abstract? Let's use a concrete example. Imagine that you select one row from the *book_sample* table. That produces what is essentially a list of key:value pairs, just like the *Workspace Data* table:

```
book_sample_id        ->    mother
input_bam             ->    gs://path/to/mother.bam
input_bam_index       ->    gs://path/to/mother.bai
```

You can therefore refer to each value based on the column name, which has been reduced to a simple key by the process of isolating that one row. That is what the `this.input_bam` syntax does. It instructs the system: for every sample row that you run the workflow on, use this row's `input_bam` value as an input for the `input_bam` variable, and so on for every variable where we use this syntax. (And again, the fact that the names match is not a requirement, though we do it intentionally for consistency.)

Alright, that was a long explanation, but we know this system trips up a lot of newcomers to Terra, so hopefully it's been worth it. The feature itself certainly is, because it's going to allow you to launch workflows on arbitrary numbers of samples at the click of a button. If that weren't enough, the data table system has another benefit: when you have your workflow wired up to use the data table as input, you can start the process directly from the data table itself.

How would you like to try it out? Go ahead and use the checkboxes to select two of the samples, as shown in Figure 11-19.

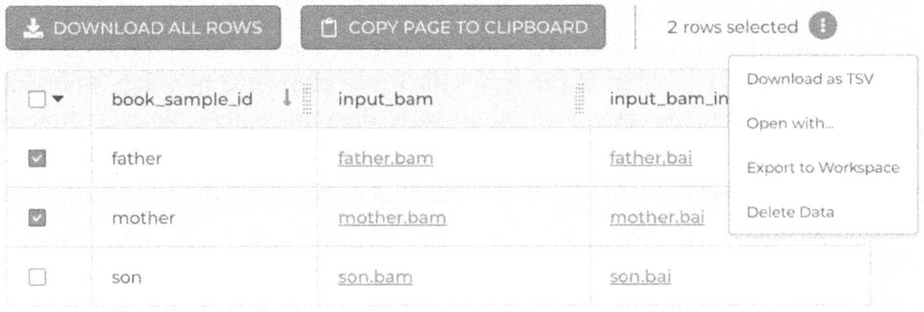

Figure 11-19. Initiating an analysis directly on a subset of data.

Having selected the samples, click the round symbol with three vertical dots located next to the count of selected rows. Click "Open with…" to open the menu of options and select Workflow as shown in Figure 11-20. Options that are not suitable for your data selection are grayed out. When you click Workflow, you're presented with a list of available workflow configurations. Be sure to pick the one that is set up to use the data table; the system does not automatically filter out workflows that are configured with direct file paths.

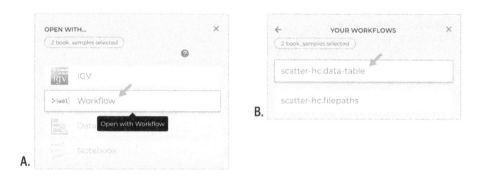

Figure 11-20. Specifying a workflow to run on the selected data.

After you pick a workflow configuration, you are brought back in the workflows section. Looking at the configuration summary, you'll see that it's now set to run on the subset of data that you selected, as shown in Figure 11-21.

Figure 11-21. Configuration updated with data selection.

When you submit this workflow execution request for the first time, the system saves a list of the samples you selected and stores the list as a row in a new data table called *sample_set*. After that, the system will add rows to the *sample_set* table every time you run a workflow on samples in this way. Note that you can also create sample sets yourself; you'll see an example of that in Chapter 13.

 Selecting rows manually is something that we mostly do for testing purposes; it wouldn't scale terribly well. For a more scalable approach, you can opt to run the workflow on all rows in a table, or you can create sets ahead of time. To access these options, simply click the Select Data link adjacent to the table selection menu. Speaking of which: yes, you can have multiple data tables in a workspace, although you cannot launch a workflow on multiple tables at the same time.

Finally, there's nothing left to do but click the Run Analysis button and then confirm the launch. But this time, when you land in the Job History section, stick around so that we can take a look at how your first workflow submission fared.

Monitoring Workflow Execution

When you're taken to the Job History page after launching a workflow, what you see is the summary page for the submission you just created. If you wandered off in the meantime and then came back to the Job History page on your own steam, what you'll see is a list of submissions, as shown in Figure 11-22. If so, you'll need to click one of them to open it in order to follow the instructions that follow.

Submission (click for details)	Data entity	No. of Workflo...	Status	Actions	Submitted	Submission...
scatter-hc.data-table Submitted by genomics.book@gmail.com	scatter-hc-data-table_2...	2	↻ Submitted	ABORT WORKFLOWS	Today	21dccf11-c...
scatter-hc.filepaths Submitted by genomics.book@gmail.com		1	✓ Done		Today	d48d9fb5-...

Figure 11-22. List of submissions in the Job History.

Workflows and Submissions

The terminology here can become a little ambiguous because the term *workflow* carries several meanings. At the most fundamental level, a workflow is the sequence of actions that must be applied to the data. But it's also the script, written in WDL, that implements that series of actions. Finally, we also use *workflow* to refer to the actual instantiation—the thing that runs in practice. The term *job*, which has similar ambiguity, can be used to refer to the overall submission, to a single workflow instance, or to individual tasks within the workflow. *Submissions*, meanwhile, refer to sets of one or more workflow runs that were launched at the same time.

The submission summary page, shown in Figure 11-23, includes a link back to the workflow configuration, information about the data that the workflow was launched on, and a table listing individual workflow executions.

Each row in the table corresponds to an individual run of the workflow. If you used the data table approach to launch the workflow, the Data Entity column shows the identifier of the corresponding row in the data table as well as the name of the table. If you ran the workflow directly on file paths, as we did in the first exercise, that column is left blank. The *Workflow ID* is a unique tracking number assigned by the system and links to the location of the execution logs in GCS; if you click it, it will take you to the GCP console. Most of the time, you'll want to ignore that link and use the View link instead, in the leftmost column, to view detailed status information about the workflow run.

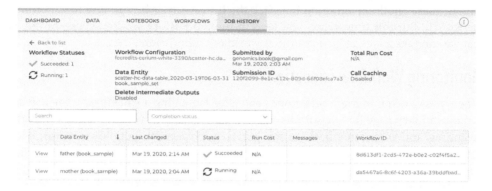

Figure 11-23. *The workflow submission summary page.*

Go ahead and click the View link in one of the rows to open up the details for that workflow run. The workflow run details page packs in a lot of information, so let's go through it one piece at a time.

 As of this writing, the workflow run details open on a new page titled Job Manager because it uses a service that is not yet fully integrated into the main Terra portal. You might need to sign in with your Google credentials to access the workflow run details page.

First, have a look at the overall status summary pane, shown in Figure 11-24. If everything goes well, you'll see your workflow change from Running to Succeeded

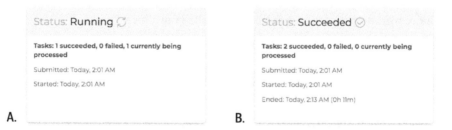

Figure 11-24. *Workflow in A) Running state and, B) Succeeded state.*

In the unfortunate event that your workflow run failed, the page will display additional details, including any error messages and links to logs, as shown in Figure 11-25.

Figure 11-25. A workflow in Failed state with ERRORS summary and Failure Message.

When everything's working, the panel below the summary pane, shown in Figure 11-26, is your go-to for everything you need to monitor the step-by-step execution of your workflow. The list of tasks will be updated as the workflow progresses and will be populated with useful resources like links to each task's logs, execution directory, inputs, and outputs. To be clear, tasks won't be listed until the work on them actually starts, so don't panic when you see only a subset of the tasks in your workflow if you check on its status while it's still running.

Figure 11-26. List of tasks and related resources.

You might notice that in Figure 11-26, the `HaplotypeCallerGVCF` task is represented a bit differently compared to the `MergeVCFs` task: its name is underlined and there is a symbol on the side that looks like a stack of paper. This is how scattered tasks are represented. As shown in Figure 11-27, you can hover over the stack symbol to see a status summary for all of the *shards* generated at that step; that is, the individual tasks resulting from the scatter.

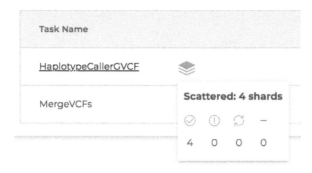

Task Name	
HaplotypeCallerGVCF	
MergeVCFs	

Scattered: 4 shards

Figure 11-27. Viewing the status of shards for a scattered task.

The panel shown in Figure 11-26 has multiple tabs, so take a minute to explore those, as well. Our favorite tab in this panel is Timing Diagram, shown in Figure 11-28, which shows you how much time each task took to run. This includes not just the time spent in the User Action stage—running the analysis command (for example, `HaplotypeCaller`), but also the time spent on getting the VM set up, pulling the container image, localizing files, and so on.

Figure 11-28. A timing diagram showing the breakdown of runtime per stage of execution for each task call.

If you recall the discussion about workflow efficiency in Chapter 10, some amount of overhead is associated with each stage. The timing diagram is a great resource for examining this. It can help you identify any bottlenecks in your workflow that you need to address; for example, if your list of genomic intervals is not well balanced or if a particular command is taking far more time than it should.

Want to see it in action? Try hovering your mouse over the colored segments, and you'll see tooltips appear that indicate the stage you're looking at and the amount of time it took. For example, in Figure 11-28, we were looking at the yellow block on the right side, which shows that `HaplotypeCaller` took just 21 seconds to run on the very short interval we gave it for testing purposes. In comparison, it took 101 seconds to pull the container image!

Finally, you might notice that the name of the task in the pop-up window includes *attempt 1*. This is because the workflow is configured to use preemptible instances, which we also discussed in the section on optimizations in Chapter 10. If one of your tasks is preempted and restarted, each run attempt will be represented by a separate line in the diagram, as shown in Figure 11-29 (which comes from an unrelated workflow that we ran separately).

Figure 11-29. A timing diagram showing preempted calls (green bars, at lines 2, 12, and 13 from the top).

The bars representing preempted jobs are displayed in a single solid color because the different stages of operation are not reported. As a result, you can usually see fairly clearly whether a pattern of serial preemptions exists; for example, in Figure 11-29 a job was started and then interrupted by preemption (line 12 from the top), restarted and interrupted again (line 13 from the top), and then restarted again and this time was successful. You can see from this example, in which even the shortest preempted job ran for almost four hours, that if you run into several preemptions in a row for a job that constitutes a bottleneck in your pipeline, the overall runtime can increase dramatically as a result. This highlights the importance of evaluating carefully whether the cost savings that you can reap from using preemptible instances is worth the risk of your pipelines taking much longer to complete. Our rule of thumb is that if our large-scale data-processing results are due next week, bring on the preemptibles. But if we're frantically trying to finish a demo for a conference in two days, turn them off and suck up the extra cost.

Locating Workflow Outputs in the Data Table

As we noted earlier, the workflow details shown in Figure 11-26 include pointers to the location of output files. That's a pretty decent way to locate outputs for a particular workflow run, but that's not going to scale well when you're working on hundreds or thousands of samples. Not to mention hundreds *of* thousands of samples. (Yep, some people are at that point—isn't it an exciting time to be alive?)

There is a much better way to do it if you used the data table, which you would need to do if you're working at that scale. Here it is: with a tiny configuration trick, you can get the system to automatically add the workflow outputs to the data table. And, of course, we enabled that in the workflow configuration we made you run, so let's go have a look at the data table, shown in Figure 11-30, to see whether it worked.

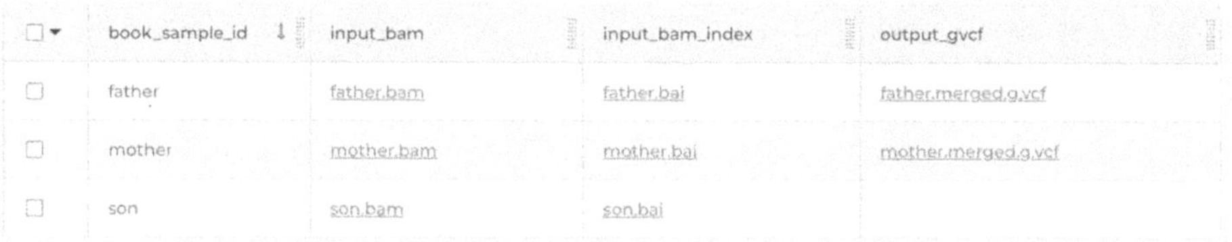

☐ ▼	book_sample_id ↓	input_bam	input_bam_index	output_gvcf
☐	father	father.bam	father.bai	father.merged.g.vcf
☐	mother	mother.bam	mother.bai	mother.merged.g.vcf
☐	son	son.bam	son.bai	

Figure 11-30. The data table showing the newly generated output_gvcf column.

And there it is! As you can see in Figure 11-30, a new column has appeared in the *book_sample* table, named *output_gvcf*, and all the samples we ran the workflow on now have file paths corresponding to the GVCFs produced by the workflow runs.

So, where does the name of the new column come from? Let's return to the `scatter-hc.data-table` workflow configuration and take a look at the Output tab, which we ignored earlier. As you can see in Figure 11-31, we specified the output name and set it to be attached to the row data using the `this.` syntax, which we described earlier. You can pick an arbitrary name (for example, you could instead use *this.my_gvcf*), or you can click "Use defaults" to automatically use the name of the variable as it is specified in the workflow script, as we did here.

Figure 11-31. The workflow outputs configuration panel.

Keep in mind that this option is available only if you're using the data table to define the workflow inputs. In either case, however, the location where the files are written is the same. By default, the system stores all outputs produced by workflows in a bucket that is tightly associated with the workspace. The bucket is created automatically when you create a workspace, and it has the same ownership permissions as the workspace, so if you share the workspace with someone (same menu as for cloning), they will also be able to access the data in your workspace. One important restriction, however, is that you cannot modify the bucket's permissions outside of Terra.

 If you delete a workspace, its bucket will also be deleted, along with all its contents. In addition, when you clone a workspace, the data is not copied, and any paths in metadata tables will continue to point to the data's original location. This is great because it allows you to avoid paying storage fees for multiple copies of the same data that you would otherwise generate. But if you thought you could save a copy of your data by cloning a workspace before deleting the original, well, you can't.

Finally, you might have noticed on the Data page that there is a Files link in the left-hand menu, which is visible in Figure 11-30 among others. If you go back to that page and click the Files link, it will open a filesystem-like interface that you can use to browse the contents of the bucket without having to leave Terra, as shown in Figure 11-32.

TABLES	⊕	Files / 120f2099-8e1c-412e-809d-66f08efca7a3 / ScatterHaplotypeCallerGVCF / 8d613df1-2cd5-472e-b0e2-c02f4f5a2043 / call-HaplotypeCallerGVCF / shard-1 /		
🎛 book_sample (3)				
🎛 book_sample_set (1)		Name	Size	Last modified
		pipelines-logs/		
REFERENCE DATA	⊕	HaplotypeCallerGVCF-1.log	11 KB	Today
		father.scatter.g.vcf	120 KB	Today
OTHER DATA		gcs_delocalization.sh	2 KB	Today
		gcs_localization.sh	2 KB	Today
🎛 Workspace Data		gcs_transfer.sh	13 KB	Today
		rc	2 B	Today
📁 Files		script	1 KB	Today
		stderr	7 KB	Today
		stdout	0 B	Today

Figure 11-32. The file browser interface showing workflow outputs in the workspace bucket.

You can even add files to your bucket by dragging and dropping them from your desktop into the file browser. That being said, if you prefer, you can always access the bucket through the GCS console and interact with its contents through `gsutil`.

Running the Same Workflow Again to Demonstrate Call Caching

Did you think we were done? Not quite; we couldn't possibly let you move on without experiencing the wonder that is Cromwell's call caching feature. We've already gone over its purpose several times—make it so you can avoid running duplicate jobs, and resume a workflow from the last point of failure, interruption or modification—so now let's see it in action. Go ahead and use the launching process you just learned to run the workflow again on one of the samples that you already ran on earlier. When you land on the Job History page, click through to the workflow monitoring details when the View link becomes available. It might take a couple of minutes, so feel free to go grab yourself a cup of tea, coffee, or other beverage of choice—just don't spill it on your laptop when you return. And remember to refresh the page; the workflow status doesn't refresh on its own.

When you're on the workflow details page, navigate to the Timing Diagram and hover over the widest bar for one of the lines. You should see something like Figure 11-33, reporting on a stage named `CallCacheReading` that took about 10 seconds to run.

Figure 11-33. A timing diagram showing CallCacheReading stage run time.

If you hover over the other bars, you'll also see that you can't find any of the stages related to pulling the container image or localizing inputs. This shows you that when call caching kicks in for a task, the system doesn't even go to the trouble of setting up VMs. In fact, Cromwell doesn't even dispatch anything about that task to PAPI.

So what exactly does happen? Good question; let's talk about how call caching works in practice. First, you should know that whenever the Cromwell server runs a task call successfully, it stores a detailed record of that call in its database. This includes the name of the task and all of its input values: files, parameters, container image version, everything—as well as a link to the output file. Then, the next time you send it a workflow to run, it will check each task against its database of past calls before sending it out for execution. If it finds any perfect matches, it will skip execution for that task and simply output a copy of the output file that it had linked to in the last successful execution. Figure 11-34 illustrates this process.

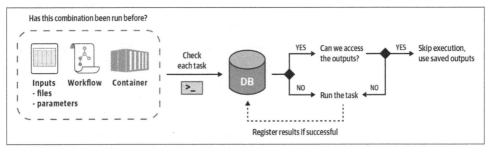

Figure 11-34. Overview of Cromwell's call caching mechanism.

In this example, all task calls in the workflow matched the call cache; in other words, *call-cached* (yep, you can use it as a verb), so nothing actually was run except a few file copying operations done by the Cromwell server itself. If you were to modify the `MergeVCFs` call even slightly and run this again, the `HaplotypeCaller` calls would call-cache, but `MergeVCFs` would not, so Cromwell would send that call to PAPI for execution. You can try this out by changing the version of the container image used by `MergeVCFs` (try 4.1.4.1) and then have a look at the Timing Diagram.

 You can disable call caching by clearing the checkbox in the workflow configuration, but why would you even want to do that? Call caching is fantastic. Call...ka-ching! (Because it saves you money.) #dadjokes.

Alright, time to take a break. How are you feeling? It's normal to feel a little lightheaded at this point; after all, you just got superpowers. Seriously, you are now capable of running real, sophisticated genomics workflows on as many samples as you can get your hands on. That's no small feat. Yet the proof, as they say, is in the pudding, so grab a spoon and let's go bite off something big.

Running a Real GATK Best Practices Pipeline at Full Scale

Do you remember Mystery Workflow #2 from Chapter 9, which turned out to be the Broad Institute's whole genome analysis pipeline? It's probably not the biggest genomics pipeline in the world (though it is plenty big), and it's not the fastest pipeline in the world either (because it needs to be inexpensive). But it might just be the pipeline that has processed the largest number of human whole genome samples in the history of genomics (so far). If you're looking to learn how to do something fairly standard that will come in handy at some point in the future, this pipeline would be a reasonable choice.

In this last section of the chapter, we're going to show you where to find it, test it, and run it on a full-scale human whole genome sample.

The good news is that it's going to be pretty straightforward. We already mentioned previously that the GATK team makes its Best Practices workflows available in a Git-Hub repository (*https://oreil.ly/D0Ofp*), but that's not all; it also provides fully loaded Terra workspaces for all of them. These Best Practices workspaces contain data tables populated with appropriate example data samples (typically a small one and a full-scale one), and workflow configurations that are already wired up to run on the example data. All you need to do is clone the workspace. Then, you can follow the procedure that you learned in this chapter to run the workflows right out of the box. (We show you how to bring in data from other sources in Chapter 13.)

Finding and Cloning the GATK Best Practices Workspace for Germline Short Variant Discovery

You already visited the Terra Library showcase (*https://oreil.ly/nUMy2*) earlier in this chapter to find the tutorial workspace we created for the book. Let's go back there, but this time you're going to look for an actual Best Practices workspace published by the GATK team. As of this writing, the featured germline short variants workspace is called Whole-Genome-Analysis-Pipeline (*https://oreil.ly/x4Pzc*), as shown in Figure 11-35. This name might change in the future because the team has plans to adapt how they name and package these resources in light of the expanding scope covered by its tools. If you're having trouble finding the right workspace, check the GATK website's Best Practices section, which hosts a list of relevant resources that includes the Terra workspaces. Be sure to also check our book's blog (*https://oreil.ly/genomics-blog*), where we'll provide updates over time.

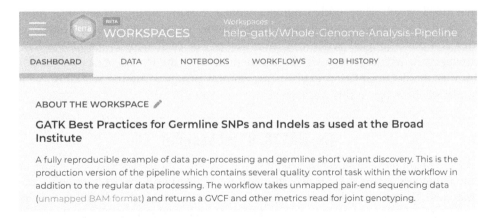

Figure 11-35. Summary information for the Whole-Genome-Analysis-Pipeline workspace (https://oreil.ly/x4Pzc).

When you've found the workspace, clone it as described earlier in this chapter. You'll notice that the Dashboard provides a lot of information about the workflow, the example data and how to use these resources effectively. It even includes a summary of how long it takes and how much it costs to run the workflow on the various samples in the example data table. Speaking of which, let's go see what's in there.

Examining the Preloaded Data

In your clone workspace, navigate to the Data section. Again, this might change, but as of this writing, this workspace contains two main data tables as well as the Workspace Data table. The latter lists the full-scale version of the genome reference files and other resources used in the whole genome pipeline. Note that these resources are all based on the hg38 build (more properly known as GRCh38) and would therefore not be compatible with the data in the tutorial workspace that we were using earlier.

The two main data tables are called *participant* and *sample*, as shown in Figure 11-36.

Figure 11-36. A list of tables and detailed view of the sample table.

The *sample* table should look familiar to you because it's the same kind of table as the *book_sample* table that you encountered earlier in this chapter, with a few additional columns. As you can see in Figure 11-35, two samples are listed there: NA12878 and

NAA12878_small. They both originate from the same study participant, dubbed NA12878. The former is a full-size whole genome sequencing dataset, whereas the latter is a downsampled version of that dataset. For each sample, we have a list of unmapped BAM files (which will be the main input to the workflow) as well as other files that the documentation explains are outputs produced by the workflow, which has already been run in this workspace.

The NA12878 Test Samples

As you might recall from earlier chapters, the NA12878 individual was the mother in the family trio whose data we've been using for testing purposes. That family's samples, and NA12878 in particular, have been sequenced and analyzed many times by many groups as part of a wide variety of research and development efforts. We might not know their names, but we all owe them a debt of gratitude for the advances that their data made possible.

If you're not familiar with the term *downsampling*, it refers to a process of selectivly removing a proportion of the data (sequencing reads, in this case) in order for it to take less storage space and/or less time for processing. We sometimes refer to the downsampled data as the *plumbing sample*, because it is useful for testing if the pipeline runs to completion but contains too little data to test the scientific validity of the pipeline.

The *participant* table, on the other hand, is probably new to you. It lists the study participants, though in this case, that's just a single person. If you looked carefully at the sample table, you might have noticed that one of its columns is *participant*, which is an identifier that points to an entry in the participant table (*participant_id* in that table). The purpose of the participant table is to provide a higher level of organization for your data, which is useful when you have multiple study participants, and each of them can have multiple samples associated with them, either corresponding to different data types, different assays, or both. With this setup, you can use the data table system to do things like run a workflow on all the samples that belong to a subset of participants, for example.

As another example, the Mutect2 somatic analysis pipeline (described in detail in Chapter 7) expects to see a tumor sample and a matched normal sample from each patient, which are formally described as pairs of samples. If you check out the corresponding GATK Best Practices workspace (this one (*https://oreil.ly/a8ksp*), at the time of writing), you'll see it has four data tables organizing the data into participants, samples, pairs of samples, and sets of samples (for the normals that are used in the PoN, described in Chapter 7). The data tables in that workspace conform to a *data model* defined by the TCGA cancer research program: The Cancer Genome Atlas.

More generally, you can use any number of tables to organize your data and describe the relationships between *data entities* like participants, samples, and others in a structured data model. In Chapter 13, we discuss options for building your data model in Terra and using it effectively to save yourself time and effort.

Selecting Data and Configuring the Full-Scale Workflow

As we've described previously, you can either head straight for the workflows page in the workspace or you can start the process from the Data page by selecting one or both samples in the *sample* table. (We recommend selecting both so you can experience the runtime difference of the plumbing test and the full-scale run for yourself.) If you choose to follow the same procedure as previously, click "Open with…" to choose the workflow option, and select the one workflow that is preconfigured in this workspace. Unless a lot has changed since the book came out, this should be the same, or practically the same, as the workflow that we examined in detail in Chapter 9. However, you'll notice that on the workflow page, you can view only the main WDL script, not any of the additional WDL files that contain the subworkflows and task libraries. This is a limitation of the current interface that will be addressed in future work.

We didn't look at the inputs to this workflow in much detail when we dissected it in Chapter 9, because at the time we were focused on understanding its internal plumbing. Now that we're looking at them through the Terra interface, in the context of the workflow inputs configuration form, it's pretty striking that it seems to have only four required inputs, which is fewer than the much simpler workflow we've been working with so far. However, this is mostly a distortion of reality; in fact, two of those four inputs represent a larger number of inputs that are bundled together using struct variables, which we encountered in Chapter 9. These two structs represent the two most typical categories of inputs that you will frequently encounter in genomics workflows. One is a bundle of reference data, grouping the genome reference sequence, associated index files, and known variant resources for validation. The other groups the files that hold the actual data that you're looking to process.

Most of the optional inputs to this workflow are task-level runtime parameters and conditional checks, which are readily recognized by their type, Boolean. You might recall that in the first workflow we examined in Chapter 9, we found a lot of conditional statements that defined settings like default runtime resource allocation. The optional inputs we see here are set up so that you can override those default settings when you configure the workflow; for example, if you have reason to believe that the defaults won't be appropriate for your use case. Yet if you don't know the first thing about what the runtime resource allocations should be, you can just leave those fields blank and trust that the presets will be good enough.

It's also worth taking a quick peek at the Output tab, where you'll be reminded that this workflow produces a ginormous number of outputs, which are mostly quality control-related metrics and reports. This is where we really appreciate being able to have the paths to the output files written to the data tables, which makes it a lot easier to find outputs than if you had to go rooting around in the Cromwell execution directories for each file. For the record, the outputs that you're most likely to care about are the trio of output_bam, output_cram, and above all, output_vcf, which is the set of variant calls that you'll want to use in downstream analysis.

That being said, all of this should mostly be prefilled and ready to go, but on the Output tab, click "Use defaults" to set up the mapping of outputs to the sample table.

Launching the Full-Scale Workflow and Monitoring Execution

Enough talk, let's run this thing! As previously, click Run Analysis and then Launch to submit the workflow for execution. You can follow the same steps as we described earlier to monitor the progress of your pipeline, but you should expect this to take quite a bit longer! As mentioned earlier, the Dashboard summary includes typical runtimes for the workflow running on both example datasets, so be sure to use that as a guide to gauge when to go check how it's going. You can also browse the Job History of the original workspace (the one you cloned your copy from); it includes past executions, so you can look up the timing diagram there.

One new thing you might notice is that the List view of the workflow details page collapses subworkflows and displays their name in underlined font, as shown in Figure 11-37. This is similar to the representation of scattered tasks but lacks the little stack icon on the side.

Figure 11-37. The List View of the task calls in the master workflow.

Have a look at the timing diagram provided at this level: you essentially get a high-level summary of the component segments of the workflow, as shown in Figure 11-38. If you hover over the various lines, you'll see that the longest segment on the left corresponds to *UnmappedBamToAlignedBam*, the data processing sub-workflow that takes in the raw unmapped data and outputs an analysis-ready BAM file. The next three lines consist of the main quality control subworkflow, *Aggregated-BamQC*, and two metrics collection tasks that are not bundled into subworkflows. Next down is the `BamToGvcf` variant-calling subworkflow, which produces the final output of the per-ample pipeline, the GVCF file of variants. You'll notice that those four segments all started at the same time, when the very first segment completed because they are independent of one another. This is parallelism in action! Finally, the last segment is the *BamToCram* workflow, which produces a CRAM file version of the processed sequencing data for archival purposes. The timing of that one might seem odd until you realize that it starts immediately after the *AggregatedBamQC* workflow finishes.

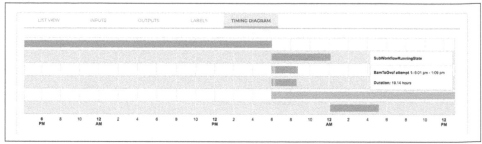

Figure 11-38. The timing diagram for the master workflow showing subworkflows (solid red bars) and individual tasks that are not bundled into subworkflows (multicolor bars).

Now go back to the List View and click through one of the subworkflows; for example, the `BamToGvcf` variant calling subworkflow. You'll see it open up on its own page as if it were a standalone workflow, with its own list of tasks, timing diagram, and so on, as shown in Figure 11-39.

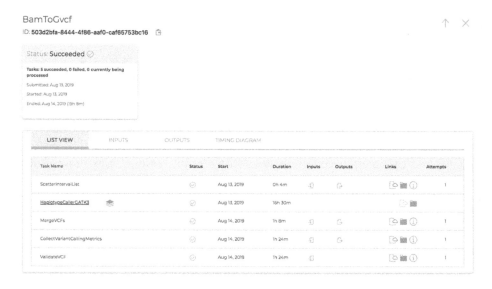

Figure 11-39. The workflow details page for the BamToGvcf subworkflow.

There is theoretically no limit to how deeply you can nest WDL subworkflows, as long as you don't create loops. In practice, though, we have not yet seen workflows with more than three levels of nesting (including task libraries). In any case, you can navigate back up to the parent workflow level by clicking the upward-pointing arrow in the upper-right corner.

Finally, you can check that the outputs were produced as expected, both in the Job History view and in the *sample* data table, as we've described earlier in the chapter.

Options for Downloading Output Data—or Not

Whether you're browsing the outputs of your workflow through the Job History page, or through either the data tables or the Files browser on the Data page, you can download any file by simply clicking its name or path. This brings up a small window that includes a preview of the file (if it's in a readable text format), its size, and an estimate of the cost of the download, which is due to egress fees. In Figure 11-40, we have examples of such download windows for two kinds of files: the list of unmapped BAM files that served as input for the pipeline, which is a plain-text file and can therefore be previewed, and the GVCF that is the final data output of the pipeline, which is originally plain text, too, but here, was gzipped to require less storage space, and therefore cannot be previewed. That would also be the case for other compressed file formats like BAM and CRAM files, which also cannot be previewed.

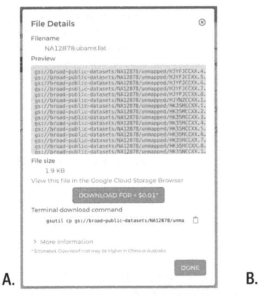

Figure 11-40. File download windows showing A) the list of unmapped BAM files, and B) the final GVCF output.

The download window offers three ways to download the file of interest. You can follow the link to the GCS console, where you can select multiple files and download them through your web browser. You can also simply click the blue Download button for that one file, which will also be downloaded by the web browser. Alternatively, you can copy the terminal download command, which includes the full path to the file's location in GCS, to download the file with the command-line tool `gsutil`.

The direct download option is fine for individual files that you want to look at quickly, especially if they're small. For example, the 1.9 KB text file in Figure 11-40 is predicted to cost less than a penny to retrieve. If you need to retrieve multiple small to medium files like the 185 MB GVCF in Figure 11-40 (just two cents? sounds reasonable), you'll usually be better off using either the GCP console or better yet, gsutil. But if you find yourself needing to retrieve large files (many gigabytes, many dollars), it might be worth pausing to rethink whether you really can't do what you need to do without a local copy of the file. For example, were you planning to look at the BAM and VCF files in IGV to check some variant calls by eye? Remember, you can do that by connecting IGV to GCS. Or did you want to run another kind of analysis that you don't have a workflow for, or that requires an interactive process? Ah, that's fair…but maybe you could continue that work in the cloud rather than falling back to a local environment. Wouldn't it be amazing if you could do *all* of your work in the cloud, from the raw data all the way to making the figures that are going to be in your paper?

That is not a pipe dream. It's already possible today—but you're going to need something better than just the VM and something different from the workflow system. You will need an integrated environment for interactive analysis on the cloud, which Terra provides as well.

 The filename produced by the workflow uses a plain *.vcf* extension instead of *.g.vcf*, which is optional but recommended for signifying that a file is, in fact, a GVCF rather than a regular VCF file. This highlights the fact that you can rarely rely on filenames and extensions to know for sure what a file contains and how it was produced. Data management frameworks like the workspace data model can mitigate such problems by helping us keep track of the relationships between pieces of data.

Wrap-Up and Next Steps

For the past four chapters, including this one, we've been living and breathing for workflows. We started out small in Chapter 8, running canned WDL scripts on a single VM. Then, in Chapter 9 we dissected real genomics workflows to learn what they did and how they were wired up. This led us to a better understanding of some of the requirements of such workflows, which are designed to take advantage of various cloud features including parallelism and burst capability.

But that made us realize that we couldn't continue running workflows on a single VM, even if we beefed up its resource allocations, so we moved to dispatching jobs to PAPI for execution in Chapter 10. We experimented with two approaches for doing so: directly with Cromwell and through the WDL Runner wrapper, which showed us the power of PAPI but also showed us the limitations of launching a single workflow at a time.

Finally, we moved to Terra in this chapter to use its built-in, full-featured Cromwell server. We were able to take advantage of some of the Cromwell server's coolest features, like the vaunted call caching mechanism and the timing diagram, without worrying for a second about server maintenance. Along the way, we also encountered unexpected benefits of using Terra, mainly thanks to the integration of data management with workflow execution. Having learned to launch a workflow on rows in a data table, we're now able to process arbitrary amounts of data without breaking a sweat. And we know how to get the outputs to be added to the table automatically, so we don't need to go digging for them.

While explaining all this, we might have implied that applying the GATK Best Practices to your own data should now "simply" be a matter of running the workflows as demonstrated. However, in the course of a real genomic study, you'll typically need to perform a variety of peripheral tasks that aren't so conveniently packaged into workflows. Whether for testing commands, evaluating the quality of results, or troubleshooting errors, you'll need to be able to interact with the data in a more immediate way. That will be even more the case if your work responsibilities extend beyond what we defined very narrowly as genomics (i.e., variant discovery) to include developing and testing hypotheses to answer specific scientific or clinical questions.

In Chapter 12, we use a popular environment for interactive analysis that also happens to do wonders for reproducible science, called Jupyter. We'll still be working within the Terra platform, which hosts a scalable system for serving Jupyter in the cloud.

Interactive Analysis in Jupyter Notebook

With this chapter, we round out the experience of working in the cloud. We started out in Chapter 4 running individual commands in a shell, and built up proficiency with GATK tools in Chapter 5 through Chapter 7. Then, in Chapter 8, you learned about scripted workflows and discovered progressively better ways to run them in subsequent chapters.

Yet we're now coming back to the inescapable fact that not everything in genomics can (or should) be done as a scripted workflow. Sometimes, you just want to interact directly with the data, maybe generate a couple of plots and determine what your next step should be based on what the plots look like. You might be in an early exploratory phase of your project, stuck midway through with some failed samples to trouble-shoot, or moving on to digging into the genetics of a group of people. In any case, you need to be able to try out ideas quickly, keep track of what each attempt produces, and share your work with others.

In this chapter, we show you how to use Jupyter Notebook in Terra to achieve these objectives. We kick it off with a brief introduction to Jupyter, in case you're not already familiar with the concept and tooling. We spend a bit more time on describing how Jupyter works in Terra, focusing on the capabilities and behaviors that are more specific to Terra and the cloud environment. Then, in the hands-on portion of this chapter, we guide you through an example notebook that demonstrates three types of interactive analyses, with direct connections to topics and exercises covered in earlier chapters.

Introduction to Jupyter in Terra

If you're reading this book in its intended order, we introduced you to Terra in Chapter 11, primarily so that you could experience working with its built-in Cromwell server to run workflows efficiently at scale. You learned to clone a workspace, read a workflow configuration, and launch the workflow on part or all of a preset dataset. However, Terra is not only built for running workflows, but also includes tooling for performing interactive analysis, including a Jupyter service. In this chapter, we're going to show you how to use Jupyter in Terra to interact with data and perform analysis in real time. You'll still work within the same workspace, but this time you'll be going to the Notebooks tab instead of the Workflows tab.

In the first part of this introduction, we aim to provide enough context and fundamentals so that if you've never heard of Jupyter before, you'll be able to complete the exercises that follow. If you're already familiar with Jupyter, feel free to skip this part. In the second part, we talk specifically about how Jupyter works in Terra, focusing mostly on what is different compared to typical local installations. We strongly recommend that you read it even if you're familiar with Jupyter, in general because it will help you to better understand key points of the exercises.

Jupyter Notebooks in General

In a nutshell, *Jupyter* is an application that creates a special kind of document that combines static content (like text and images) with executable code and even interactive elements. For example, the tutorial notebook that you'll work with in this chapter has sections of plain text that explain briefly what's going on as well as *code cells* that include fully functional tool commands (which you can execute to actually run GATK on real data). It includes an integrated IGV module that allows you to view the results of the commands. To run the contents of a code cell, you simply click the cell and then press Shift+Enter on your keyboard, or, on the menu bar, click the Run icon. As the command runs, the output log of the command appears directly below the code cell, as shown in Figure 12-1. When you send to collaborators a copy of a notebook with code cells that have been run, they can view your results embedded within the document.

1.1 Hello Python

Let's try a basic Hello World example in Python.

```
In [1]:  print ("Hello World")

         Hello World
```

```
In [ ]:  # Now you try adding a variable
         greeting =
```

Figure 12-1. Doc text, code cell, and execution output in a Jupyter notebook.

The basic idea is to combine analysis methods and findings in a single place, in a form that anyone can easily distribute. In a way, this is a logical evolution of the traditional scientific paper, but much better because it dramatically shortens the path between reading how an analysis was done and actually being able to reproduce it. It's difficult to overstate how powerful this concept is and what a dramatic impact it can have on the reusability and reproducibility of findings in the computational sciences.

So what kind of code can you run in such an interactive notebook? The original concept had been developed under the name IPython (*https://ipython.org*) to run Python code specifically. The Jupyter project came out of IPython with the goal of extending the concept to other languages, starting with Julia, Python, and R, which are reflected in the name Jupyter (hence, the *py* spelling of Jupyter, as opposed to Jupiter, the Roman god of thunder). There are now Jupyter *kernels* for other popular languages (*https://oreil.ly/Pf8Or*), such as Ruby, for example. In this chapter's exercises, we work with a Python kernel that supports including R code, as well as running pretty much anything you could run in a shell environment, thanks to a neat set of features called *Python magic methods*. In our opinion, it's like having the best of both worlds, but for many worlds.

 In general computing, the *kernel* is the program at the core of an operating system. In the Jupyter context, the kernel is the program that interprets code cells, passes those instructions to the actual operating system of the machine that the notebook is running on, and retrieves results to display them in the notebook.

Jupyter notebooks are backed by a server application that has fairly simple requirements and can be run on almost any kind of computing infrastructure, including your laptop. In addition, a Jupyter notebook enables researchers to encapsulate the information necessary to re-create the software environment used by the analysis that it describes, making it a great vector for distributing reproducible code. It's also an increasingly popular teaching tool, for reasons that will hopefully become evident as you work through the exercises in this chapter.

The growing popularity of Jupyter has spawned a rich ecosystem of add-on tools and services. For example, GCP operates a service called Colaboratory (*https://oreil.ly/3Tr2r*) that offers free access to cloud-based notebooks, with tutorial materials that are heavily geared toward machine learning applications. The Google Cloud AI Platform, meanwhile, offers a paid service that offers preconfigured VMs for running notebooks integrated with other Google Cloud services. Another example is Binder (*https://mybinder.org*), an open source community-driven project that can take any Jupyter notebook in a GitHub repository and open it in an interactive environment. These free services usually have limitations on the amount of computing power associated with the environments they provide, but they can be extremely convenient nonetheless for sharing working code, tutorials, and so on.

That being said, it's practically impossible for a single tool to satisfy the full spectrum of needs and preferences of people in computational sciences, and we recognize that Jupyter notebooks do have some shortcomings that limit their appeal to certain audiences. For example, people with advanced programming experience typically criticize the lack of development features that are standard on most modern programming tools, like syntax highlighting and code introspection. In addition, data scientists who are used to exploratory analysis interfaces like RStudio (*https://oreil.ly/1M5Jh*) tend to find the primary Jupyter interface too basic and lacking in assistive features. The JupyterLab (*https://oreil.ly/RCfhB*) project aims to remedy that limitation by providing a richer interface that is closer in concept to RStudio. Given the surge in economic investment in data sciences that we've seen in recent years, we expect that the tooling options will only improve over time, and we look forward to seeing what the next generation of interfaces will look like.

In the meantime, we choose to use Jupyter for its accessibility to newcomers and its as-yet unparalleled support of portability and reproducibility.

How Jupyter Notebooks Work in Terra

Before we get into the details, you should know that Terra uses a standard Jupyter server implementation, so the interface and core capabilities that you will be working with are all basically the same as those you would see in any other setting. As a result, you can take advantage of the wealth of documentation and tutorials available on the internet for learning how to use the various menu options, widgets, and so on that we're not going to cover in detail here.

The one thing that is truly different about how Jupyter notebooks work in Terra compared to typical local installations is the way the computing environment is set up. Let's go over that now given that it will have important consequences for you on several fronts; for example, the amount of flexibility you have to customize the environment, and the way you will access data and then save your analysis results.

Overview

In Terra, notebook documents live in your workspace's storage bucket. When you open a notebook for the first time, Terra requests a VM in GCP, spins up a container with a Jupyter server on it, and loads your notebook within that container environment, as illustrated in Figure 12-2.

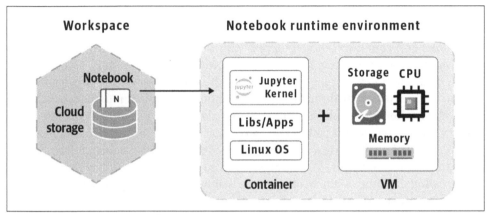

Figure 12-2. An overview of the Jupyter service in Terra.

From that point on, any code that you run in the notebook will be executed in the container on that VM. You can even run commands that install software packages or load libraries on the VM to customize the environment on the fly. Conceptually, it's similar to working with the VM you set up through the GCP console in Chapter 4 and subsequently used in every chapter up through Chapter 8, except it is originally created by Terra, and it provides you with the Jupyter interface instead of the bare shell terminal.

It is possible to pull up a terminal interface to the notebook, which allows you to perform actions like listing files and installing packages without having to put these actions into code cells. However, we recommend using this capability sparingly, largely because it runs counter to the underlying purpose of Jupyter, which is to capture every meaningful action taken during an analysis. Listing directory contents might not be meaningful in that context, but installing a package or importing data, on the other hand, can play an essential role. Omitting such actions from the notebook record could create missing links that break the reproducibility of the work.

The cost model is the same, as well; GCP will charge you for the amount of time that the VM is in a running state, even if it's not actively doing anything. The good news is that Terra has an automatic feature to detect inactivity; more on that in a minute. The base rate depends on the VM configuration you choose to use; by default, Terra provides a basic configuration that accommodates common performance needs, but you can dial it up or down based on your specific needs. We get back to that shortly.

The overall computing environment constituted by the VM plus the container and all the software it contains is called the *notebook runtime*. It is strictly personal to you; no one else can access it even if you share your workspace with them. We talk about sharing and collaboration in the last section of this introduction, when you'll have a better sense of how it all works.

Accessing data

Your notebook runtime comes equipped with local storage space. Cloud subtleties notwithstanding, it's basically like a hard drive with a filesystem on it. You can interact with the filesystem through code cells in the notebook, via the Jupyter built-in graphical file explorer, or through the aforementioned terminal interface, which supports classic commands like ls and cd. This allows you to run commands in the notebook using regular file paths as you would on a regular local system, such as your laptop. In addition, there are a few ways to run commands on data without first copying it to the local filesystem. We're not going to go through the laundry list of options for accessing data from a notebook, because that would be boring. Instead, here's a list of those we use most commonly:

- Upload files from your desktop to your notebook's local storage through the notebook's graphical file explorer.
- Copy data from a GCS bucket to your notebook's local storage using gsutil cp.
- Run tools that support streaming (like GATK4) directly on files in GCS.
- Import tabular data from tables on the DATA tab of the workspace by using a programmatic interface (API).
- Import tabular data from Google's BigQuery datastore service.

We show you how to use the second and third options in practice in the next section, and we will provide pointers to additional resources for learning the fourth and fifth on the book's companion blog.

Saving, stopping, and restarting

While you work in the notebook, the system will regularly save changes back to the original notebook document in your workspace bucket. However, aside from the notebook itself, Terra does not automatically save any files from the notebook's local storage back to the workspace. For reasons that we discuss shortly, you can't depend on the notebook runtime for permanent storage, so you must take action to save a copy of any output files you care about, preferably to the workspace storage bucket.

We show you an easy way to do this in "Setting Up a Sandbox and Saving Output Files to the Workspace Bucket" on page 352.

When you're done working and close the notebook, Terra instructs GCP to stop the notebook runtime but save its state, which includes the state of the Jupyter container, with any modifications that you might have made by installing packages, for example, and any files present on its local storage partition. That way, you can resume working at any time with minimal effort: when you reopen the notebook, Terra restarts the VM and restores the notebook runtime to its saved state. The restart process can take up to two minutes on the GCP side, but while you wait, Terra gives you a read-only view of the notebook contents based on the document's latest saved state. Finally, as mentioned in the previous paragraph, Terra is able to detect when you are no longer actively working in your notebook (based on how long the VM has been idle) and will automatically save the notebook and stop the VM to limit your costs.

Customizing your notebook's computing environment

You might want to customize your notebook runtime in two main ways: modify the VM resource allocations (how many CPUs, how much memory, etc.) and/or modify which software is preinstalled in the container.

As noted earlier in passing, you can readily modify the VM resources allocated to your notebook runtime; for example, if you need more CPUs or memory than is included in the default configuration, you can adjust the notebook runtime configuration accordingly. You can even request a Spark cluster instead of a single VM if you're planning to use tools that are Spark enabled. Conveniently, Terra includes a notebook runtime configuration panel that is much simpler than the equivalent GCP interface, which you might remember from Chapter 4. But what's really cool is that you can do it at any time, even if you've already started working in the notebook. You'll just pull up the configuration panel, specify what you want, and let the system regenerate your notebook runtime with the new specifications. As a result, you don't need to try too hard to guess up front the kind of resources you're going to need to do your work. You can start working with minimal settings and then dial them up if you run into limitations.

However—and this is a big caveat—the new runtime *will be a blank slate* because the regeneration process provisions a new VM with a fresh install of the original container image and an empty storage partition. If the work you're doing in your notebook includes mostly short-running commands that don't amount to much computation cost, this isn't a big problem: the Jupyter menu includes an option to rerun all code cells (or all up to a certain point) so that you can simply regenerate the previous state. However, if some of your work involves massive computations that would not be trivial to rerun, you might need a better strategy. For example, you could explicitly save the outputs generated so far to the workspace bucket and then set your notebook to take those saved outputs as inputs for the next section of the work.

The second customization point, modifying the software in the container when the notebook runtime is being generated, is a little more complicated but worth taking a few minutes to discuss. First, why would you want to do that? Suppose that your analysis is going to require software packages that are not part of the default notebook runtime configuration. You could start the notebook with some installation steps, but that can turn into a maintenance headache if you have multiple notebooks that require the same configuration commands. It would be much easier to move some of those software installation steps out of the notebooks and into the environment configuration proper. You can do exactly that in Terra in two ways: you can specify a setup script that Terra will run in the container when it is creating or regenerating the notebook runtime, or you can supply a custom container image to the runtime service. Or you can, in fact, combine both: specify a custom container and a startup script to modify its setup. Figure 12-3 illustrates the difference between these options.

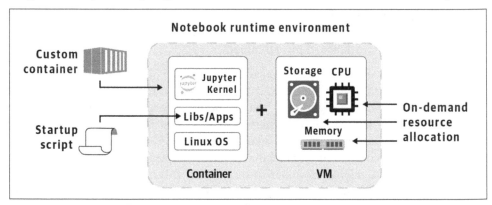

Figure 12-3. Options for customizing the software installed in the notebook runtime.

In the upcoming exercises, we show you how to use the startup script option because it's a good compromise of power and ease of use: it's not difficult to make your own, and you can readily distribute the script for others who are also using the same kernel in Terra. The custom container image option is technically more powerful and more portable, but it's a bit more complicated. If you're interested in learning how that works, check out the Terra documentation on custom containers (*https://oreil.ly/R-Grd*).

Now that you know how much power you have to customize your notebook runtime, we need to address a crucial question that we've been careful to avoid bringing up until now: how many notebook runtimes do you get to work with? Do you use the same runtime for everything, one per workspace, one per notebook, or can you create new runtimes willy-nilly like you can spin up VMs in the GCP console? To be frank, the answer is likely to evolve over time as the platform matures further and the product development team collects more data regarding what researchers actually want (so feel free to give them your opinion!).

As of this writing, the Terra Notebooks service provides you with a single notebook runtime for all workspaces within a particular billing project. To be clear, this means that if you open or create another notebook in any workspace attached to the same billing project as your first notebook, the new notebook will open in the same runtime environment. This can be very convenient if you have a lot of overlap between the data, resources, and software that you plan to use in both notebooks. However, it can be a source of major complications if you are working on different projects with very different configuration needs. In that case, you might want to consider developing notebooks with incompatible requirements under different billing projects, because that will give you a completely separate notebook runtime for each one.

Sharing and collaboration

With that, you've reached the last section of this introduction, and now it's time to talk about how to play nice with others. As we noted earlier, the notebook runtime is personal to you: only you have access to that machine, container, and Jupyter server. If you share your workspace with a collaborator who then opens up the same notebook, it will open in their own runtime environment. As a result, any work they do in the notebook will not affect the state of your runtime environment. However, the system will automatically save any changes they make to the shared document in the workspace, so it's important to set expectations clearly with your collaborators about whether it's OK for them to modify the notebook or whether they should work in a separate copy.

In addition, be aware that Terra will *lock* the notebook document in the workspace whenever someone is actively working with it, to avoid having multiple people making conflicting changes at the same time. When this happens, your collaborator can open the notebook in the read-only preview mode, or they can open it in a special *playground mode* that allows them to make changes and run code in their own runtime environment but does not save any changes to the original file, as illustrated in Figure 12-4. This falls a bit short of the ideal collaborative experience that you could envision based on Google Docs, for example, but it provides a reasonable compromise given the constraints at play.

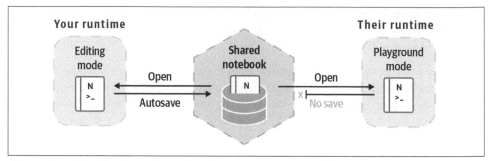

Figure 12-4. Notebooks in shared workspaces are protected from overwriting when two people open them concurrently.

If you're having a difficult time envisioning how this all works in practice, don't worry! First, that's totally OK; if anything is unclear, it's our fault, not yours. Second, good news: you've reached the end of the wall of theory, and now it's time to work through practical exercises in a real-life notebook.

Getting Started with Jupyter in Terra

We're going to work with a prewritten notebook, so you'll mostly just need to run cells, though whenever possible we tried to include hints for additional things that you can try in order to reinforce the learning. In Chapter 13, we talk about creating your own notebooks from scratch or importing existing notebooks from external sources.

To get started, go back to the workspace that you created by cloning the original book workspace in Chapter 11. If you don't have the URL on hand, you should be able to find it in Your Workspaces (*https://oreil.ly/bKWll*); or if you deleted it, you can clone the original again by following the relevant instructions at the beginning of Chapter 11. After you've found your workspace, open it and go to the Notebooks tab. There, you'll see two notebooks listed, as shown in Figure 12-5. Those are actually two copies of the same notebook: one that has never been run, and the other where we ran everything so that you can see the expected output.

Figure 12-5. The Notebooks tab showing two copies of the notebook: one already executed and another without any previous results.

We recommend that you open the second only in preview mode to preserve its contents as a reference in case you encounter anything surprising in the other one, which you will use to run through the upcoming exercise.

However, before you open anything, we're going to walk you through customizing your runtime configuration. If you already started opening one of the notebooks, don't panic; you'll still be able to reconfigure the notebook runtime. We just want to save you a little bit of time given that it takes a few minutes to get a new runtime up and running, and we know that we're going to want more than what the default configuration has to offer.

Inspecting and Customizing the Notebook Runtime Configuration

As noted earlier, the runtime environment that Terra creates for you by default is set up with basic resource allocations and a set of standard software packages. You can view this configuration at any time without having to open a notebook, as long as you're in a workspace under the appropriate billing project. To do so, look for the Notebook Runtime status widget, which, as of this writing, is displayed in the upper-right corner of almost all workspace pages, as demonstrated in Figure 12-6.

Figure 12-6. The Notebook Runtime status widget.

 We have heard rumblings from the Terra product development team that the display of the Notebook Runtime status widget might change in the near future, in which case you'll need to poke around to find it, or if that fails, consult the documentation in the book's repository on GitHub (*https://oreil.ly/genomics-repo*).

Click the gear icon on the right to bring up the runtime configuration page. If you haven't made any customizations to the runtime under your current billing project, the form should display all default settings, as depicted in Figure 12-7.

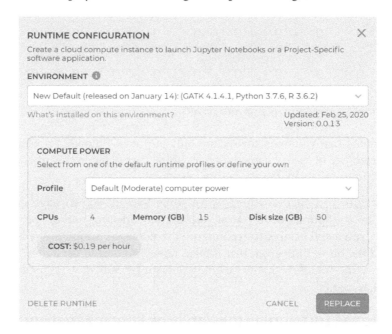

Figure 12-7. The default Notebook Runtime configuration settings.

You can see a small list of default environments to choose from, identified by the packages that are considered most important. For full details on what is installed on each, select the environment you're interested in and click "What's installed on this environment?" to bring up the detailed view. As shown in Figure 12-8, this detailed view is further broken into categories such as Python, R, and Tools. Selecting either Python or R brings up the full list of packages of the corresponding language that are included in the runtime environment. Selecting Tools will bring up the list of command-line executable tools also included in the runtime environment.

In fact, if you select Tools, you'll see that this default set actually includes GATK, which is a nice touch for those of us in the genomics field, considering Terra has a much wider audience than just genomics. That being said, for the purposes of this book, we are using a different version than the one included in the default configuration as of this writing, so we need to customize this. In addition, we're going to want to use a Python library that (spoiler alert) makes it possible to embed an IGV browser window within the notebook (which is so cool). We could install both from within the notebook itself, but as noted in the introduction, we prefer to use a startup script that will install them in the Jupyter container during the notebook runtime creation process. We provide a closer look at the script in question in the accompanying side-bar in case you're curious, but feel free to skip it if it's not your cup of tea or if you're eager to get started with the notebook itself.

Figure 12-8. Detailed view of the packages installed on the default runtime environment.

A Closer Look at the Startup Script that Installs GATK and IGV on the Notebook Runtime

This startup script was originally provided by the GATK team for its workshops, which use Terra as a teaching platform. You can find it with a collection of other startup scripts here (*https://oreil.ly/rUPmm*), and we've also included a copy in the book repository on GitHub. Let's walk through it briefly to highlight the key parts.

This line declares that it's a Bash script:

```
#!/bin/bash
```

The next block of lines is much more interesting because it contains the installation commands for the IGV library. First the script installs the `igv-jupyter` Python package by using the Python package management tool, `pip` (specifically its Python 3 version, `pip3`). Then, it enables Jupyter Notebook extensions that are necessary for the IGV browser to be properly interactive in the context of the notebook:

```
pip3 install igv-jupyter

jupyter serverextension enable --py igv --sys-prefix
jupyter nbextension install --py igv --sys-prefix
jupyter nbextension enable --py igv --sys-prefix
```

This next block contains accessory packages that are useful but not particularly interesting, so we won't go into the details:

```
pip3 install rpy2==3.0.4
pip3 install singledispatch
pip3 install tzlocal

echo
"install.packages(c(\"optparse\",\"data.table\"),repos=\"
http://cran.us.r-project.org\")" | R --no-save
```

And, finally, this longer block contains the commands to install the GATK components that we want. First it deletes any preinstalled GATK package, retrieves the zipped archive for the specific release of GATK that we want from GitHub, and then unzips it and sets up a symbolic link to make it callable from the notebook. That takes care of the main GATK package. There's also a secondary package of Python code, which the script installs with pip:

```
set -e

GATK_VERSION=4.1.3.0
GATK_ZIP_PATH=/tmp/gatk-$GATK_VERSION.zip

# remove pre-existing GATK version
rm -rf /bin/gatk

# download the gatk zip if it doesn't already exist

if ! [ -f $GATK_ZIP_PATH ]; then
  # curl and follow redirects and output to a temp file
  curl -L -o $GATK_ZIP_PATH
https://github.com/broadinstitute/gatk/releases/download/$G
ATK_VERSION/gatk-$GATK_VERSION.zip
fi

# unzip with forced overwrite (if necessary) to /bin
unzip -o $GATK_ZIP_PATH -d /etc/

# make a symlink to gatk right inside bin so it's available from the existing
# PATH
ln -s /etc/gatk-$GATK_VERSION/gatk /bin/gatk

pip3 install /etc/gatk-$GATK_VERSION/gatkPythonPackageArchive.zip

export PATH=$PATH:/home/jupyter-user/.local/bin
```

If you're familiar with installation scripts written in Bash, you can see that there is nothing specific to Terra about this particular script. You could provide it to anyone who wants to set up the same environment as your notebook runtime outside of Terra. That contributes to making the "outsourcing" of custom environment setup instructions to a startup script (as opposed to keeping them within the notebook itself) a really solid option for portability and reproducibility.

So where should we specify the startup script? You might be tempted to look in the environments menu on the notebook runtime customization page, maybe even select the Custom Environment option, and you would be so close to being right, conceptually—but in practice, you'd be wrong. That is where you would go to specify a custom Docker image to substitute for the built-in one. Instead, you need to look a little farther down at the Compute Power section (shown earlier in Figure 12-7), which allows you to modify the VM resource allocations. This section includes a menu that allows you to choose from three preset configurations, designated as providing Moderate, Increased, or High computer power, or provide your own under the label Custom.

 To be frank, we wouldn't be surprised if this part of the interface also evolved a bit in the near future because it's not super logical to find a software customization option grouped with hardware allocations. Not to mention the names of the preset configurations, which are about as helpful as Starbucks cup sizes: once you get used to them, they kind of make sense, but the first time you set foot in a Starbucks, you're just happy that the prices tell you which one is bigger.

Feel free to select each preset and look at how their respective resource allocations differ. When you're ready to move on, select the Custom option to bring up an editable configuration window. You should see a new field named "Startup script" appear that wasn't available earlier (Figure 12-9). You can finally input the path to the startup script there, in the text box labeled URI (for uniform resource identifier (*https://oreil.ly/Ltao0*), a close cousin of URL (*https://oreil.ly/jUtVr*), the uniform resource locator). We included a copy of the script in the book bucket, so you can use this path:

```
gs://genomics-in-the-cloud/v1/scripts/install_GATK_4130_with_igv.sh
```

The rest of the resource allocations will be fine with the default (Moderate) values of 4 CPUs, 15 GB of memory (RAM), and 50 GB of disk (storage space). Figure 12-9 demonstrates what this looks like.

As a reminder, this startup script installs GATK version 4.1.3.0 in the runtime environment, as well as an IGV integration module that will make it possible to view genomic data using IGV from within the notebook itself.

Figure 12-9. The Compute Power section allows you to specify a startup script if you choose the Custom profile.

Click the Create button when you are done (labeled Replace if you had already created a runtime earlier), and Terra begins to create a new runtime environment with your settings. You can go grab yourself a cup of something nice or continue on to the next set of instructions, as you prefer. You will probably need to wait a few minutes in either case while Terra communicates with GCP to provision your shiny new runtime environment.

Opening Notebook in Edit Mode and Checking the Kernel

You don't need to wait for the runtime to be ready in order to take a peek at the notebook, so go ahead and open the never-been-run copy of the tutorial notebook. Whether your runtime is ready or not, Terra will initially open the notebook in preview mode, in which the notebook is read-only. As Figure 12-10 illustrates, a menu at the top of the preview panel offers you the two options for opening the notebook in an interactive mode: Edit, to work with the notebook normally; or Playground Mode, to experiment without saving anything, as we briefly discussed in the introduction.

Figure 12-10. Menu on the notebook preview page displaying the main options: Preview, Edit, and Playground Mode.

Click Edit and wait for the runtime to be ready. You'll recognize when the transition to Edit mode happens by the appearance of the standard Jupyter menu bar, shown in Figure 12-11. A more subtle but also important sign is the little white box toward the right of the menu bar showing the label Edit Mode. If you had opened the notebook in Playground Mode by mistake, the Jupyter menu bar would also be displayed, but instead of the white Edit Mode label, you'd see an orange "Playground Mode (Edits not saved)" label.

Figure 12-11. The standard Jupyter menu bar.

Further to the right in that area of the menu bar is the Python 3 label that identifies the active kernel as well as the Python logo. You might remember that the *kernel* of the notebook is the computational engine that interprets the code in the notebook and initiates execution of each cell that you run. For this particular notebook, we decided to use a Python 3 kernel so we can execute terminal commands (including to programs like gsutil and GATK) using Python magic methods—which is an actual technical term, we swear. You'll see it in action shortly: it is truly magical.

In case you're wondering, yes you can switch the kernel while the notebook is running using the Kernel menu, but we really don't recommend doing that. It might sound useful in principle—for example, to run code in a different language in a subset of cells—but in practice it's dangerous and can mess you up if you're not careful. If you need to use both Python and R code within the same notebook, you're much better off using the Python 3 kernel and magic methods commands as we demonstrate in this notebook.

And with that, it's time to run some actual code cells!

Running the Hello World Cells

Let's run through a few simple examples so you can get a feel for what it's like to work in a Jupyter notebook if you've never done so before. We're going to run three types of commands to demonstrate the syntax for running Python code as well as R code and a command-line tool from the Python context.

Python Hello World

This is the classic Hello World in Python, using the print() function and giving it the string Hello World. As you may recall from the Hello World we did in WDL in Chapter 8, this is the equivalent of the echo "Hello World" command that we used at the time. To run the cell, click anywhere in the gray area to select the cell and then press Shift+Enter on your keyboard, or, at the top of the page, on the toolbar, use the Run menu to run the cell:

```
In [1]  print("Hello World!")

Out [1] Hello World!
```

You can use a variable for the greeting if you want to give yourself some flexibility. Double-click the cell to edit it and modify the code, as follows, and then run it again:

```
In [2]  greeting = "Hello World!"
        print(greeting)

Out [2] Hello World!
```

Incidentally, this shows you that when you have a cell containing multiple lines, running it executes all the code in the cell. Sometimes, it makes sense to group multiple commands in a single cell because you're always going to want to run them all. However, you could also choose to divide them into separate cells. Try doing that now. You can add a new cell to the notebook either by clicking the "+" icon on the toolbar at the top of the page or by going to the Insert menu. The former automatically creates the new cell below the one that is currently active, whereas the Insert menu gives you an explicit choice to add it either above or below the active cell.

```
In [3]  greeting = "Hello World!"
In [4]  print(greeting)

Out [4] Hello World!
```

This allows you to decouple the assignment of the variable and the execution of the task itself. In this simple example, it doesn't make much difference, but when you are performing more complex operations, the choice of which commands to group versus which to break apart becomes more important.

Order and State in Jupyter

Even though the commands are organized linearly on the page to accommodate our puny human brains, you can go back and run code cells in any order you want. That's why the Jupyter server adds a number in the little brackets in the left margin; it gives you an indication of when the cells were last run relative to each other.

This system is not perfect—for example, running the same code cell again makes the previous number disappear, so if you go back and forth multiple times, you could lose track of what was run when. This is one of the shortcomings of Jupyter that is most often criticized by experienced programmers, because it can lead to risky situations in which the underlying state of the notebook is not clear; for instance, which variables are live and what their values are. If you get into a situation like that, the safest course of action might be to reset the notebook state by restarting the kernel and clearing all outputs, which you can do from the Kernel menu on the Jupyter toolbar.

Now that you have the basic mechanics down, let's have a look at how we can run some R code within this Python notebook.

R Hello World using Python magic methods

This one requires a tiny bit of setup before we can dive into the Hello World exercise itself. Remember the startup script we used to customize the notebook runtime? One of the steps in that script installs the `rpy2` package, which handles the interpretation of R code within the Python notebook, into the runtime environment. It's one of the magic methods features we mentioned earlier in the chapter. To activate it, you first need to import the `rpy` package and activate the corresponding notebook extension:

```
In [5]  import rpy2
        %load_ext rpy2.ipython
```

This can take a few seconds, during which time the server displays an asterisk (*) in the brackets to the left of the cell to indicate that it's working on it. After the package is loaded, you have two ways to invoke the magic methods: use `%R` for a single line of code or `%%R` for an entire cell.

Here's the same basic Hello World as we ran earlier but in R this time, using the single-line magic methods invocation:

```
In [6]  %R print ("Hello World!")

Out [6] [1] "Hello World"
```

Yes, that's the same code as we ran in the Python example, because R also has a `print()` function. You can recognize that it's the R version because the output shows up as an array with the greeting as a single string element, whereas the Python version just returned the text of the greeting as a string by itself.

That being said, modifying it to use a variable assignment makes it a little more obvious that it's R code as opposed to Python. This time, use `%%R` to apply the magic methods to the whole cell:

```
In [7]  %%R
        greeting <- "Hello World!"
        print(greeting)

Out [7] [1] "Hello World"
```

There you go: you're running R code in a Python notebook. This is going to be really handy when we get to the "serious" exercises, because we need to use Python in order to embed an IGV browser (coming up real soon!), but we also want to use an existing R script later for plotting. Now we have access to the best of both worlds. Just remember that when you start using this in your own notebooks, you'll need to include the cell that imports `rpy2` and activates the extension.

This brings us to the third type of command that we're going to want to run in our notebook: command-line tools like `ls`, `gsutil`, and GATK.

Command-line tool Hello World using Python magic methods

For this one, there's nothing to load; the Hello World case works out of the box. We'll use the classic echo command that we used in our WDL Hello World example. Simply prepend an exclamation mark to the command and then run the cell:

```
In [8]  ! echo "Hello World!"
```

```
Out [8] Hello World!
```

You can use all the classic shell commands in this way; for example, if you want to list the contents of the working directory, type **! ls** in a cell and run it. Similarly, you can run any command-line tool installed in the notebook runtime environment in this way. All preset environments available in Terra include the gsutil package, so you can use those tools in any Terra notebook. We walk you through specific examples of that in the next section. Later, we also run commands to run GATK, which our startup script installed, using the same basic syntax.

We've been focusing on code cells so far, but keep in mind all of the descriptive text cells are also editable, of course. Feel free to double-click a few and see how their appearance changes to show that they are in editing mode. Try making some edits and then, when you're done, "run" the cell (just as you would run a code cell) to exit the text-editing mode. The descriptive text cells use a simple formatting markup language called *Markdown*, so you can set header levels, make bullet-point lists, and so on. For more on working with Markdown, see this helpful page (*https://oreil.ly/07KtL*) in the Jupyter project documentation. When you create new cells in your notebook, the system makes them code cells by default, but you can switch them to Markdown via the Cell menu, by choosing Cell Type > Markdown.

Now that you have a firm grip on the fundamentals, it's time to do some work that is more specific to the cloud environment and the genomics subject matter that interests us.

Using gsutil to Interact with Google Cloud Storage Buckets

Most of the time, the data we want to work with in the notebook resides in GCS buckets, so the first thing you need to learn is how to access that data. Here's some good news: you can use gsutil commands to do all the same things we've previously shown you in Chapter 4 and beyond. For example, use gsutil ls to list the contents of the book bucket:[1]

---

1 From this point on, we don't show the output of the cells. You can check your outputs against the copy of the notebook included in the Terra workspace. We also provide an html version of the pre-run notebook in the book's Github repository (*https://oreil.ly/genomics-repo*).

```
In [9]   ! gsutil ls gs://genomics-in-the-cloud/
```

Similarly, you can use `gsutil cp` to *localize* files; that is, copy them from a bucket to the notebook's local storage space. For example, use the following command to copy a file from the book bucket to the *sandbox* directory in the notebook runtime:

```
In [10]   ! gsutil cp gs://genomics-in-the-cloud/hello.txt .
```

Then, you can run `cat` to read the contents of the localized file. You could write this in Python instead because this is a Python notebook, but it's hard to beat the brevity of `cat`!

```
In [11]   ! cat hello.txt
```

As you can see, these are essentially the same commands we used in Chapter 4, except that we added the `!` in front to signal the Python interpreter to execute that line as a terminal command instead of reading it as Python code.

Setting Up a Variable Pointing to the Germline Data in the Book Bucket

The next few exercises are going to make use of the germline data that we provide in the book's data bundle. Because the path to the example data is rather long, let's set up a variable to store it in the notebook in a more concise form, much as we did with an environmental variable back in Chapter 5:

```
In [12]   GERM_DATA = "gs://genomics-in-the-cloud/v1/data/germline"
```

This is a Python variable, so when you use it in a shell command, you'll need to wrap it up in curly braces. For example, a command using `gsutil` to list the contents of that directory would look like this:

```
In [13]   ! gsutil ls {GERM_DATA}
```

Take a moment to think about how we're using curly braces around this variable instead of calling it as `$GERM_DATA`, which you might have expected based on the Bash environment variables that we've been using quite a bit so far. The key point to remember here is that we set up the bucket shortcut as a Python variable, not a shell environment variable. It is possible to use shell environment variables in the notebook, but that's a topic for another time.

You can also compose paths based on this variable in order to list subdirectories; for example, to get a list of the BAM files, or perform operations on specific files. For each of the following commands, try to infer what its function is, then run it in the notebook and evaluate the result against your expectation.

```
In [14]   ! gsutil ls {GERM_DATA}/bams
```

```
In [15]   ! gsutil cp {GERM_DATA}/bams/mother.ba* .
```

```
In [16]  ! ls .
```

Later in this chapter, we also show you how to use this same GERM_DATA variable in the context of some Python code.

Setting Up a Sandbox and Saving Output Files to the Workspace Bucket

When you run commands in the notebook that produce output files, by default those files will be saved on the notebook's local storage space. However, the local storage associated with your notebook is temporary, so you'll need to copy any outputs that you care about to a GCS bucket. You can use any bucket to which you have write access for that purpose, but we recommend using the workspace's dedicated bucket.

To streamline the process of saving outputs to the bucket, we like to do two things: create a sandbox to house the output files that we're going to produce, and set up a variable pointing to the workspace bucket.

Let's begin with the *sandbox* directory; go ahead and create a new directory, and then move some files there for demonstration purposes. Once again, we include the commands here but don't detail their purpose or results. We do provide additional detail in the notebook. Try to infer the function of each command before you look at its full description and run it in the notebook:

```
In [17]  ! mkdir -p sandbox/

In [18]  ! mv mother.ba* sandbox/

In [19]  ! ls sandbox
```

When you have your sandbox ready, let's tackle the workspace bucket. The workspace bucket name is a long machine-generated sequence of letters and numbers, which is annoying to work with. Fortunately, we can import it programmatically (rather than looking it up manually in the workspace dashboard) because Terra makes it available to the notebook as a system variable. And to make it even easier, we're going to create a Python variable from that system variable as follows:

```
In [20]  import os
         WS_BUCKET = os.environ['WORKSPACE_BUCKET']

In [21]  print(WS_BUCKET)
Out [21] 'gs://fc-46207b9c-d593-4e7a-9057-7aca3bb5c9a7'
```

Here, we use Python commands to set variables at the Python level. The import os command allows us to interact with the operating system from within Python code, and the os.environ['WORKSPACE_BUCKET'] call uses that to access the value of the environment variable `'WORKSPACE_BUCKET'`, which was originally set for you by the Terra Notebooks service.

From here on, you'll be able to refer to the workspace bucket as `WS_BUCKET`. For example, you can use `gsutil ls` to list its contents:

```
In [22]  ! gsutil ls -r {WS_BUCKET}
```

Notice that we wrapped the `WS_BUCKET` variable in curly braces, just as we did for the germline data variable in the previous section.

After that's set up, you can simply run the same `gsutil cp` command on the *sandbox* directory whenever you want to save your outputs to the workspace bucket:

```
In [23]  ! gsutil -m cp -r sandbox {WS_BUCKET}
```

```
In [24]  ! gsutil ls {WS_BUCKET}/sandbox
```

This copies the entire *sandbox* directory from the notebook's local storage to the workspace bucket. Keep in mind that this approach might not scale very well if you're producing many large files; in that case, you might want to consider dividing and managing your sandbox in separate subdirectories. It is admittedly not ideal that you must synchronize files manually to the bucket; we look forward to seeing improvements to the experience of using these tools as the technology develops further. At least the notebook file itself is saved automatically, as we discussed earlier.

Visualizing Genomic Data in an Embedded IGV Window

Now that our notebook is all set up and ready to roll, let's use it to try out a cool trick: visualizing genomic data with IGV within the context of the notebook. In previous chapters, we had you work with the desktop version of IGV, which you had set up to pull data from GCS. This time, we're going to do something a little different: we're going to use a special IGV package called *IGV.js* that allows us to embed an IGV browser in the notebook. This is especially convenient when you want to include the data visualization within a tutorial for students, or within a report to communicate results to collaborators, in such a way that they don't need to resort to using a separate program.

 The *IGV.js* package has limitations that we describe further in just a moment. In a nutshell, it's very cool when it works, so we do think it's worth showing you how to use it in a Jupyter notebook, but it doesn't always work seamlessly. If you experience any difficulties while using it, we recommend that you fall back to using Desktop IGV, as previously described.

In this exercise, we're going to load two BAM files: the WGS mother sample, which has been our go-to test file for most of the book, and an exome sample from the same person for comparison. You might recall that in Chapter 2 (a lifetime ago), we touched on the differences between several library design strategies, including WGS and exome sequencing. In particular, we discussed how their coverage profiles have very different shapes: the WGS tends to look like a distant mountain range, whereas the exome sample looks more like a series of volcanic islands scattered in the ocean. This is an opportunity for you to see that for yourself, essentially replicating Figure 2-18 in an interactive form, while trying out the *IGV.js* integration.

Setting Up the Embedded IGV Browser

The good news is that there's not much you need to do here. Remember the startup script that you ran as part of the runtime environment customization? That included instructions to install all the prerequisite software to run *IGV.js*, which the system executed when it created your customized runtime environment. As a result, we just need to do a one-time import to activate the `igv` Python package. After that, it's just a matter of creating an IGV browser wherever you want one to appear, as shown in the code cell that follows. This is all Python code that follows the guidelines documented by the IGV team in the IGV-Jupyter repository (*https://oreil.ly/JgOtt*) on GitHub:

```
In [25]  import igv

In [26]  IGV_Explore = igv.Browser(
             {"genome": "hg19",
             "locus": "chr20:10,025,584-10,036,143"
             }
         )

In [27]  IGV_Explore.show()
```

The name we give the browser (here, `IGV_Explore`) is completely arbitrary. You could provide other parameters to initialize the browser, but the only one that is absolutely required is the genome reference; everything else is optional. That being said, we usually specify some coordinates (or the name of a gene of interest) for the browser to zoom in on straight away.

When you run the cell, you should see an embedded IGV browser that includes the reference genome and a RefSeq gene track, but no actual data tracks, as shown in Figure 12-12.

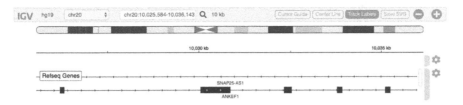

Figure 12-12. A newly created IGV browser.

Next, let's add the two sample BAM files that we want to compare.

Adding Data to the IGV Browser

For each track that we want to load, we need to provide this same set of metadata: a name for the track, the path to where the file lives in GCS, the format, and the path to the corresponding index file. We provide this information to IGV using the `load_track()` function:

```
In [28]  IGV_Explore.load_track(
            {
                "name": "Mother WGS",
                "url": GERM_DATA + "/bams/mother.bam",
                "indexURL": GERM_DATA + "/bams/mother.bai",
                "format": "bam"
            })
```

```
In [29]  IGV_Explore.load_track(
            {
                "name": "Mother Exome",
                "url": GERM_DATA + "/bams/motherNEX.bam",
                "indexURL": GERM_DATA + "/bams/motherNEX.bai",
                "format": "bam"
            })
```

Because this is all Python code, we can use the Python variable that we set up for the germline data simply by referring to its name, `GERM_DATA`.

Take a moment to note the Python syntax, which explicitly uses a + operator to concatenate the variable and the subdirectory strings in order to compose the full address pointing to where the data files reside. This is in contrast to the shell syntax that we used earlier in the `gsutil` command, `"{GERM_DATA}/bams"`, which is a more implicit instruction. If you're not very familiar with Python, know that this exemplifies one of the cardinal rules of Python programming: explicit is better than implicit.

After you've run both cells and each of them returns OK as a result, scroll up to the browser, where you should see spinning symbols that indicate the data is loading. When the spinners go away and the data displays, you should have two data tracks in your IGV browser: the WGS and exome versions of the mother sample, respectively, as shown in Figure 12-13.

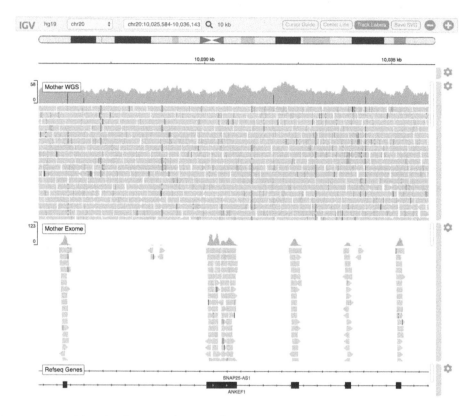

Figure 12-13. The IGV browser showing the two sequence data tracks.

Try zooming in and out, and drag the sequence left and right to pan the view and get a sense of how the data is distributed in these two samples. You'll observe the classic "mountain range versus volcanic islands" difference in coverage profile, which you can use from here onward to identify the library design type of any sequencing sample on sight.

It is technically possible to load VCF and BAM files in the embedded IGV browser without specifying an index file. To do so, omit the `"indexURL"` line and replace it with `"indexed": False`. Be aware, however, that doing so will cause IGV to take much longer to load the data. It can take a couple of minutes for the data to load, and you might see a pop-up window stating that the page is unresponsive. If so, dismiss the alert and give it another minute. If it takes much longer than that, you might need to fall back to using the desktop version of IGV.

We hope you'll agree this is a neat way to include a view of the genomic data within an analysis log or report, even though it suffers from a few limitations. One limitation is the lag that you might experience when you originally load the data; another is the fact that not all display customization options are available compared to the desktop version of IGV. Authentication can be a major stumbling block if you don't have proper guidance: if you want to access data in private buckets (which includes your workspace bucket!), you need to jump through additional hoops that involve access credentials. You might recall that in Chapter 4 you had to enable a Google login option in the desktop version of IGV in order to view files from a private bucket. Here, we're going to do something similar, except instead of a point-and-click process, it will consist of a few lines of code.

Setting Up an Access Token to View Private Data

In the previous example, we were reading data from completely public buckets, so we didn't need to do any authentication. However, you'll eventually want to view files in private buckets. To do that, you need to set up an access token that IGV can use to access data in your private bucket.

First, let's use `gcloud auth` to generate an access token and save it to a file:

```
In [30]   ! gcloud auth print-access-token > token.txt
```

As long as this file is saved only to your notebook's local storage, it is secure because your runtime environment is strictly personal to you and cannot be accessed by others, even if you share your workspace or your notebook with them. But don't save this file to your workspace bucket! Saving it to the bucket would make it visible to anyone with whom you share the workspace.

Next, read the contents of the token file into a Python variable. Because the token consists of a single line of text, we can use the `readline()` function, which reads the first line of a file into a string:

```
In [31]   token_file = open("token.txt","r")
          token = token_file.readline()
```

At this point, you have the `token` variable stored and ready to use with IGV whenever you want to load a file that resides in a private bucket.

For example, recall that in the previous section, we had you copy the *mother.bam* file and its index to your workspace bucket. Even though you own that bucket, the IGV process that is running in your notebook doesn't "know" that you're allowed to access it. You must instruct it explicitly by providing the token that you just set up when you make the call to `load_track()` function, as follows:

```
In [32] IGV_Explore.load_track(
            {
                "name": "Workspace bucket copy of Mother WGS",
                "url": WS_BUCKET + "/sandbox/mother.bam",
                "indexURL": WS_BUCKET + "/sandbox/mother.bai",
                "format": "bam",
                "oauthToken": token
            })
```

As you can see, we copied the same code we used earlier to load BAM files, except this time we provided the path to the files in the workspace bucket and added the token that we generated earlier. If you're curious to see what would happen if you didn't provide the token, feel free to try it out by deleting that line (as well as the comma that ends the previous line). Note that it is also possible to access data that resides in private buckets managed outside of Terra. As we'll see in Chapter 13, this requires giving access to the bucket to your proxy group service account.

If you were to have multiple private files to load in the IGV browser, you would include the token in each track definition. The IGV documentation (*https://oreil.ly/ RWDqg*) states that it is possible to set a global IGV configuration variable, `igv.set GoogleOauthToken(accessToken)`, that would apply to all tracks, but as of this writing, that did not work within our notebook.

Running GATK Commands to Learn, Test, or Troubleshoot

It's all well and good that we can visualize the sequencing data from within a notebook, but we were already able to achieve the equivalent result with the desktop version of IGV. What's really cool about the notebook concept is that we can run analysis commands and then visualize the output, all within the same environment.

What kind of analyses can you run, you ask? Well, just about anything you want. As you saw when we had you run `gsutil` commands, you're not constrained to running only Python code in a Python notebook. You can run pretty much anything that you can install and run in the shell environment.

Here we're relying on the startup script that you used to initialize your notebook runtime environment at the start of this chapter's exercises. That script includes instructions to download the GATK package and make it available for command-line invocation, which were executed when the environment was created for you, so you don't need to do any of it yourself.

In this section, we show you how to run GATK commands and visualize the results in IGV, both from within the notebook. We find that this provides a much more integrated and seamless experience than the "split-screen" approach we took in earlier chapters in which we were running GATK commands in a VM and then visualizing results in the desktop version of IGV. We put you through all that because we're sadists, and also because it gave you the opportunity to build up foundational skills. Through that process, you gained a measure of familiarity with the underlying components of cloud computing that should help you conceptualize what's happening behind the scenes when you run a workflow or work in a notebook. And perhaps having gone through that will enhance your appreciation of the notebook-based approach, if only for purposes like teaching, testing, and troubleshooting.

To that end, we're going to revisit exercises that you previously worked through in Chapter 5 so you can focus your attention on *how* you are doing the work rather than what the analysis means.

Running a Basic GATK Command: HaplotypeCaller

Let's begin by running the `HaplotypeCaller` tool on the same sample we've been using throughout the book. You should recognize this command, which we copied almost verbatim from Chapter 5:

```
In [33]   ! gatk HaplotypeCaller \
              -R {GERM_DATA}/ref/ref.fasta \
              -I {GERM_DATA}/bams/mother.bam \
              -O sandbox/mother_variants.200k.vcf.gz \
              -L 20:10,000,000-10,200,000
```

What are the differences in this command compared to how we ran it in Chapter 5? By now you should recognize the ! that precedes the GATK command as the signal to bypass the Python interpreter and run it as a shell command. The reference to the file path variable is also a little different since we're using curly braces instead of $, as noted earlier. We're also writing the output VCF file in the compressed *gzip* form, which is a requirement for IGV in the next step.

Another difference is a bit hidden by our use of a Python variable to store the common part of the input file paths, *gs://genomics-in-the-cloud/v1/data/germline*. This time, we're using the paths to the files in GCS instead of pointing to a local copy. We can do that because, as we've noted several times by now, GATK tools are capable of streaming most types of file inputs directly from GCS. In practice this behavior kicks in whenever the GATK command-line parser identifies that an eligible input file path starts with *gs://*. This is great because it allows us to avoid localizing the relevant files to the notebook's local storage. Incidentally, this also works for writing output files directly to GCS, though we don't demonstrate it here.

In Chapter 5 through Chapter 7, we had you localize the full data bundle and run all GATK commands with local file inputs. We could have had you run most of the commands with the bucket paths instead and relied on GATK's data streaming capabilities; it would have worked just fine on your VM. However, we felt that introducing those aspects so early would overcomplicate what might well be your first experience with the cloud. We chose instead to have you work with localized files in the hope that the ensuing experience would provide enough familiarity to make you feel more comfortable. We bring this up now in case you choose to go back to working in a VM environment, so that you know that you can still take advantage of the streaming feature. And, as you might remember from the discussion on optimizations in Chapter 10, this also works in the context of WDL workflows.

When you run the command, you should see the log output being written to the notebook below the cell. This is a really nice touch in terms of keeping all the information about the analysis together in a single place—it's one of the key benefits of the Jupyter concept. On the downside, if you're running a tool that's particularly verbose (as GATK can occasionally be), you can end up with pages and pages of a log in the middle of your notebook. That's where it really helps to use clear section headers in Markdown cells to demarcate the different parts of your analysis, especially in combination with the notebook widget that automatically creates a table of contents and sidebar navigation menu.

After the `HaplotypeCaller`'s run is complete, let's list the sandbox contents to confirm that the command worked and that the VCF of variant calls was created as expected:

```
In [34]  ! ls sandbox/
```

Yep, there it is, along with its index file. Let's look at it in IGV.

Loading the Data (BAM and VCF) into IGV

Suppose that we want to open the output VCF with IGV in our notebook, mainly to do a visual check and compare it to the BAM file. We could use the IGV browser that we created earlier to look at the different BAM files, but because this is a separate exercise with a different purpose—and we're too lazy to scroll up a bunch of pages—we're going to create a new one.

The first bit of code is essentially the same as what we used earlier except that we're using a different name for the browser object:

```
In [35]   IGV_InspectCalls = igv.Browser(
              {"genome": "hg19",
              "locus": "chr20:10,002,294-10,002,623"
              }
          )

          IGV_InspectCalls.show()
```

This creates a new browser below the cell, zoomed in on intervals of interest but without any data. So, let's load the variant data from the VCF file that we produced with the `HaplotypeCaller` command, which resides in the *sandbox* directory on the notebook's local storage space:

```
In [36]   IGV_InspectCalls.load_track(
              {
                  "name": "Mother variants",
                  "url": "files/sandbox/mother_variants.200k.vcf.gz",
                  "indexURL": "files/sandbox/mother_variants.200k.vcf.gz.tbi",
                  "format": "vcf"
              })
```

This is the same code that we used earlier to load BAM files, except this time we changed the track `format` property to `vcf` instead of `bam`, and the file paths (`url` and `indexURL`) are pointing to local files instead of pointing to locations in GCS.

 Pay attention to those file paths: you should notice that they're not exactly the file paths you would expect based on the directory structure of the notebook's local storage space. Do you see it? The `files/` part does not refer to a real directory! It's a prefix that we add for IGV's benefit, as instructed in the IGV-Jupyter project documentation (*https://oreil.ly/JgOtt*).

Alternatively, you could run the `gsutil cp` command to copy the sandbox to the workspace bucket and then use the paths to the workspace bucket copy to load the VCF track. However, if you do that, don't forget to include the access token as explained in the previous section.

Finally, let's load the BAM file and its index from the original germline data bundle. These files are located in a public bucket and therefore do not require specifying the access token (but if you do include it, nothing bad will happen):

```
In [37]   IGV_InspectCalls.load_track(
              {
                  "name": "Mother WGS",
                  "url": GERM_DATA + "/bams/mother.bam",
                  "indexURL": GERM_DATA + "/bams/mother.bai",
                  "format": "bam"
              })
```

The resulting view, shown in Figure 12-14, should look essentially the same as what you produced in Chapter 5, with a few differences in appearance between the desktop version and the *IGV.js* version of the visual rendering.

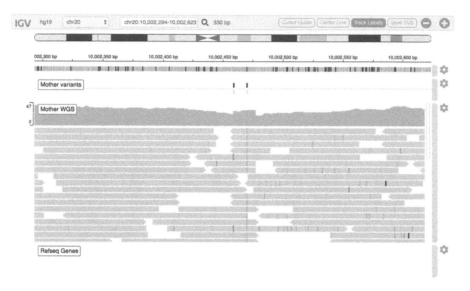

Figure 12-14. IGV.js rendering of the sequencing data ("Mother WGS" track) and output variants produced by HaplotypeCaller ("Mother variants" track).

You can click elements of data (e.g., reads or variants) in the viewer to bring up additional details, just as we did in Chapter 5. The visual display is a little different, but it's basically the same functionality, except that you can't switch it to show details "on hover."

One difference that's not obvious here is that the embedded IGV window organizes tracks a little differently compared to the way the desktop version of IGV does it. In the desktop version, variant tracks are automatically displayed above sequence data tracks, regardless of the order in which they are loaded. You could load a BAM file first and then a VCF file, yet the variant track will always be on top. In contrast, the embedded IGV window displays tracks in whatever order they are added. So, if you load the BAM file first, that's what will be on top, even if you load a VCF file afterward.

Hopefully this gives you a good sense of how you can use the embedded IGV within a notebook. Let's work through one more exercise from the original Chapter 5 curriculum to practice using this tooling and cover a few more minor options.

Troubleshooting a Questionable Variant Call in the Embedded IGV Browser

You might recall that in Chapter 5 we took a closer look at the homozygous variant insertion of three T bases that appears in the variant track in this region. At first glance, we were skeptical of `HaplotypeCaller`'s decision because the call didn't seem to be supported by the sequencing data. Do you remember the first thing we did to investigate? That's right, we turned on the display of soft clips, those bits of sequence data tagged as "unusable" by the mapper that are normally hidden by default. Let's do that now in the IGV window in the notebook.

As you can see in Figure 12-15, you can bring up track-viewing options by clicking on the gear icon to the right of the track of interest. Do that now for the Mother WGS sequence data track and select "Show soft clips;" then, in the upper-right corner of the menu, click the X to close it.

Figure 12-15. Menu of display options for the Mother WGS sequence data track.

You should see the entire area light up in the bright glare of multitudes of mismatches, as shown in Figure 12-16.

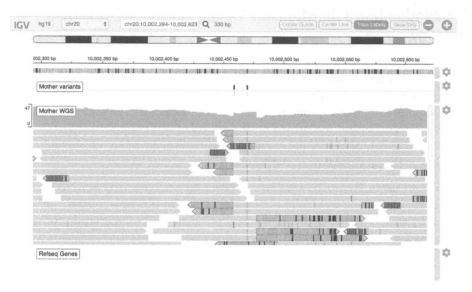

Figure 12-16. Display of soft clips.

You know what to do now, right? Questionable indel call, tons of soft clips…Yes, it's time to generate a bamout to see what `HaplotypeCaller` was thinking when it made that call:

```
In [38]  ! gatk HaplotypeCaller \
            -R {GERM_DATA}/ref/ref.fasta \
            -I {GERM_DATA}/bams/mother.bam \
            -O sandbox/motherHCdebug.vcf.gz \
            -bamout sandbox/motherHCdebug.bam \
            -L 20:10,002,000-10,003,000
```

That should run very quickly and produce the key output we're interested in, which is the BAM file that shows how `HaplotypeCaller` has realigned the read data, as explained in Chapter 5. Let's add that file to our IGV browser:

```
In [39]  IGV_InspectCalls.load_track(
            {
                "name": "Mother HC bamout",
                "url": "files/sandbox/motherHCdebug.bam",
                "indexURL": "files/sandbox/motherHCdebug.bai",
                "height": 500,
                "format": "bam"
            })
```

Again, this should produce a view that is equivalent, though not identical, to the one we encountered in Chapter 5. As previously, we can conclude that `HaplotypeCaller`'s call of an indel was reasonable, given the realigned data.

Incidentally, you might notice that in this one we specified the height of the track with `"height": 500`. This can be useful when we're trying to showcase a specific view of the data in a way that minimizes scrolling, for example. Feel free to experiment with setting the height of different tracks.

What do you think of this approach to running and examining GATK commands? We could continue mirroring all of the material that we covered in Chapter 5 through Chapter 7 in this way, and in fact, there are several such GATK tutorial notebooks in public Terra workspaces, which the GATK team uses in its popular series of international workshops. We encourage you to check those out for further study.

However, for the purposes of this book, and this chapter in particular, we want to focus on covering the most useful aspects of Jupyter notebooks in relation to the types of interactions you would typically have with genomic data. We have a few more that we're excited to show you, so we need to move on.

The next logical step is to plot variant data. There are many aspects of variant data that you might want to explore visually, but we can't cover them all—in fact, we can really cover only one. So, let's tackle the topic of visualizing how variant context annotation values are distributed, which can be helpful for understanding variant filtering methods, as we discussed in Chapter 5.

Visualizing Variant Context Annotation Data

You might recall that in Chapter 5 we described using an annotation (`callsets`) derived from the GiaB truth set in order to understand how the distributions of variant context annotations can inform us about the quality of our variant calls. We used a visual approach to making that assessment, which involved plotting variant context annotation values in a couple of ways (density plots and scatter plots). If that doesn't ring a bell or if you're feeling fuzzy on the details, please take a few minutes to read through that section again to refresh your memory. At the time, we focused on the concepts and outlined the procedure only in general terms, so here we're going to take the opportunity to show you how to apply key steps to reproduce the plots shown in Figure 5-8 through Figure 5-11.

Exporting Annotations of Interest with VariantsToTable

We start with the VCF of SNPs called from the Mother WGS sample that we've previously annotated with information for the GiaB truth set. For a tutorial showing how to perform the full procedure, including subsetting and annotation steps, see this GATK tutorial workspace (*https://oreil.ly/WGncb*).

The VCF file format is rather painful to work with directly, so for this exercise, we're going to make life easier on ourselves and export the information we care about from the VCF file into a tab-delimited table, to make it easier to parse in R. To that end, we run the GATK tool `VariantsToTable` on the annotated input VCF file, providing it with the list of annotations we're interested in. We use the `-F` argument for INFO (site-level) annotations and `-GF` for FORMAT (sample-level) annotations, where the *F* stands for field, and *GF* for genotype field, respectively:

```
In [40]  ! gatk VariantsToTable \
                -V {GERM_DATA}/vcfs/motherSNP.giab.vcf.gz \
                -F CHROM -F POS -F QUAL \
                -F BaseQRankSum -F MQRankSum -F ReadPosRankSum \
                -F DP -F FS -F MQ -F QD -F SOR \
                -F giab.callsets \
                -GF GQ \
                -O sandbox/motherSNP.giab.txt
```

The `VariantsToTable` command should run very quickly to produce the output file, *motherSNP.giab.txt*. This is a plain-text file, so we can view a snippet of it using `cat`:

```
In [41]  ! cat sandbox/motherSNP.giab.txt | head -n300
```

As you can see, the tool produced a table in which each line represents a variant record from the VCF, and each column represents an annotation that we specified in the export command. Wherever a requested annotation was not present (for example, homozygous sites do not have `RankSum` annotations, because that annotation can be calculated only for heterozygous sites), the value was replaced by `NA`. With this plain-text table in hand, we can easily load the full set of variant calls and their annotation values into an R DataFrame.

To load the table contents into an R DataFrame, we call the `readr` library and use its `read_delim` function to load the *motherSNP.giab.txt* table into the `motherSNP.giab` DataFrame object. Notice that the R command is preceded by the `%%R` symbol, which as we learned earlier instructs the notebook kernel that all of the code in this cell should be interpreted in R:

```
In [42]  %%R
         library(readr)
         motherSNP.giab <- read_delim("sandbox/motherSNP.giab.txt","\t",
                    escape_double = FALSE,
                    col_types = cols(giab.callsets = col_character()),
                    trim_ws = TRUE)
```

When the DataFrame is ready, you can manipulate it using your favorite R functions. And, as it happens, we have some handy plotting functions all lined up for you.

Loading R Script to Make Plotting Functions Available

We're going to take advantage of an existing R script provided by the GATK support team. The script, which is available in the book repository and in the bucket, defines three plotting functions that utilize a fantastic R library called `ggplot2` to visualize the distribution of variant annotation values.

To make these functions available in the notebook, we could simply copy the contents of the R script into a code cell and run it. However, because this is a script that we might want to run in multiple notebooks, and we don't want to have to maintain separate copies, let's use a smarter way to import the code. You're going to copy the R script to the notebook's local storage, and then use the source() function in R to load the R script code into the notebook:

```
In [43]   ! gsutil cp gs://genomics-in-the-cloud/v1/scripts/plotting.R .

          %R source("plotting.R")
```

This outputs about a page's worth of log, which we're not showing here. The log output is displayed on a red background, which is a tad alarming, but don't worry about it unless the next steps fail. If you do encounter issues, check whether your output is different from what is shown in the prerun copy of the notebook (which also contains solutions to the do-it-yourself exercises). If everything works as it should, you'll now have a few new R packages installed and loaded, and the plotting functions will be available.

Let's try that out, shall we? First up, the density plot.

Making Density Plots for QUAL by Using makeDensityPlot

The makeDensityPlot function takes a DataFrame and an annotation of interest to generate a density plot, which is basically a smoothed version of the histogram, representing the distribution of values for that annotation. Here's how we can use it to reproduce Figures 5-8 and 5-9. In each of the following cells, the first line creates the plot and then the second line calling its name displays it:

```
In [44]   %%R
          QUAL_density = makeDensityPlot(motherSNP.giab, "QUAL")
          QUAL_density
```

The QUAL distribution shown in Figure 12-17 has a very long tail on the right, so let's zoom in by restricting the x-axis to a reasonable maximum value by using the optional xmax argument, with the result presented in Figure 12-18:

```
In [45]   %%R
          QUAL_density_zoom = makeDensityPlot(motherSNP.giab, "QUAL", xmax=10000)
          QUAL_density_zoom
```

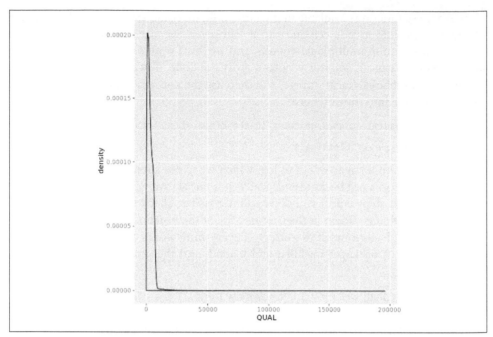

Figure 12-17. QUAL distribution.

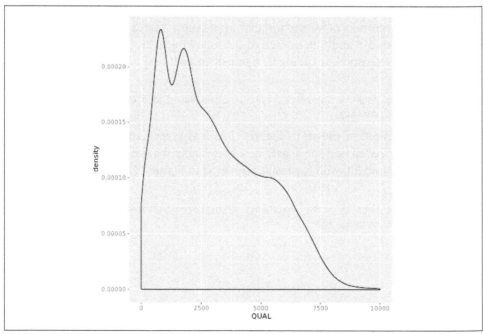

Figure 12-18. QUAL density plot.

We can also specify an annotation to use for organizing the data into subsets and have the function generate a separate density curve for each subset of data. Here, we use the `giab.callsets` annotation, which refers to the number of callsets in the GiaB truth set called the same variant. The higher the number, the more we can trust the variant call:

```
In [46]  %%R
         QUAL_density_split = makeDensityPlot(motherSNP.giab, "QUAL", xmax=10000,
         split="giab.callsets")
         QUAL_density_split
```

Figure 12-19 shows the result.

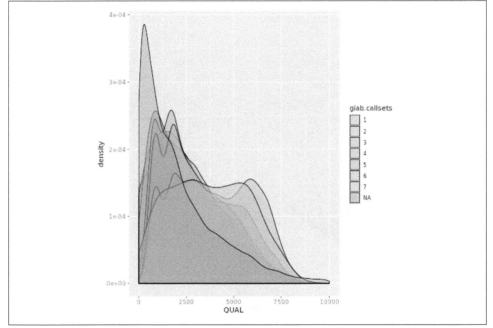

Figure 12-19. QUAL density plots by callsets from GiaB.

After you have that working, try generating the same kind of plots for other annotations. As an example, we used similar commands to generate Figure 5-10 for the `Qual ByDepth` (QD) annotation.

Now let's try making scatter plots. Everyone loves a good scatter plot, right?

Making a Scatter Plot of QUAL Versus DP

The `makeScatterPlot` function takes a DataFrame and two annotations of interest to generate a 2D scatter plot of the two annotations in which each data point is an individual variant call. Here's how we can use it to reproduce Figure 5-8 from Chapter 5, with Figure 12-20 displaying the result:

```
In [47]   %%R
          QUAL_DP_scatterplot = makeScatterPlot(motherSNP.giab, "QUAL", "DP")
          QUAL_DP_scatterplot
```

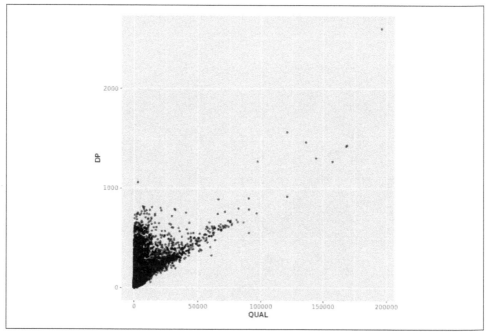

Figure 12-20. Scatter plot QUAL versus DP.

This function accepts the same `xmax` argument as `makeDensityPlot` for limiting the range of values on the x-axis, and a new `ymax` argument to limit values on the y-axis. Feel free to experiment with these arguments to zoom in on subsets of data.

You can also use the same `split` argument for splitting the data into subsets, with the effect of coloring the points based on the subset they belong to. Try doing that now based on what you learned in the previous exercise and then try applying the same principles to plot other annotations.

Finally, in our last plotting exercise, we're going to combine both the scatter and the density plotting.

Making a Scatter Plot Flanked by Marginal Density Plots

The `makeScatterPlotWithMarginalDensity` function takes a DataFrame and two annotations, combining the other two functions to generate a scatter plot flanked horizontally and vertically by the annotations' respective density plots. Here's how we can use it to reproduce Figure 5-11 from Chapter 5, with the result shown in Figure 12-21:

```
In [48]  %%R
         QUAL_DP_comboplot = makeScatterPlotWithMarginalDensity(motherSNP.giab,
         "QUAL",
         "DP", split="giab.callsets", xmax=10000, ymax=100, ptSize=0.5,
         ptAlpha=0.05)
         QUAL_DP_comboplot
```

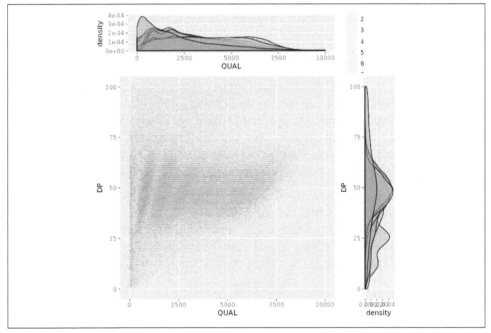

Figure 12-21. A scatter plot along with density plots.

As previously, we specify `giab.callsets` as the variable to use for splitting the variant data points into groups according to how much we trust them. We also set optional parameters (`xmax` and `ymax`) to limit the axes to display a subset of values, and we tweak the display of the data to optimize readability (`ptSize` and `ptAlpha`).

Go ahead and try applying this to other pairs of annotations. Note that some annotations can have negative values, so be aware that the plotting functions also accept `xmin` and `ymin` arguments to limit the range of negative values to display.

To be clear, there are a lot of other ways to manipulate and plot variant data from within a notebook. In fact, this particular method would not scale well for larger datasets, because it involves reading a potentially very large table directly into memory. We chose it for this tutorial because it has the advantage of being approachable for newcomers, and our primary goal was to give you a sense of the possibilities and familiarize you with the basic mechanics involved. However, for full-scale work, you'll probably want to use more robust methods. We recommend checking out Hail (*https://hail.is*), a Python-based, genetics-focused toolkit that is extraordinarily scalable and includes a suite of variant quality control functions, among other capabilities. Like some of the more recent GATK tools, Hail is capable of using Spark to parallelize analysis, and has been used to perform genome-wide analysis studies (GWAS) on massive datasets like the UK Biobank (*https://oreil.ly/mes1R*). The Terra Library has a few workspaces that feature Hail, including a set of tutorial notebooks (*https://oreil.ly/-h7Zj*) and a complete GWAS example (*https://oreil.ly/Q-LJD*).

Wrap-Up and Next Steps

In this chapter, you learned how to use Jupyter in Terra to interact with your data. You began by learning the basic mechanics of using notebooks on the cloud, setting up your computing environment, opening an example notebook, and running code cells. With those foundations in place, you worked through three types of interactive analysis: visualizing genomic data in an embedded IGV browser, running and troubleshooting GATK commands, and plotting variant context annotation data in R.

This was by no means an exhaustive catalog of what you can do in this environment; if anything, we barely scratched the surface of what is possible. However, you now have enough grounding in the technology and tooling to start adapting your own analyses to work within the Terra framework. In Chapter 13, we show you how to assemble your own workspaces from component elements: data, tools, and code from various origins.

Assembling Your Own Workspace in Terra

In Chapters 11 and 12, you learned how to use workflows and interactive notebooks in Terra using prebaked workspaces. Now, it's time for you to learn to bake your own so you can build your own analyses within the Terra framework. This is an area provides a lot of options and multiple valid approaches, so rather than attempting to provide a one-size-fits-all path, we're going to walk through three scenarios.

In the first scenario, we re-create the book tutorial workspace (*https://oreil.ly/n7oOr*) from its base components to demonstrate the key mechanisms involved in assembling a workspace from the ground up. In the second and third scenarios, we show you how to take advantage of existing workspaces to minimize the amount of work you have to do when starting a new project. In one case, we explain how to add data to an existing workspace that is already set up for a particular analysis, such as the official GATK Best Practices workspaces. In the other, we demonstrate how to build an analysis around data exported from the Terra Data Library. However, before we dive into those three scenarios, let's explore the data management strategy that we're applying in all three cases.

Managing Data Inside and Outside of Workspaces

One of the most important aspects of moving your work to the cloud is designing a data management strategy that will be sustainable for the long term, especially if you expect to work with large datasets that will serve as input for multiple projects. It's a complex enough topic that a full discussion would be beyond the scope of this book, and entire books cover that subject alone. However, it's worth taking some time to talk through a few key considerations that apply specifically in the context of Terra and shape how we decide where data should reside.

The Workspace Bucket as Data Repository

As you learned in Chapter 11, each Terra workspace is created with a dedicated GCS bucket. You can store any data that you want in the workspace bucket, and all notebooks as well as logs and outputs from workflows that you run in that workspace will be stored there. However, there's no obligation to store your *input* data in the workspace bucket. You can compute on data that resides anywhere in GCS as long as you can provide the relevant pointers to the files, assuming that you are able to grant access to that data to your Terra account (more on that in a minute).

This is important to note for a couple of reasons. First, if you're going to use the same data as input for multiple projects, you don't want to have to maintain copies of the data in each workspace because you'll get charged storage costs for each bucket. Instead, you can put the data in one location and point to that location from wherever you need to use it. Second, be aware that the workspace bucket lives and dies with your workspace: if you delete the workspace, the bucket and its contents will also be deleted. Related to this, when you clone a workspace, the only bucket content that the clone inherits from its parent is the notebooks directory. The clone will inherit a copy of the parent's data tables, retaining links to the data in the parent's bucket, but not the files themselves. This is called making a *shallow copy* of the workspace contents. If you then delete the parent workspace, the links in the clone's data tables will break, and you will no longer be able to run analyses on the affected data.

As a result, we generally recommend storing datasets in dedicated *master* workspaces that are kept separate from analysis workspaces, with more narrow permissions restricting the number of people authorized to modify or delete them. Alternatively, you could also store the data in buckets that you manage outside of Terra, as we did for the example data provided with this book. The advantage of storing the data outside of Terra is that you (or whoever owns the billing account for the bucket) have full administrative control over it. That gives you the freedom to do things like setting granular per-file permissions or making contents fully public, which are not currently possible in Terra workspace buckets for security reasons. It also doesn't hurt that you can choose meaningful names for the buckets you create yourself, whereas Terra assigns long, abstract names that are not very human-friendly. However, if you decide to follow that route, you'll need to enable Terra to access any private data that you manage yourself outside of Terra, as explained in the next section.

Accessing Private Data That You Manage Outside of Terra

So far, we've worked with data that's either in public buckets or in the workspace's own bucket, which is managed by Terra. However, you will eventually want to access data that resides in a private bucket that is not managed by Terra. At that point, you will encounter an unexpected complication: even if you own that private bucket, you will need to go through an additional authentication step that involves the GCP console and a *proxy group account*. Here's what you need to know and do to get past this hurdle.

Whenever you issue an instruction to do something in GCP through a Terra service, the Terra system doesn't actually use your individual account in its request to GCP. Instead, it uses a service account that Terra creates for you. In fact, you can have multiple service accounts created for you by Terra because you get one for each billing project to which you have access. All of your service accounts are collected into a *proxy group*, which Terra uses to manage your credentials to various resources. This has benefits for both security and convenience under the hood.

Most of the time, you don't need to know about this because any resources that you create or manage within Terra (like a workspace and its bucket, or a notebook) are automatically shared with your proxy group account. In addition, whenever someone shares something with you in Terra, the same thing happens: your proxy group account is automatically included in the fun, so you can seamlessly work with those resources from within Terra. However, when you start connecting to resources that are not managed by Terra, the Google system that controls access permissions doesn't automatically know that your proxy group account is allowed to act as your proxy. So, you need to identify your proxy group account in the Google permissions system, and specify what it should be allowed to do.

Fortunately, this is not too difficult if you know what to do, and we're about to walk you through the process for the most common case: accessing a GCS bucket that is not controlled by Terra. For other resources, the process would be essentially the same.

First, you need to find out your proxy group account identifier. There are several ways to do that. The simplest is to look it up in your Terra user profile (*https://app.terra.bio/#profile*), where it is displayed as shown in Figure 13-1.

Figure 13-1. The proxy group identifier displayed in the user profile.

With that in hand, you can head over to the GCS console and look up your *external bucket*; that is, the bucket you originally created for the exercises in Chapter 4. Go to the Bucket details page and find the Permissions panel, which lists all accounts authorized to access the bucket in some capacity, as shown in Figure 13-2.

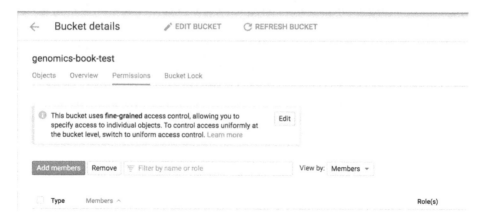

Figure 13-2. The bucket permissions panel showing accounts with access to the bucket.

Click the "Add members" button to open the relevant page and add your pet service account identifier in the "New members" field. In the "Select a role" drop-down menu, scroll if needed to select the Storage service in the left column and then further select the Storage Object Admin role in the right column, as shown in Figure 13-3. Click Save to confirm.

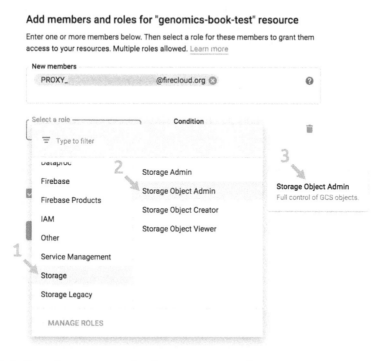

Figure 13-3. Granting access to a bucket to a new member.

After you've done this, you will be able to access external GCP resources such as buckets not managed by Terra from within your notebooks and workflows. For any buckets that you don't administer yourself, you'll need to ask an administrator to do this procedure on your behalf.

Accessing Data in the Terra Data Library

As you might recall from the context-setting discussion in Chapter 1, in which we originally introduced the platform, Terra includes a Data Library (*https://oreil.ly/VD6cJ*) that provides connections to datasets hosted in GCP by various organizations. Some of those datasets are simply provided in Terra workspaces, like the 1000 Genomes High Coverage dataset that you'll use in the second and third scenarios in this chapter. You'll have the opportunity to try out a few ways to take advantage of that type of data repository.

Other datasets are hosted in independent repositories that you access through dedicated interfaces called *data explorers*, which enable you to select subsets of the data based on metadata properties and then export them to a regular workspace. We won't cover those in detail but encourage you to explore the library and try retrieving data from the ENCODE (*https://oreil.ly/lBDFN*) or NeMo (*https://oreil.ly/qCWuS*) repositories, for example. Both of these are fully public and have very different interfaces, so they make for interesting navigation practice.

Exploring ENCODE and NeMo

For ENCODE, try selecting a few metadata properties to narrow the data selection. Then, look for the Export button to become active in the upper-right corner of the data explorer window. For NeMo, on the top menu bar, click the Data button, select a few properties in the menu on the left, and then go to the shopping cart icon in the upper right (interesting design choice, indeed). In the cart, click the Download options menu and select "Export to Terra." In both cases, you'll have the choice to create a new workspace or send the data to an existing workspace.

Most of the datasets in the Terra Data Library are restricted to authorized researchers because they contain protected health information, and the access modalities depend on the project host organization. Some, like TCGA (*https://oreil.ly/Dsnm4*), are mediated through dbGAP credentials, which you can link in your Terra user profile to gain automatic access if you are already authorized. Others are managed outside of Terra and require interaction with the project maintainers. If you are interested in learning more about any of the datasets in the library, click through to their project page, where you will typically find a description of access requirements.

Given the current heterogeneity of the data repositories included in the library, it is not yet trivial to use these resources effectively, especially if you intend to cross-analyze multiple datasets—which is unfortunate because that is one of the key attractions of moving to the cloud. This is an area of ongoing development, with many organizations actively collaborating to improve the level of interoperability and usability of these resources. We are already seeing early adopters successfully harness these resources in their research, and we are optimistic that upcoming improvements will make it easier for a wider range of investigators to do so as well.

For now, however, let's dial back our ambition by a couple of notches and focus on getting the basics in place. Back to work!

Re-Creating the Tutorial Workspace from Base Components

Our most immediate goal in this chapter is to equip you with the knowledge and skills to assemble a complete yet basic workspace on your own. We've chosen to do this by having you re-create the tutorial workspace from Chapters 11 and 12 given that you've been using it extensively in previous chapters and it contains all the basic elements of a complete workspace in a fairly simple form. As we guide you through the process, we're going to focus on the most commonly used mechanisms for pulling in the various components (data, code, etc.), and to avoid overwhelming you, we're intentionally not going to discuss every option that is available on the platform.

Are you ready to get started? Because we'll have you check the model workspace multiple times during the course of this section, we recommend that you open two separate browser windows (or tabs): one for the model workspace and another for the workspace that you are about to create.

Creating a New Workspace

In your second browser tab or window, navigate to the page that lists workspaces to which you have access, either by selecting Your Workspaces (*https://oreil.ly/3fL7H*) in the collapsible navigation menu or View Workspaces from the Terra landing page. You should see plus sign next to the Workspaces header in the top left of the page. Click that now to create a new blank workspace.

 By default, this page lists only private workspaces that either you have created yourself or that were shared with you under the "My workspaces" tab, excluding workspaces that are publicly accessible to everyone. You can view public workspaces by selecting one of the other workspace category tabs.

In the workspace creation dialog that pops up, give your new workspace a name and select a billing project, just as you've done when cloning workspaces in previous chapters. Because this is a brand-new workspace, the Name and Description fields will initially be blank. The name of your workspace must be unique within your billing project, but aside from that, you have a lot of freedom in naming, including using spaces in the name, as you can see in the example in Figure 13-4.

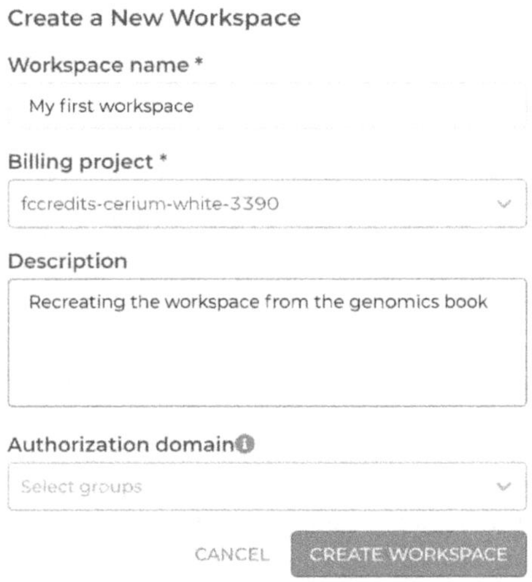

Create a New Workspace

Workspace name *

My first workspace

Billing project *

fccredits-cerium-white-3390

Description

Recreating the workspace from the genomics book

Authorization domain ⓘ

Select groups

CANCEL CREATE WORKSPACE

Figure 13-4. The Create a New Workspace dialog box.

The Description field is what will show up in the Dashboard of your new workspace. It's good practice to enter something informative for future reference, even if it's optional. You'll have the opportunity to edit it later, though, so don't sweat the details.

The Authorization domain field allows you to restrict access to the workspace and all its content to a specific group of users, which you must define separately. This is a useful feature for securing private information, but we're not going to demonstrate its use here, so leave this field blank. See the article in the Terra user guide (*https://oreil.ly/ikgz9*) if you're interested in learning more about this feature.

Click the Create button; you're then directed to the Dashboard page of your brand-new workspace. Feel free to click through the various tabs to inspect the contents, but as you'll quickly notice, there's not much to see there, with the exception of the description you provided (if you provided one) and a link to the dedicated GCS bucket that Terra created for your workspace. You might also notice the small widget showing the estimated cost per month of the workspace, which corresponds to the cost of storing the bucket contents. That cost estimate does not include the charges resulting from the analyses you might perform in the workspace. Right now, the cost estimate is zero because there's nothing in there. Yay?

Not that we want you to spend money, but this empty workspace is begging for some content, so let's figure out how to load it up. Because this is supposed to be a re-creation of the tutorial from Chapter 11, we're going to follow the same order of operations. Our first stop, therefore, will be the Workflows panel to set up the `Haplo typeCaller` workflow.

Adding the Workflow to the Methods Repository and Importing It into the Workspace

As you might recall, the workflow that you used in Chapter 11 was the same `Haploty peCaller` workflow that you worked with in previous chapters. For the purposes of the tutorial, we had already deposited the workflow in Terra's internal Methods Repository, so technically you could just go look it up and import it into your workspace. However, we want to empower you to bring in your own private workflows, so we're going to have you deposit your own copy of the `HaplotypeCaller` WDL and import that into your workspace.

In your blank workspace, navigate to the Workflows panel and click the large box labeled "Find a workflow." This opens a dialog that lists a selection of example workflows as well as two workflow repositories that contain additional workflows: Dockstore and the Broad Methods Repository. Click the latter to access the Methods Repository.

Expectations for Working with the Broad Methods Repository

As of this writing, the Broad Methods Repository has not yet been fully integrated into Terra and is technically considered a separate service. If that is still the case when you follow these instructions, you'll see a blue navigation theme and references to something called FireCloud, which was the precursor to Terra and lives on as a project powered by Terra and funded by the National Cancer Institute (*https://fire cloud.terra.bio*). If you see a green theme, you live far enough in the future for the integration to be complete. We hope the planet still has ice caps and the following instructions are still valid enough to guide you successfully through the next few steps. We tried to predict what might change and minimize our dependence on screenshots, but if you find yourself stuck, check out the changelog in the book repository for updated instructions.

You may need to sign into your Google account again and accept a set of terms and conditions. Find the Create New Method button (which might be renamed to Create New Workflow) to open the workflow creation page. Provide the information as shown in Figure 13-5, substituting your own namespace, which can be your username or another identifier that is likely to be unique to you. You can get the original WDL file rom the book GitHub repository (*https://oreil.ly/GGMoF*); either open the text file and copy the contents into the WDL text box, or use the "Load from file" link to upload it to the repository. By default, the optional Documentation field will be populated using the block of comment text at the top of the WDL file. The Synopsis box allows you to add a one-line summary of what the workflow does, whereas the Snapshot Comment box is intended to summarize what has changed if you upload new versions of the same workflow. You can leave the latter blank.

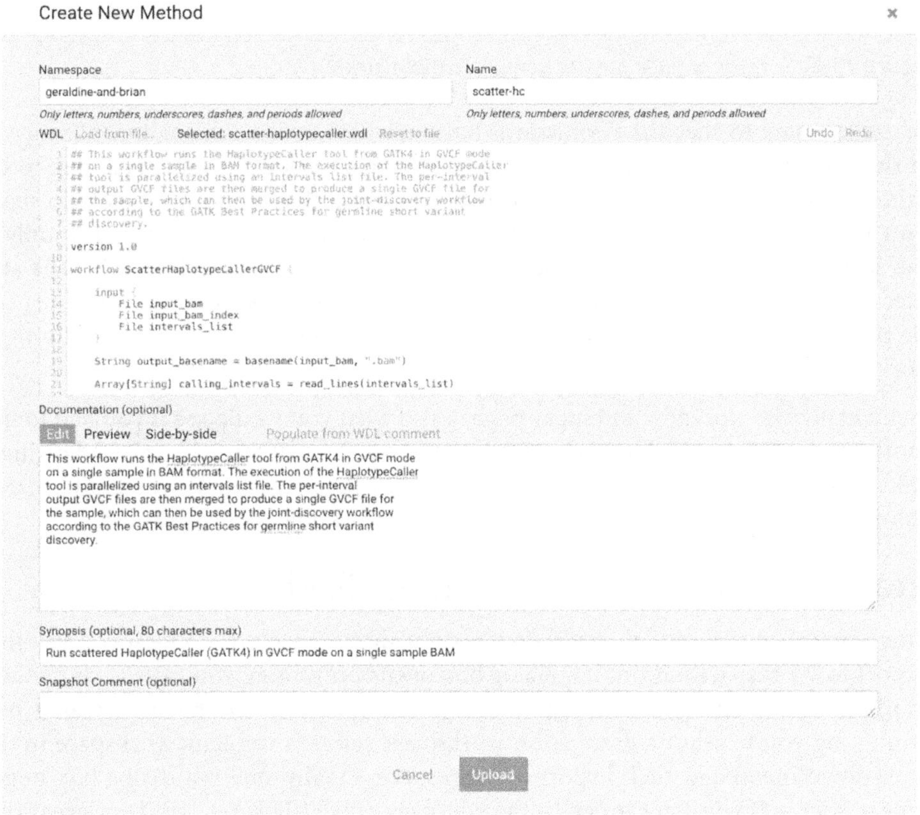

Figure 13-5. The Create New Method page in the Broad Methods Repository.

When you click the Upload button, the system validates the syntax of your WDL (using Womtool under the hood). Assuming that everything is fine, it will create a new workflow entry in the repository under the namespace you provided and present you with a summary page, as shown in Figure 13-6.

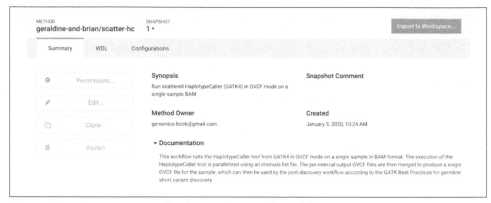

Figure 13-6. Summary page for the newly created workflow.

Take a moment to click the Permissions button, which opens a dialog that allows you to share your workflow with specific people, or make it fully public. Feel free to do either of those things as you prefer; just keep in mind that if and when you do share your workspace with others, they will be able to view and run your workflow only if you have done this. That being said, you will be able to come back to this page at a later date to do so if needed. We just bring this up now because we frequently see people run into errors after their collaborator forgot to share a workflow along with their workspace.

Incidentally, the workflow summary page is also what you would see if you had found another workflow in the Methods Repository, either by searching for it or following a link shared by a collaborator. As a result, the following instructions will apply regardless of whether you were the one who created the workflow in the first place.

Creating a Configuration Quickly with a JSON File

After you're done sharing the workflow (or not sharing; we're not judging), click the Export to Workspace button. If a dialog box opens prompting you to select a method configuration, click the Use Blank Configuration button. In the next dialog box prompting you to select a destination workspace, select your blank workspace in the drop-down menu and click Export to Workspace. Finally, one last dialog box might appear, asking if you want to "go to the edit page now." Click Yes—and we swear this is the last dialog box. You should now arrive back at the Workflows page in Terra, with your brand-new workflow configuration page in front of you.

Configuring the workflow is going to be fairly straightforward. First, because we're following the flow of Chapter 11, make sure to select the "Run workflow with inputs defined by file paths" option (located under the workflow documentation summary), which allows you to set up the workflow configuration with file paths rather than using the data tables. Then, turn your attention to the Inputs page, which should be lit up with little orange exclamation marks indicating that some inputs are missing. In fact, all of them are missing because you used a blank configuration and the workflow itself does not specify any default values.

 The orange exclamation marks will also show up if you have an input that is badly formatted or of the wrong variable type, like a number instead of a file. Try hovering over one of them to see the error message; all such symbols in Terra generally display me information when you hover over them and can sometimes hold the answer to your problems.

So you need to plug in the appropriate file paths and parameter values on the Inputs page. You could look up each input value individually in the original workspace and type them manually, but there's a less tedious way to do it. Can you guess what it is? That's right, make Jason do all the work. Er, we mean the JSON *inputs* file, of course. (Yes, that's a lame joke, but if it helps you remember that you can upload a JSON file of inputs, it will have been worth the shame we're feeling right now.)

Because we previously ran this same workflow directly through Cromwell in Chapter 10, we have a JSON file (*https://oreil.ly/I5Yzl*) that specifies all the necessary inputs here in the book bucket. You just need to retrieve a local copy of the file and then use the "Drag or click to upload json" option on the Inputs page (right side next to the Search Inputs box) to add the input definitions it contains into your configuration. This should populate all of the fields on the page. Click the Save button and confirm that there are no longer any orange exclamation marks.

At this point, your workflow should be fully configured and ready to run. Feel free to test it by clicking the "Run analysis" button and following the rest of the procedure for monitoring execution as you did previously in Chapter 11.

With that, you have successfully imported a WDL workflow into Terra by way of the Broad Methods Repository, and configured it to run on a single sample using the direct file paths configuration option. That's great because it means that you're now able to take any WDL workflow you find out in the world and test it in Terra, assuming that you have the correct test data and an example JSON file available.

However, what happens when you've successfully tested your workflow and you want to launch it on multiple samples at once? As you learned in Chapter 11, that's where the data table comes in. You just reproduced the first exercise in that chapter, which demonstrated that you can run a workflow using paths to the files entered directly in the inputs configuration form. However, to launch the workflow on multiple samples, you need to set up a data table that lists the samples and the corresponding file paths on the workspace's Data page. Then, you can configure the workflow to run on rows in the data table as you did in the second exercise in that chapter. In the next section, we show you how to set up your data table.

Adding the Data Table

Remember that our tutorial workspace used an example dataset that resides in a public bucket in GCS, which we manage outside of Terra. The *sample* table on the Data page listed each sample, identified by a name that is unique within that table, along with the file paths to the corresponding BAM file and its associated index file, each in a column of its own.

So the question is, how do you re-create the *sample* data table in our blank workspace? What does Terra expect from you? In a nutshell, you need to create a *load file* in a plain-text format with tab-separated values. A Terra documentation article (*https://oreil.ly/lONJL*) goes into more detail, but basically, it's a spreadsheet that you save in TSV format. It's reasonably straightforward except for a small twist that often trips people up on their first attempt: the file needs to have a header containing column names, and the first column must be the unique identifier for each row. Here's a pro tip: rather than trying to create a load file from scratch, you can download an existing one from any workspace that has data tables and use that as a starting point. Here we're going to be extra lazy and download the table TSV file from the original tutorial workspace, look at it and discuss a few key points, and then upload it as is to our blank workspace.

In the browser tab or window where you opened the original tutorial workspace, navigate to the Data page and click the blue Download All Rows button. This should trigger the download in your browser, and because it's a very small text file, the transfer should complete immediately. Find the downloaded file and open it in your preferred spreadsheet editor; we're using Google Sheets.

 You can also use a plain-text editor for viewing the file contents, but we recommend using a spreadsheet editor whenever you modify TSV load tables because it reduces the risk of messing up the tab-delimited format.

As you can see in Figure 13-7, this file contains the raw content from the sample data table in the original workspace, so instead of seeing just *father.bam* with a hyperlink, for example, we have the full path to where the BAM file resides in the storage bucket. In addition, this table has a header row that contains the names of the columns as displayed in the workspace, with one exception: the name of the first column is preceded by *entity:* in the file, which is not the case in the workspace. That little detail marks the most important takeaway of this whole section: the column that contains the unique identifier for each row must be the first listed in the load file, and its name must end with the _id suffix. When you upload the load file into Terra, the system will derive the name of the data table itself from the name of that column, by chopping off both the *entity*: prefix and _id suffix. If you don't include a column named following that pattern, Terra will reject your table. To be clear, the name that you give to the load file has no effect on the name of the table because it will be created in Terra.

	A	B	C	D	E	F
1	entity:book_sample_id	input_bam	input_bam_index			
2	mother	gs://genomics-in-t|	gs://genomics-in-the-cloud/v1/data/germline/bams/mother.bai			
3	father	gs://genomics-in-t|	gs://genomics-in-the-cloud/v1/data/germline/bams/father.bai			
4	son	gs://genomics-in-t|	gs://genomics-in-the-cloud/v1/data/germline/bams/son.bai			
5						

Figure 13-7. A sample data table from the tutorial workspace, viewed in Google Sheets.

Let's try this out in practice. In your blank workspace, navigate to the Data page and click the "+" icon adjacent to TABLES on the DATA menu to bring up the TSV load file import dialog box, as shown in Figure 13-8. You can drag and drop the file or click the link to open a file browser; use either method as you prefer to upload the TSV file that we were just looking at.

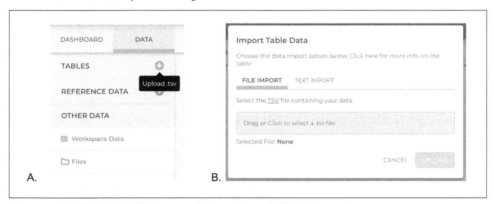

Figure 13-8. TSV load file import A) button, and B) dialog.

If the upload completes successfully, you should now see the sample table listed in the data menu. Click its name to view its content, and compare it to the original: if you didn't make any changes, it should be identical.

Feel free to experiment with this sample load file to test the behavior of the TSV import function. For example, try modifying the name of the first column and see what happens. You can also try adding more columns with different kinds of content, like plain text, numbers, file paths, and lists of elements (use array formatting, with square brackets and commas to separate elements). And try modifying rows and creating more rows, either by adding rows to the same file or by making a new file with different rows. Just remember that if you're editing the load file in a spreadsheet editor, as we recommend, you'll need to make sure to save the file in TSV format.

 Many spreadsheet editors (including Microsoft Excel) call this format *Tab-delimited Text* and will save the file with a *.txt* extension instead of *.tsv*, which is absolutely fine; Terra will not care about the extension as long as the contents are formatted correctly.

If your data page ends up getting a bit messy as a result of this experimentation, don't worry: you can delete rows or even entire tables; just select the relevant checkboxes (use the checkbox in the header to select an entire table) and click the trashcan icon that appears in the lower right of the window. The one limitation is that it's not currently possible to delete columns from tables, so if you want to get rid of unwanted columns, you'll need to delete the entire table and redo the upload procedure. For that reason, it can be a good idea to store versioned copies of what you consider the "good" states of your tables when you're working on an actual project, especially if you're planning to add data over time. There isn't yet any built-in functionality to do that in Terra, so it's a manual process of downloading the TSV and storing it somewhere (for example, in the workspace bucket), and in another location outside the workspace if you want to be extra cautious.

After you're done having fun with the data tables, move on to the resource data in the Workspace Data table.

Filling in the Workspace Resource Data Table

As you might recall, the purpose of this table is to hold variables that we might want to use in multiple workflow configurations, like the genome reference sequence file, for example, or the GATK Docker container. This allows you to configure it only once, and just point to the variable in any configuration that needs it. Not only do you not need to ever look up the file path again, but if you decide to update the version or location of the file, you need to do it in only one place. That being said, you might not always want updates to be propagated to every use of a given resource file or parameter setting. When you go to set up your own analysis, be sure to think carefully about how you want to manage common resources and "default" parameters.

In practice, this data table has a very simple structure because it's all just key:value pairs, and it comes with an easy form-like interface for adding and modifying elements. You can simply use that interface to copy over the contents of the table from the original workspace, if you don't mind the manual process. Alternatively, you can download the contents of the original in TSV format by clicking the Download All Rows link, and upload that without modification in your workspace using the Upload

TSV link. You can even drag the file from your local filesystem into the table area if you prefer. Feel free to experiment and choose the method that suits you best. With that, you're done setting up the data, so it's time to get back to the workflow.

Creating a Workflow Configuration That Uses the Data Tables

In your not-so-blank-anymore workspace, navigate back to the Workflows page and look at the configuration listed there, but don't open it. Click the circle with three vertically stacked dots, which you might by now recognize as the symbol that Terra uses to provide action options for a particular asset, be it a workspace, workflow, or notebook. On the action menu that opens, select "Duplicate workflow" and then give the copy a name that will indicate that it is going to use the data table, as we did in the tutorial workspace.

 It's currently not possible to simply rename a workflow configuration in place. If you want to give the "file paths" version of your workflow configuration a name that is equally explicit as we did in the tutorial workspace, you'll need to create another duplicate with the name you want and then delete the original.

Open the copy of the configuration that you want to modify to use the data table, and this time, switch the configuration option to "Process multiple workflows." On the drop-down menu, select the *sample* table and choose whether to select a subset of the data or go with the default behavior, which is to run the workflow on all items in the table. Now comes the more tedious part of this exercise, which is to connect the input assignments on the Inputs page to specific columns in the data table or variables in the workspace resource data table. Unfortunately, this time you don't have a prefilled JSON available as a shortcut, but on the bright side, the Inputs page has an assistive autocomplete feature that speeds up the process considerably. For each variable that needs to be connected (excluding a couple of runtime parameters that we simply leave hardcoded, which you can look up in the original workspace), start typing either **workspace** or **this** in the relevant text box. This brings up a contextual menu listing all options from the relevant table: this points to whatever table you selected on the configuration's drop-down menu, and workspace always points to the workspace resource data, which lists the reference genome sequence, Docker images, and so on.

Another way to use the assistive autocomplete feature on the Inputs page is to start typing part of the input variable name in the corresponding input text box. If the relevant data table column or workspace resource variable was set up with the same name as the input variable, as is the case in our tutorial workspace, this will typically pull up a much shorter list of matching values. For example, typing **ref** in one of the input fields would bring up only the workspace reference sequence, its index file, and it dictionary file. This approach can be blissfully faster when you're dealing with a large number of variables, but it is not guaranteed to work for every configuration because it relies on the names matching well enough, which will not always be the case. Some people just want to watch the world burn.

Go ahead and fill out the Inputs page, referring to the original workspace if you become stuck at any point. When you're done, make sure to click the Save button and confirm that there are no more exclamation marks left, as occurred previously.

What do you think, is this workflow ready to launch or what? What. Meaning, you do have one more configuration task left, which was not applicable in the "file paths" round. Pop over to the Outputs page, which requires you to indicate what you want to do with the workflow outputs. To be clear, this won't change where the files are written; that is determined for you by the built-in Cromwell server, as you learned in Chapter 11. What this does is determine under what *column name* the outputs will be added to the data table. You can click the "Use defaults" link to simply use the output variable names, or you can specify something different—either a column that already exists, or a new name that the system will use to create a new column. In our workspace, we chose to use the defaults.

Considerations for Naming Outputs

Terra will happily overwrite any existing element in the table; the underlying data from any previous runs will still exist in its original location, but the link in the table will be updated to point to the latest result. To compare results from different configurations, you'll probably want to give your outputs names that indicate which is which. Similarly, watch out for variable names that are overly generic; for example, if you have multiple workflows that produce qualitatively different outputs which are nevertheless all called output_vcf, you're going to have a bad time.

Alright, now you're done. Go ahead and launch the workflow to try it out and then sit back and delight in the knowledge that you've just leveled up in a big way. Figuring out how to use the data table effectively is generally considered one of the most challenging aspects of using Terra, and here you are. You're not an expert yet—that will come when you master the art of using multiple linked data tables, like participants and sample sets—but you're most of the way there. We cover that in the next scenario, in which you'll import data from the Terra Data Library into a GATK Best Practices workspace.

Before we get to that, however, we still need to tackle the Jupyter Notebook portion of the tutorial workspace, which we used in Chapter 12. But don't worry, it's going to be mercifully brief—if you take the easy road, that is.

Adding the Notebook and Checking the Runtime Environment

In your increasingly well-equipped workspace, navigate to the Notebooks page, and take a guess at what should be your next step. Technically, you have a choice here. You could create a blank new notebook and re-create the original by typing in all the cells. To do that, click the box labeled Create New Notebook, and then, in the dialog that opens, give it a name and select Python 3 as the language. Then, you can work through Chapter 12 a second time, typing the commands yourself instead of just running the cells in the prebaked version as you did last time. Alternatively, you can take the easy road and simply upload a copy of the original notebook, which like everything else is available in the GitHub repository. Specifically, the cleared copy and the previously run copy of the notebook are here (*https://oreil.ly/jbM8T*). Retrieve a local copy of the file and use the "Drag or Click to Add an ipynb File" option on the Notebooks page to upload it to your workspace.

Whichever road you choose to take, you don't need to customize the runtime environment again if you created your workspace under the same billing project as the one you used to work through Chapter 12, because your runtime is the same across all workspaces within a billing project. Convenient, isn't it? Perhaps, but it's also a bit of a cop-out; if you were truly building this workspace from scratch, you would need to go back and follow the instructions in Chapter 12 again to customize the runtime environment. So, enjoy the opportunity to be lazy, but keep in mind that in a different context, you might still need to care about the runtime environment.

When you're done playing with the notebook(s) in your workspace, take a step back and check that you have successfully re-created all functional aspects of the tutorial workspace. Does it all work? Well done, you! You now know all the basic mechanisms involved in assembling your own workspace from individual components.

Importing Notebooks from Outside of Terra

Speaking of different context, we'd be remiss if we didn't address the thus-far-unspoken question: what does it take to adapt an existing notebook from a source outside of Terra? Here we're taking advantage of importing a notebook that is already set up to use the workspace bucket and all that jazz; but what if we were just grabbing a notebook out of someone's Binder of Python tutorials? It's a great question, and one that you're very likely to face in practice at some point. Again, there is no single one-size-fits-all answer, but you can use the exercises in Chapter 12 to guide you.

Review the key steps that you had to go through in the beginning that involved cloud- or Terra-specific components, like accessing data in GCS using gsutil and setting up an environment variable to easily access the workspace bucket. The steps that have to do with accessing data and saving results are typically the parts of a new-to-you notebook that you need to tweak to work smoothly in the Terra context. For the rest, the work of tweaking the runtime environment to make sure you have the right versions

---

of software packages and libraries is very similar to what you would need to do on a local system, except in Terra you need to decide between making those tweaks from within the notebook, using a startup script, or using a custom container image (which we do not cover in detail in this book).

Documenting Your Workspace and Sharing It

The one thing you haven't done yet to match (or outdo) the original tutorial workspace is to fill out the workspace description in the Dashboard, beyond whatever short placeholder you put in there when you created the workspace. If you have a little bit of steam left in you, we encourage you to do that now while your memory is fresh. All it takes is clicking the editing symbol that looks like a pencil on the Dashboard page, which opens up a Markdown editor that includes a graphical toolbar and a split-screen preview mode.

We don't usually like to say anything is self-explanatory, because that kind of qualifier just makes you feel even worse if you end up struggling with the thing in question, but this one is as close as it gets to deserving that label. Take it for a spin, click all the buttons, and see what happens. Messing around with the workspace description is one of the only things you can do on the cloud that is absolutely free, and in a private clone, it's completely harmless, so enjoy it. This is also a great opportunity to write up some notes about what you learned in the process of working through the Terra chapters, perhaps even some warnings to your future self about things you struggled with and need to watch out for.

Finally, consider sharing your workspace with a friend or colleague. To do so, click the same circle-with-dots symbol you used to clone the tutorial workspace in Chapter 11. This opens a small dialog box in which you can add their email address and set the privileges you want to give them. Click Add User and then be sure to click Save in the lower-right corner. If your intended recipient doesn't yet have a Terra account, they will receive an email inviting them to join, along with a link to your workspace. They don't need to have their own billing project if they just want to look at the workspace contents.

Starting from a GATK Best Practices Workspace

As you just experienced, setting up a new workspace from its base components takes a nontrivial amount of effort. Not that it's necessarily *difficult*—you probably perform far more complex operations on a daily basis as part of your work—but there are a lot of little steps to follow, so until you've done it a bunch of times, you'll probably need to refer to these instructions or to your own notes throughout the process.

The good news is that there are various opportunities to take shortcuts. For example, if you simply want to run the official GATK Best Practices workflows as provided in the featured workspaces by the GATK support team, you can skip an entire part of this process outright by starting from the relevant workspace. Those workspaces already contain example data and resource data as well as the workflows themselves, fully configured and ready to run. All you really need to do is clone the workspace of interest and add the data that you want to run the workflow(s) on to the data tables.

In this section, we run through that scenario so you can get a sense of what that entails in practice, the potential complications, and your options for customizing the analysis. The lessons from this scenario will generally apply beyond the GATK workspaces, to any case for which you have access to a workspace that is already populated with the basic components of an analysis. We think of this as the "just add water" scenario, which is an attractive model for tool developers to enable researchers to use their tools appropriately and with minimal effort. It's also a promising model for boosting the computational reproducibility of published papers given that such a workspace constitutes the ultimate methods supplement. We discuss the mechanics and implications of this in more detail in Chapter 14. For now, let's get to work on that GATK Best Practices workspace.

Cloning a GATK Best Practices Workspace

We're going to use the Germline-SNPs-Indels-GATK4-hg38 workspace (*https:// oreil.ly/2I8RE*), which showcases the GATK Best Practices for short variant discovery implemented as three separate workflows. The first workflow, named 1_Process ing_for_Variant_Discovery, takes in raw sequencing data and outputs an analysis BAM file for a single sample. The second, 2_HaplotypeCaller_GVCF, takes that BAM file and runs variant calling to produce a single-sample GVCF file. Finally, the third, 3_Joint_Discovery, takes multiple GVCF files and applies joint calling to produce a multisample callset for the cohort of interest.

Navigate to that workspace now and clone it as you have done previously with other workspaces. When you're in your clone, have a quick look around to become acquainted with its contents. You'll find a detailed description in the Dashboard, three data tables and a set of predefined resources on the Data page, and the three fully configured workflows mentioned earlier on the Workflows page.

Examining GATK Workspace Data Tables to Understand How the Data Is Structured

Let's take a closer look at the three data tables on the Data page. One table is named *participant*. It contains a header line showing the name of the single column, *participant_id*, and a single row, populated by our beloved NA12878, or *mother* as we call her in this book:

participant_id
NA12878

This identifies her as a study *participant*; in other words, the actual person from whom biological samples were originally collected to produce the data that we will ultimately analyze.

The `participant_id` attribute is the unique identifier for the participant and is used to index the table. If you were to add attributes to the table (for example, some phenotype information like the participant's height, weight, and health status), they would be added as columns to the right of the identifier.

The second table is called *sample* and contains multiple columns as well as two sample rows below the header. This table is more complicated than the previous one largely because in addition to the minimum inputs required to run the workflow, it contains outputs of previous runs of the workflow. Here is a minimal version of the table without those outputs:

sample_id	flowcell_unmapped_bams_list	participant _id
NA12878	NA12878.ubams.list	NA12878
NA12878_small	NA12878_24RG_small.txt	NA12878

In this table, you see that the leftmost column is *sample_id*, the unique identifier of each sample, which is used as index value for that table. Jumping briefly to the other end of the table, the rightmost column is *participant_id*. Can you guess what's happening here? That's right, this is how we indicate which sample belongs to which participant. The *sample* table is linking back to the *participant* table. This might seem boring, but it's actually quite important, because by establishing this relationship, we make it possible for the system to follow those references. Using a similar relationship between two tables, we'll be able to do things like configure a workflow to run on a set of samples to analyze them jointly instead of launching the workflow individually on each sample. If you find that confusing, don't worry about it now; we discuss it in more detail further down. For now, just remember that these tables are connected to each other.

The middle column, *flowcell_unmapped_bams_list*, points to sequencing data that has been generated from a biological sample. Specifically, this data is provided in the form of a list of unmapped BAM files that contain sequencing data from individual read groups. As you might recall from Chapter 2, the read groups are subsets of sequence data generated from the same sample in different lanes of the flowcell. The data processing portion of the workflow expects these subsets of data to be provided in separate BAM files so that it can process them separately for the first few steps and then merge the subsets in a single file containing all the data for that sample.

For a given sample, the list of unmapped BAMs will be the primary input to the *Processing_for_Variant_Discovery* workflow. You can verify this by taking a look at the workflow configuration, where the input variable `flowcell_unmapped_bams` is set to `this.flowcell_unmapped_bams_list`. Recall from our Chapter 11 forays into launching workflows on data tables that syntax reads out as "for each row in the table, give the value in the *flowcell_unmapped_bams_list* column to this variable." The main output of that workflow will be added under the *analysis_ready_bam* column, which you can see in the full table in the workspace. That column in turn will serve as the input for the *HaplotypeCaller_GVCF* workflow, which will then output its contents into the *gvcf* column.

The two sample rows correspond to two versions of the original whole genome dataset derived from participant *NA12878*. In the first row, the NA12878 sample is the full-scale dataset, whereas the *NA12878_small* sample in the second row is a downsampled version, meaning that it contains only a subset of the original data. The purpose of the downsampled version is to run tests more quickly and cheaply.

Finally, the third table is called *sample_set*, which might or might not sound familiar, depending on how much you paid attention in Chapter 11. Do you remember what happened when you ran a workflow on rows in the data table? The system automatically created a row in a new *sample_set* table, listing the samples on which you had launched the workflow. In this workspace, you can see a sample set called *one_sample*, with a link labeled *1 entity* under the *samples* column, as shown here simplified:

sample_set_id	samples
one_sample	1 entity

If you click that, a window opens and displays a list referencing the NA12878 sample. This is another example of a connection between tables, and we're going to take advantage of this one specifically in a very short while.

The system doesn't actually show how this sample set was created, but from what we know about the workflows, we can deduce that running either the first or the second workflow in this workspace on the NA12878 sample would replicate the creation of the sample set. Mind you, it's also possible to create sample sets manually; we show how that works in a little bit.

What's interesting here is that the *one_sample* row in this sample set has output files associated with it, which was not the case when we ran workflows in Chapter 11. Can you guess what that might mean? We'll give you a hint: the *output_vcf* column contains a VCF of variant calls. Got it? Yes, this is the result of running the *Joint_Discovery* workflow on the sample set: it took as input the list of samples referenced in the *samples* column of the sample set, and attached its VCF output to the sample set. This is a somewhat artificial example given that the sample set contains only one sample, so we'll run this again on a more realistic sample set later in this section in order to give you a more meaningful experience of this logic.

Now that we've dissected the data tables in this way, we can summarize the structure of the dataset and the relationships between its component entities. The result, illustrated in Figure 13-9, is what we call the *data model*: Participant NA12878 has two associated Samples, NA12878 and NA12878_small, and one Sample Set references the NA12878 Sample.

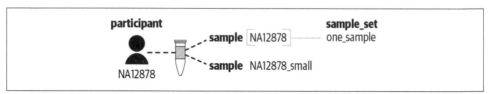

Figure 13-9. The data model—the structure of the example dataset.

Note that having a sample identifier that is identical to a participant identifier normally happens only when there is just one sample per participant. In a realistic research context in which a participant has multiple samples associated with it, you would expect to have different identifiers. You would also expect the two sample datasets to have originated from different biological samples or to have been generated using different assays, whereas we know that in this case they originated from the same one. Again, this is due to the somewhat artificial nature of the example data. In the next step of this scenario, we're going to look at some data that follows more realistic expectations.

Getting to Know the 1000 Genomes High Coverage Dataset

As we discussed earlier, the Data Library provides access to a collection of datasets hosted by various organizations in several kinds of data repositories. Accordingly, the protocol for retrieving data varies depending on the repository. In this exercise, we retrieve data from a repository that hosts a whole-genome dataset of the 2,504 samples from Phase 3 of the 1000 Genomes Project, which were recently resequenced by the New York Genome Center as part of the AnVIL project (*https://oreil.ly/Z2vfO*). This particular repository simply consists of a public workspace (*https://oreil.ly/195yq*) containing only data tables, which serves as a simple but effective form of data repository at this scale.

Go to the 1000 Genomes data workspace now, either by clicking the direct link above or by browsing the Data Library. If you choose the latter route, be aware that there is another 1000 Genomes Project data repository, as shown in Figure 13-10, but it's quite different: that one contains the Low Coverage sequence data for the full 3,500 study participants, whereas we want the new High Coverage sequence data from the 2,504 participants who were included in Phase 3 of the project.

Figure 13-10. The Terra Data Library contains two repositories of data from the 1000 Genomes Project.

When you're in the workspace, head over to the Data page to check out the data tables. You'll see the same three tables as we described in the GATK workspace: *sample*, *participant*, and *sample_set*. As expected, the *participant* table lists the 2,504 participants in the dataset. Similarly, the *sample* table lists the paths to the location of the corresponding sequence data files, which are provided in CRAM format as well as a GVCF file produced by running a HaplotypeCaller pipeline on the sequence data. In addition, both contain a lot of metadata fields that were not present in the GATK workspace, including information about the data generation stage (type of instrumentation, library preparation protocol, etc.) and the population of origin of the study participant. The presence of all this metadata is a big indicator that this is a more realistic dataset, compared to the toy example data in the GATK workspace. Meanwhile, the *sample_set* table contains a list of all 2,504 samples associated with the sample set name *1000G-high-coverage-2019-all*.

We can summarize the data model for this dataset as follows: each Participant has a single associated Sample, and there is one Sample Set that references all the Samples, as shown in Figure 13-11.

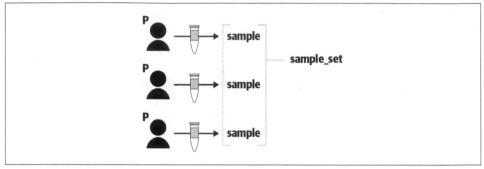

Figure 13-11. The data model for the 1000 Genomes High Coverage dataset.

At this point, we have a confession to make: all this time we've spent looking at the data tables of the two workspaces wasn't mere data tourism. It all had a very specific purpose, to answer the unspoken question that you might or might not have seen coming: will their data models be compatible?

And the good news is yes, they seem to be compatible, meaning that you can combine their data tables without causing a conflict, and you should be able to run a *federated data analysis* across samples from both datasets. Admittedly the degree of federation here is not really meaningful because only one person's genomic information is in the GATK workspace, but the underlying principle will apply equally to other datasets that you might want to bring together into a single workspace.

Now that we're satisfied that we should be able to use the data in our workspace, let's figure out how to actually perform the transfer of information.

Copying Data Tables from the 1000 Genomes Workspace

There are two main ways to do this: a point-and-click approach, and a load file–based approach. Let's begin with the point-and-click approach because it abstracts away some of the complexity involved in using load files.

On the Data page of the 1000 Genomes High Coverage workspace, click the *sample* table and choose a few samples by selecting their corresponding checkboxes on the left. Locate and click the three-dots symbol above the table to open the action menu (as you did in Chapter 11 to configure data inputs for the workflow) and select "Export to Workspace" as shown in Figure 13-12.

Figure 13-12. The Copy Data to Workspace dialog box.

This should open a dialog box with the list of samples and a workspace selection menu. Select your clone of the GATK workspace and then click Copy. Keep in mind that despite what the buttons seem to imply, we're not going to copy over any files; we just want to copy over the metadata in the data tables, which include the paths to the locations of the files in GCS. So, rest assured you're not suddenly going to be on the hook for a big data storage bill.

You should see a confirmation message that asks you whether you want to stay where you are (in the 1000 Genomes workspace) or go to where you copied the data (the GATK workspace). Select the latter option so we can go look at what the transfer produced. If everything worked as expected, you should see the samples you selected listed in the *sample* table. However, you will not see the corresponding participants listed in the *participant* table, because the system does not automatically copy over the data from other tables to which the selected data refers. The only exceptions to this are sets, like the sample sets we've been working with. If you select a set and copy that over, the system will also copy over the contents of the set. If you also want to copy over the participants, you need to do it as a separate step, either by selecting the rows of interest using the checkboxes or by defining a participant set using a load file.

If you play around with the "copy data to workspace" functionality a little bit, you will quickly realize that this approach is fairly limiting for dealing with large datasets. Why? Because when you try to select all rows in a table, even using the checkbox in the upper-left corner of the table, the system really selects only the items that are displayed on the page. Because the maximum allowed for that is one hundred items, the set approach is your only option for copying over more than one hundred samples at a time using the point-and-click approach.

So now let's try the load file–based approach, which has some advantages if you are working with large datasets or need to tweak something before copying over the data.

Using TSV Load Files to Import Data from the 1000 Genomes Workspace

Technically, you've already done this—that's how we had you copy over the data table from the original book tutorial workspace to re-create your own earlier in this chapter. You selected the table, clicked the Download All Rows button, retrieved the downloaded file, and then uploaded it to the new workspace. Boom. All our samples are belong to you.

However, this time there's a twist: several data tables are involved, and some of them have references to the others. As a result, there's going to be an order of precedence: you must start with the table that doesn't have any references to the others because Terra can't handle references to things that it hasn't seen defined yet. For example, you can't upload a sample set before having uploaded the samples to which the sample set refers. This is an annoying limitation, and you could imagine having the system simply create a stand-in for any reference to something that it doesn't recognize. Yet we need to work with the system in its current state, so the bottom line is this: order matters; deal with it.

In practice, for this exercise, you need to download the TSVs for the three data tables from the 1000 Genomes workspace, as previously described (in any order), and then upload them in the correct order to your GATK workspace clone. What is the correct order, you ask? Here's a hint: look at the visual representation of the data model in Figure 13-11 and use the direction of the arrows to inform your decision. What do you think? That's right, you first upload the participants, then the samples, and then the sample set. Go ahead and do that now.

Oh wait, did you encounter something you didn't expect while doing that? Indeed: there are *two* load files for the sample set: one that defines the table itself and names the sample sets that it contains, identified as *entity:sample_set_id*, and one that lists the members that each sample set contains, identified as *membership:sample_set_id*. Have a look at their contents to get a sense of how they relate to each other. And relate they do; because the membership TSV refers to the name of the sample set defined in the entity TSV, you'll need to upload the entity list first, then the membership list.

 Dealing with load files and the precedence rules is one of the more awkward, and frankly tedious, parts of setting up data in Terra. We expect that future developments will address this by providing some kind of wizard-style functionality to smooth out the sharp edges.

On the bright side, after you're done with that, you should see all 2,504 samples from the 1000 Genomes dataset listed in your GATK workspace as well as their corresponding participants. However, there's a problem. Can you see it? Take a good look at the columns in the sample table. Most of the columns were different between the two original tables, of course, because the GATK workspace version mainly had output files from the workflows that are showcased in the workspace, and the 1000 Genomes workspace version mainly had metadata about the provenance of the data. However, they were supposed to have one thing in common: GVCF files. Can you find them? Yes, they do all have GVCF files—but they're in different columns.

In the original GATK workspace, the name of the column containing the GVCF files was *gvcf* in all lowercase, whereas in the 1000 Genomes workspace it was *gVCF* in mixed case. Incidentally, the columns containing their respective index files are also named with subtle differences, *gvcf_index* versus *gVCF_TBI*, where TBI refers to the extension of the index format generated by a utility called Tabix.

Well, shucks. We were going to surprise you by saying, "Look, now you can run a joint calling analysis of your NA12878 sample combined with all of the 1000 Genomes Phase 3 data. Isn't that cool?" But it's not going to work if the GVCF files are in different columns (sad face Emoji).

Chin up; this is fixable. You still have the TSV files that you used to upload the 1000 Genomes data tables, right? Just open the *sample* TSV file and rename the two columns to match the corresponding names in the GATK workspace. No need to change anything else, just those two column names. Then, upload the TSV and see what happens: now the GVCF files from the 1000 Genomes samples and their index files also show up in the right columns. Crisis averted! The old columns with the mismatching names are still there, but you can ignore them. Literally, in fact; you can hide them (and any others you don't care about) to reduce the visual clutter. Simply click the gear icon in the upper-right corner of the table to open the table display menu. You can clear the checkboxes for the column names to hide them and even reorder them to suit your preferences.

 It's admittedly a little annoying that you can't simply edit column names in place or delete unwanted columns. The import dialog box feels a bit limited, as well—we would love to be able to preview the data and get a summary count of rows and columns, for example, before confirming the import operation. We're looking forward to seeing these aspects of the interface improve as Terra matures.

Fortunately, we were able to get past this trivial little naming mismatch by editing just two column names. However, minor as it might seem, this stumbling block is symptomatic of a much larger problem: there is not enough standardization around how datasets are structured and how their attributes are named. Almost any attempt to federate datasets from different sources can quickly turn into an exercise in frustration as you find yourself battling conflicting schemas and naming conventions. There is no universal solution (yet), but when you encounter this kind of issue, it can be helpful to start by reducing the problem to the smallest set of components that need to be reconciled. For example, try to define the core data model of each dataset; that is, determine the key pieces of data and how they relate to one another. From there, you can gauge what it would take to make the datasets compatible to the extent that you need them to be.

In our case, we now have what we need: all our samples are defined in the sample table and have a GVCF file listed in the *gvcf* column as well as a corresponding index file in the *gvcf_index* column. Anything else is irrelevant to what we want to do next, which is to apply joint calling to all the samples in the workspace.

Running a Joint-Calling Analysis on the Federated Dataset

To cap off this scenario, we're going to run the *3_Joint_Discovery* workflow that is preconfigured in this workspace, which applies the GATK Best Practices for joint-calling germline short variants on a cohort of samples as described in Chapter 6. We're going to run it on a subset of the samples for testing purposes, but we'll provide pointers for scaling up the analysis in case you want to try running it on all the samples. In any case, the workflow will produce a multisample VCF containing variant calls for the set of samples that we choose to include.

As we admitted earlier, calling this a *federated dataset* is a bit of an exaggeration given that we're really adding just one sample to the 1000 Genomes Phase 3 data. However, the principles we're discussing would apply equally if you were now to add more samples to this workspace. For example, you might want to use the 1000 Genomes data to pad your analysis of a small cohort in order to maximize the benefits you can get from doing joint calling, as is recommended in the GATK documentation.

The workflow is already configured, so let's have a look at what it expects. Go to the Workflow page and click the *3_Joint_Discovery* workflow to view the configuration details. First, we're going to look at which table the workflow is configured to run on. The data selection drop-down menu shows Sample Set, which means that we'll need to provide a sample set from the *sample_set* table. That table currently holds two sample sets; one that lists the NA12878 sample alone, and the other that lists the full 1000 Genomes Phase 3 cohort that we imported into the workspace. We need a sample set that lists some samples from the 1000 Genomes cohort and the NA12878 sample from the GATK workspace, so let's create one now.

 We're going to use 25 samples because that is how much data the predefined configuration of the workflow can accommodate; we'll provide guidance for scaling up once we've completed the test at this scale.

To quickly create a test sample set, set the data table selector to *sample_set* under Step 1, and click Select Data under Step 2. In the dialog box that opens up, select the "Create a new set from selected samples" option. You can select the checkboxes of individual samples or select the checkbox at the top of the column to select all 25 samples that are displayed by default. When you've selected the samples, you can use the box labeled "Selected samples will be saved as a new set named" to specify a name for the sample set or leave the default autogenerated name as is. Press OK to confirm and return to the workflow configuration page. Note that, at this point, you haven't actually created the new sample set; you've just set up the system to create it when you press the Launch button.

But don't press the button yet! We want to show you another way to create a sample set, not just because we have a mean streak (we do), but also to equip you to deal with more-complex situations.

The interface for creating sets can be too limiting when you're working with a lot more samples than can be displayed on a single screen, so we're also going to show you how to use the manual TSV approach to do this. This is going to be a two-step process, similar to what you did earlier: create a new sample set using an *entity* load file and then provide a list of its members using a *membership* load file. The first one is really minimal because it's just the one column header, *entity:sample_set_id*, and the name of the new sample set in the next row, as shown here:

You can save this as a TSV file and upload it to your workspace as previously, or you can take advantage of the alternative option illustrated in Figure 13-13, which involves copying and pasting the two rows into a text field. Annoyingly, it's not possible to type text directly into this text box or edit what you paste in, but it does cut out a few clicks from the import process.

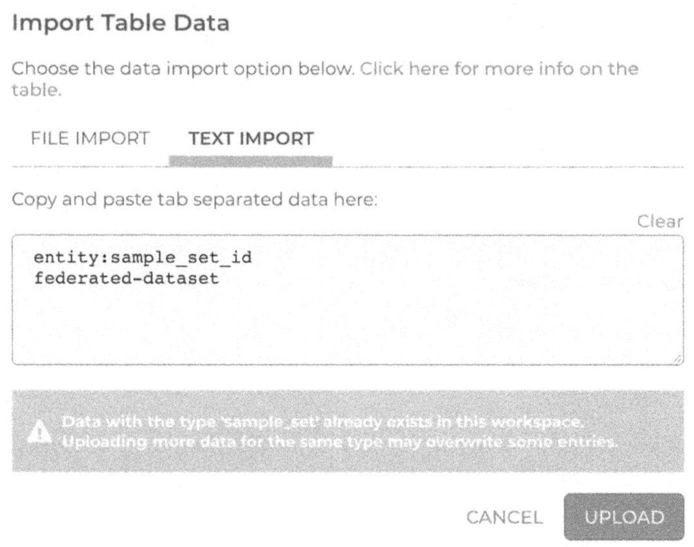

Import Table Data

Choose the data import option below. Click here for more info on the table.

FILE IMPORT   **TEXT IMPORT**

Copy and paste tab separated data here:

Clear

```
entity:sample_set_id
federated-dataset
```

⚠ Data with the type 'sample_set' already exists in this workspace. Uploading more data for the same type may overwrite some entries.

CANCEL   **UPLOAD**

Figure 13-13. Direct text import of TSV-formatted data table content.

Uploading that content creates a new row in the *sample_set* table, but the new sample set has no samples associated with it yet. To remedy that, we need to make a *membership* TSV file containing the list of samples that we want to run on. We like to start from existing data tables because it reduces the chance that we'll get something wrong, especially those finicky header lines. To do so, download the TSV files corresponding to the sample set table and open the *sample_set_membership.tsv* file in your spreadsheet editor. As shown in Figure 13-14, you should see the full list of 1000 Genomes samples in the second column, with the name of the original 1000 Genomes sample set displayed in the first column of each row. If you scroll all the way to the end of the file, you'll also see the NA12878 sample from the GATK workspace, which is assigned to the *one_sample* set.

	A	B
1	membership:sample_set_id	sample
2	1000G-high-coverage-2019-all	SRS000030
3	1000G-high-coverage-2019-all	SRS000031

...

	A	B
2505	1000G-high-coverage-2019-all	SRS000631
2506	one_sample	NA12878

Figure 13-14. Start and end rows of the membership load file sample_set_membership.tsv.

We're going to take a subset of this and transform it into a membership list that associates the samples we choose with our newly created sample set, *federated-dataset*. Start by deleting all but 24 of the samples that belong to the 1000 Genomes cohort from the list, so that you're left with 25 samples in total in the list. It doesn't matter which samples you keep. Then, use the Find and Replace or Rename function of your editor (typically found on the Edit menu) to change the sample set name in the first column to *federated-dataset* for all rows, as demonstrated in Figure 13-15. Make sure to also replace the sample set name for the NA12878 sample. Then, save and upload the file as you've done previously.

	A	B
1	membership:sample_set_id	sample
2	federated-dataset	SRS000030
3	federated-dataset	SRS000031

...

	A	B
25	federated-dataset	SRS000055
26	federated-dataset	NA12878

Figure 13-15. Updated membership load file sample_set_membership.tsv assigning 25 samples to the federated-dataset sample set.

After you've uploaded this membership list, the *sample_set* table should now list the 25 samples as belonging to the *federated-dataset* sample set, as shown in Figure 13-16.

☐ ▾	sample_set_id	↓	samples	
☐	1000G-high-coverage-2019-all		2504 entities	
☐	federated-dataset		25 entities	
☐	one_sample		1 entity	

Figure 13-16. The sample_set table showing the three sample sets.

By the way, this shows you that you can add or modify rows by uploading TSV content for just the parts of the table you want to update or augment. You don't need to reproduce the full table every time. We feel this makes up a bit for not being able to just create the sample sets in the graphical interface.

 In case you're wondering why we're including only the full-scale *NA12878* sample from the GATK workspace and not the downsampled *NA12878_small* sample, it's because they come from the same original person so it would be redundant to use both. Because the downsampled one has less data, it's less likely to produce meaningful results, hence that's the one we eliminate.

Now that we have a sample set that lists all of the samples that we want to use in our analysis, we can proceed with the next step. Select the checkbox on the left of the *federated-dataset* row in the *sample_set* table and then click the "Open with" button, select Workflow, and finally, select *3_Joint_Discovery* to initiate the workflow submission. As we've shown you previously, you can choose to kick off the process from a selection of data on the Data page like this, or you can do it from the Workflows page, as you prefer. The advantage of doing it as we just described is that now your workflow is already set to run on the right sample set; otherwise, you would need to use the Select Data menu to do it when you get to the configuration page.

At this point, all that remains for us to do is check the remainder of the configuration, starting with the Inputs. There are a lot of input fields so we're not going to look at all of them; instead, let's focus on the main file inputs. Based on what you learned about joint calling in Chapter 6, you should have an idea of what kind of input variable you're looking for: one that refers to a list of GVCFs. Sure enough, if you scroll down a bit, you'll find the workflow variable named *input_gvcfs*, as shown in Figure 13-17. The *Type* of the variable is *Array[File]*, which means the workflow expects us to provide a list of files as input, so that checks out. You also see a corresponding variable set up for the list of index files.

| JointGenotyping | input_gvcfs | Array[File] | this.samples.gvcf | {...} |
| JointGenotyping | input_gvcfs_indices | Array[File] | this.samples.gvcf_index | {...} |

Figure 13-17. Input configuration details for the input_gvcfs and input_gvcfs_indices variables.

Now if you look at the value provided in the rightmost column of the input configuration panel, you see *this.samples.gvcf.* That should seem both familiar and new at the same time. Familiar because it appears to be a variation of the *this.something* syntax that we've seen and used previously, which allows us to say, "For each row of the table that the workflow runs on, give the content of the *something* column to this input variable." We used this previously when launching the `HaplotypeCaller` workflow on multiple samples for parallel execution; this syntax allowed us to easily wire up the *input_bam* variable to take the input BAM file for each workflow invocation, without having to specify any files explicitly.

Yet it's also a bit new because it includes an extra element, forming a *this.other.something* syntax. Can you guess what's going on there? This is really cool; it's the payoff from establishing the connections between tables that we mentioned earlier in the chapter. This syntax is basically saying, "For each sample set, look up the list in its *samples* column, and then go to the *sample* table and round up all the GVCF files from the corresponding rows." You can use this *.other.* element to refer to any list of rows from another table, query them for a particular field, and return a list of the corresponding elements. And that is how we generate the list of input GVCF files based on the list of samples linked in the sample set. As a bonus, the order in which the list is made is consistent, so we can assign *this.samples.gvcf* and *this.samples.gvcf_index* to two different variables, and rest assured that the lists of GVCF files and their index files will be in the same sample order.

This is one of the key benefits of having a well-designed data model in place; you can take advantage of the relationships between data entities at different levels. You can even daisy-chain them several levels deep; there is theoretically no limit to how many *.other.* lookups you could do. For more examples of what you can do along these lines, check out the data model used in the somatic Best Practices workspaces, where each participant has a tumor sample and a normal sample. In that data model, one additional table lists the Tumor-Normal pairs and another lists sets of pairs.

Feel free to scroll through the rest of the Inputs configuration page. Then, when you feel like you have a good grasp of how the inputs are wired up, head over to the Outputs for one last check before you launch the workflow. The outputs are all set up along a *this.output_** pattern, and you should see a line that says, "References to outputs will be written to Tables / sample_set," which tells you that the final VCF and related output files will be attached to the sample set. This makes sense because by definition the results of a joint calling analysis pertain to all of the samples that are included.

Finally, go ahead and launch the workflow. You can monitor the execution and explore results the same way as in previous exercises. On 25 samples, our test run of the workflow cost $10 and took about 10 hours (with a couple of preemptions) to run to completion. The longest-running step by far was ImportGVCFs, which ran for about three hours per parallelized segment of the genome and spent half that time in the file localization stage. That's a classic indication that this step would strongly benefit from being optimized to stream data. The other long-running task that is parallelized, GenotypeGVCFs, took about one hour per parallelized segment. Finally, the variant recalibration step, which is not parallelized, took about two hours.

We recommend running this again with a few cohort sizes and comparing the timing diagram for the various runs in order to understand how the time and cost of this workflow scale, depending on the number of samples. You'll need to create new sample sets by following the same instructions as earlier, but keeping however many samples you're interested in. Be sure to give each sample set a different name and do the entity TSV step to add it to the table; otherwise, it will just update the sample set you originally created. If you choose to run the workflow on a larger number of samples than the 25 we've tested so far, you'll need to make a few configuration changes according to the instructions in the workspace dashboard, under the *3_Joint_Discovery* workflow input requirements. In a nutshell, you'll need to allocate more disk space to account for the larger number of sample GVCFs that will need to be localized, using the disk override input variables.

The truth of the matter is that this version of the joint discovery workflow suffers from key scaling inefficiencies, including that it does not use streaming, as we pointed out a moment ago, and does not automatically handle resource allocation scaling. We used it for this exercise because it's conveniently available and fully supported by the GATK support team. Frankly, it's a great way to observe the consequences of inefficiencies that might seem small individually but would cause substantial difficulties at the scale of thousands of samples. However, we do not recommend attempting to run it on the full cohort—in fact, we wouldn't advise going above a few hundred samples until you've at least run some preliminary scale testing to gauge how the workflow time and cost increase with the number of samples.

If you need something that scales better out of the box, an alternative version of this workflow (*https://oreil.ly/WI0NE*) is supposed to scale better, but as of this writing, it has not gone through the GATK support team's publishing process, so it is not officially supported. In addition, its input requirements do not allow you to run it directly on the data tables in your workspace because it expects a sample map file that lists sample names and GVCF files as shown in this example (*https://oreil.ly/mpFnK*)

(but with multiple lines, one for each sample). To get past this obstacle, you could make such a file based on the *sample* table in your workspace and add it as a property of the sample set, or if you're feeling adventurous, you could even write a short WDL task that would run before the main workflow. We leave this as the proverbial exercise for the reader.

This scenario started out with the goal of illustrating a few shortcuts that you can take when assembling a workspace to run GATK Best Practices, but in the process, you also picked up some bonus nuggets of knowledge: what to watch out for when combining data from different origins, how to set up an analysis on the resulting federated dataset, and how a well-crafted data model empowers you to pull together data across different levels. To cap it all off, you ran a complex analysis across a cohort of multiple whole-genome samples without breaking a sweat. Nicely done.

For our last scenario of the chapter, we're going to flip the order of operations and see how that affects our workspace-building process.

Building a Workspace Around a Dataset

So far, we've been taking a very tool-first approach in our scenarios, mainly because our primary focus is on teaching you how to work with a certain range of tools within the cloud computing framework offered by GCP and Terra. Accordingly, our guiding pattern has been to set up tools and then bring in data. However, we recognize that in practice, many of you will follow a data-first approach: bring in data and then figure out how to apply various tools.

In this last scenario, we're going to walk through an example of what that would look like applied to the same 1000 Genomes dataset that we used in the previous scenario. This time, instead of cloning the GATK workspace and pulling in the 1000 Genomes data from the library, we're going to clone the 1000 Genomes workspace and pull in a GATK workflow from the public tool repository Dockstore.

Cloning the 1000 Genomes Data Workspace

Navigate back to the 1000 Genomes High Coverage dataset workspace in the Data Library and clone it as you've done previously, specifying a name and a billing project. As we discussed at the beginning of this chapter, *cloning* a workspace makes a shallow copy of its contents, which means that the data files in the bucket will not be copied to the clone. The clone's data tables will simply point to the original file locations; you can assure yourself that this is true by looking up the file locations in the clone and in the original. This is equivalent to the result of the copy operation we performed in the previous scenario, first by using the data copy option in the interface, and then by repurposing the original workspace's TSV load files.

In your clone, feel free to replace some or all of the description in the Dashboard with a link to the original to make space for your own notes.

Importing a Workflow from Dockstore

Now the question becomes how do we get some workflows in to analyze this data? As usual in Terra, you have a few options, depending on what you're trying to achieve. First, you could simply use the internal Methods Repository as you did previously—either put in your own workflow or browse the public section to see if you can find a workflow you like. However, the internal Methods Repository is used only by people who do their work in Terra, so it's probably missing a lot of interesting workflows that are developed by others in the wider biomedical community.

Another option is to simply copy over a workflow from another workspace; for example, you could grab the *3_Joint_Discovery* workflow from the GATK Best Practices workspace (either the original or the clone you used in the previous scenario) by using the "Copy to another workspace" option in the workspace actions menu (circle icon with stacked dots). However, that again limits you to workflows that have already been brought into Terra. In practice, you will likely find some interesting workflows outside of Terra, so we're going to show you how to import workflows from Dockstore, an increasingly popular registry targeted at the biomedical community that supports workflows written in WDL, CWL, and Nextflow.

Introducing the Dockstore Repository

Dockstore is a public workflow registry developed and operated by the University of California, Santa Cruz, and Ontario Institute for Cancer Research (OICR) intended as a place for workflow authors to publish their workflows for use by the wider community. This is similar to the way code is released on GitHub or a Docker image is published on DockerHub; in fact, Dockstore uses both under the hood. In a nutshell, publishing a workflow in Dockstore consists mainly of linking a GitHub repository where the workflow code resides and providing metadata about the workflow and its inputs. Similarly, registering a container image for a tool involves linking a repository in DockerHub or Quay.io. To learn more about how to publish your own tool containers and workflows in Dockstore, see the extensive online documentation (*https://docs.dockstore.org*).

In your clone of the 1000 Genomes workspace, navigate to the Workflows page and click the Find a Workflow button, as you've done previously. This time, instead of choosing the internal Methods Repository, select the Dockstore option. This takes you to the Dockstore website, where you'll be able to browse a collection of WDL workflows. On the lefthand menu, you should see a search box that says "Enter search term." Type **joint discovery** and view the list of results in the righthand panel (there is no button to press), which will include several GATK workflows, as shown in Figure 13-18.

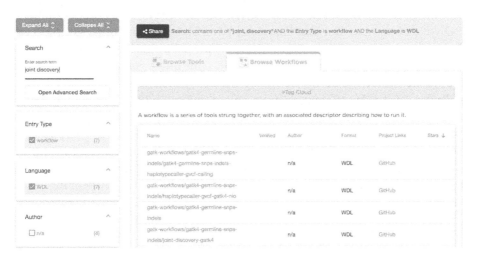

Figure 13-18. Search results for "joint discovery" in Dockstore.

Some of the workflows listed in the results are different versions of the same workflow, registered separately for historical reasons. Unfortunately, the information summarized in the list of search results does not provide a way to easily differentiate between them, so you might sometimes need to click through to the detailed descriptions to understand how they differ.

Find the workflow called *gatk-workflows/gatk4-germline-snps-indels/joint-discovery-gatk4* (*https://oreil.ly/Kfndw*) and click through to its detailed information page. This workflow's name and description should feel at least a little bit familiar because this is actually the source of the *3_Joint_Discovery* workflow that you used in the previous scenario. The code itself is in GitHub, in one of the repositories under the gatk-workflows organization used by the GATK support team to publish official GATK workflows. If you click the Versions tab on the workflow details page, you can see a list of code development branches and releases from the GitHub repository. There should be one version tagged as *default* by the person who registered the workflow, which amounts to them saying, "Unless you know otherwise, this is the version you should use."

It can sometimes be interesting to look at this to see whether the workflow you're interested in is under active development; for example, in this case, as of this writing, we see that the default was set to version 1.1.1 in May 2019, yet recent work was done in a development branch as recently as January 2020. Perhaps those changes will have been released by the time you read this, and a new default version will have been set.

You can explore the various other tabs for yourself, but we'd like to point out a couple more that we find especially useful in addition to those we've already reviewed. The first is the Files tab, which references the WDL source file under Descriptor Files, and a JSON file of inputs under Test Parameter Files. This is boring now, but it will come in handy later. The other tab we really like is the DAG tab, which stands for *directed acyclic graph* and refers to the workflow graph. This offers you an interactive visualization of the workflow akin to what you generated with Womtool in Chapter 9, though the display is slightly different, as you can see in Figure 13-19. By default, this just shows tasks in the workflow, which is a pretty neat way to get a high-level view of how the workflow is wired up without looking at the WDL code. There is also an option to switch to a different visualization tool made by a company called EPAM, which provides a much more granular view that includes inputs and outputs. This is very cool for a deep dive, but the amount of detail can be a bit overwhelming.

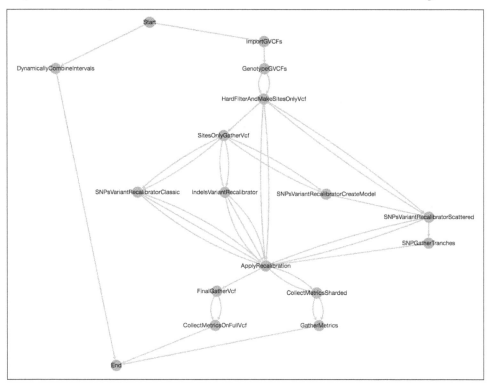

Figure 13-19. The Joint Discovery workflow provided in the DAG tab in Dockstore.

When you're done exploring the workflow details in the Dockstore interface, look for the box in the right menu labeled "Launch with," which offers you several options for running the workflow, including Terra. Click the Terra button to return to Terra with the workflow in your pocket. In the import screen that asks you which workspace you want to put it in, select your clone of the 1000 Genomes workspace. You will land on the configuration page of the newly imported workflow in your workspace.

Configuring the Workflow to Use the Data Tables

The Inputs page will be completely bare, so the first thing we recommend doing is uploading the example JSON file of inputs that was listed on the Files tab in Dockstore. Yes, this is why we mentioned that boring tab. Unfortunately, currently there is no way to have Terra import that file along with the workflow itself (obvious feature request right there, folks), so you'll need to download a copy and then upload it to your workflow configuration using the Upload a JSON option. When the upload is complete, save the configuration and check that there are no remaining error indicators (orange exclamation marks) on the Inputs page. Make sure you select the option to run the workflow directly on file paths and then try running the imported workflow to make sure that everything works out of the box.

Assuming the test run completes successfully, you're going to need to reconfigure the inputs to point to the data tables instead of direct file paths. This can be the most difficult part of the process, because you must figure out which inputs need to be plugged into the data tables, which ones make sense to move to the workspace resources table, and which one to just leave in the configuration. A good rule of thumb for figuring out the first set is to look at the workflow documentation for a description of its input requirements. There won't always be one in the workflows you find out in the field, but when there is one, it's usually highly illuminating with regard to what you should consider to be the main inputs of the workflow. In the case of this particular workflow, you should know what to wire up because we just spent a good chunk of this chapter poking at it, so go do that now. If you get stumped, remember that you can look at the original GATK workspace for a model of what you're aiming for.

 Regarding what's appropriate to move to the workspace resources table, it's really just a matter of judgment and preference because there is no real technical difference as far as Terra is concerned. You could simply decide to leave everything in the workflow until you import another workflow and realize they utilize the same resources, and plan to move the relevant resources over to the workspace data table at that time.

When you have the configuration set up, you'll need to define a sample set by following the same instructions as we gave you in the previous scenario. In fact, at this point, all the remaining work is almost exactly the same as the last two sections of the previous scenario, so we don't feel the need to walk you through every step again; we're confident that you can do it on your own. Good luck!

Finally, if you're feeling a bit short-changed because we made you import the same workflow as in the previous section, here's a suggestion for a stretch assignment to flex your new muscles on a new-to-you workflow. Sometimes, you might want to redo the GVCF calling step if enough has changed in the algorithms or the supporting resources (e.g., if there is a new reference genome build), but the sequencing data files provided with the 1000 Genomes High Coverage dataset are all in CRAM format, which the GATK team does not yet recommend using for direct data access when using streaming. As a result, you need to find either a CRAM-to-BAM conversion workflow to run as a preprocessing step, or a GATK `Haplotype-Caller` workflow that starts with a built-in CRAM-to-BAM conversion task. That should ring a bell if you paid attention in Chapter 9.

So go ahead and try finding the appropriate workflow or combination of workflows to achieve that with the 1000 Genomes data, using the various approaches and resources we've given you over the past few chapters. If you get really stuck, let us know by posting an issue in the GitHub repository (*https://oreil.ly/genomics-repo*), and we'll help you work through it. If a lot of people report difficulties with the stretch assignments, we'll consider posting step-by-step solutions in the blog.

Wrap-Up and Next Steps

In this chapter, you learned how to assemble your own workspace in Terra from data and code components originating from various sources, both internal and external to Terra. We put a lot of emphasis on thinking about how datasets are structured in order to empower you to take advantage of the power of linked data tables in Terra and to combine data from different sources in general. You practiced importing data from the Data Library and running a federated analysis that included data from the 1000 Genomes Project, setting the stage for large-scale analyses. You also learned to import workflows from Dockstore, a tool and workflow repository that can connect to several other platforms besides Terra. At this point, you have all the necessary foundations to be able to assemble and execute your own scalable analysis from publicly or privately available tools and datasets on the cloud. In Chapter 14, the final chapter, we walk through an example of a workspace that reproduces an end-to-end analysis from a published paper to illustrate current capabilities, obstacles, and perspectives for achieving optimal computational reproducibility in biomedical research.

Making a Fully Reproducible Paper

Throughout this book, you've been learning how to use an array of individual tools and components to perform specific tasks. You now know everything you need to get work done—in small pieces. In this final chapter, we walk you through a case study that demonstrates how to bring all of the pieces together into an end-to-end analysis, with the added bonus of showcasing methods for ensuring full computational reproducibility.

The challenge posed in the case study is to reproduce a published analysis in which researchers identified the contribution of a particular gene to the risk for a form of congenital heart disease. The original study was performed on controlled-access data, so the first part of the challenge is to generate a synthetic dataset that can be substituted for the original. Then, we must re-create the data processing and analysis, which include variant discovery, effect prediction, prioritization, and clustering based on the information provided in the paper. Finally, we must apply these methods to the synthetic dataset and evaluate whether we can successfully reproduce the original result. In the course of the chapter, we derive lessons from the challenges we face that should guide you in your efforts to make your own work computationally reproducible.

Overview of the Case Study

We originally conceived this case study as a basis for a couple of workshops that we had proposed to deliver at conferences, starting with the general meeting of the American Society for Human Genetics (ASHG) in October 2018. The basic premise of our workshop proposal was that we would evaluate the barriers to computational reproducibility of genomic methods as they are commonly applied to human medical genetics and teach the audience to overcome those barriers.

We started out with certain expectations about some of the key challenges that might affect both authors and readers of genomic studies, and we were aware of existing solutions for most of those challenges. Our primary goal, therefore, was to highlight real occurrences of those known challenges and demonstrate how to overcome them in practice, based on approaches and principles recommended by experts in the open-science movement. We would then develop educational materials with the ultimate goal of popularizing a set of good practices for researchers to apply when publishing their own work.

Through a series of circumstances that we describe shortly, we selected a study about genetic risk factors for a type of congenital heart disease and worked with one of its lead authors, Dr. Matthieu J. Miossec, to reproduce the computational analysis at the heart of the paper (so to speak). In the course of the project, we verified some of our assumptions, but we also encountered obstacles that we had not foreseen. As a result, we learned quite a bit more than we had expected to, which is not the worst outcome one could imagine.

In this first section of the chapter, we set the stage by discussing the principles that guided our decision making, and then we start populating that stage by describing the research study that we set out to reproduce. We discuss the challenges that we initially identified and describe the logic that we applied to tackle them. Finally, we give you an overview of our implementation plans, as a prelude to the deep-dive sections in which we'll examine the nitty-gritty details of each phase of the project.[1]

Computational Reproducibility and the FAIR Framework

Before we get into the specifics of the analysis that we sought to reproduce, it's worth restating what we mean by *reproducibility* and making sure we distinguish it from *replication*. We've seen these terms used differently, sometimes even swapped for each other, and it's not clear that any consensus exists on the ultimate correct usage. So let's define them within the scope of this book. If you encounter these terms in a different context, we encourage you to verify the author's intent.

We define *reproducibility* with a focus on the repeatability of the analysis process and results. When we say we're reproducing an analysis, we're trying to verify that if we put the same inputs through the same processing, we'll get the same results as we (or someone else) did the first time. This is something we need to be able to do in order to build on someone else's technical work, because we typically need to make sure

---

1 This case study was originally developed by the Support and Education Team in the Data Sciences Platform at the Broad Institute. Because that work was led at the time by one of us (Geraldine), we've kept the description in the first person plural, but we want to be clear that other team members and community members contributed directly to the material presented here, either toward the development and delivery of the original workshop or in later improvements of the materials.

that we're running their analysis correctly before we can begin extending it for our own purposes. As such, it's an absolutely vital accelerator of scientific progress. It's also essential for training purposes, because when we give a learner an exercise, we usually need to ensure that they can arrive at the expected result if they follow instructions to the letter—unless the point is to demonstrate nondeterministic processes. Hopefully, you've found the exercises in this book to be reproducible!

Replication, on the other hand, is all about confirming the insights derived from experimental results (with apologies to Karl Popper, legendary contrarian). To replicate the findings of a study, we typically want to apply different approaches that are not likely to be subject to the same weaknesses or artifacts, to avoid simply falling in the same traps. Ideally, we'll also want to examine data that was independently collected, to avoid confirming any biases originating at that stage. If the results still lead us to draw the same conclusions, we can say we've replicated the findings of the original work.

This difference, illustrated in Figure 14-1, comes down to *plumbing* versus *truth*. In one case, we are trying to verify that "the thing runs as expected," and in the other, "yes, this is how this small part of nature works."

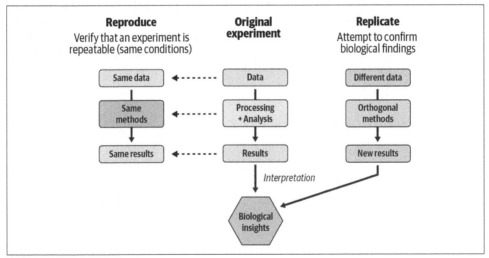

Figure 14-1. Reproducibility of an analysis versus replicability of study findings.

Those definitions shouldn't sound too outlandish given that we've already mentioned this core concept of reproducibility in earlier chapters. For example, in Chapter 8, when we introduced workflows, we noted their great value as a way to encode the instructions for performing a complex analysis in a systematic, automatable way. We also touched on similar themes when we explored Jupyter Notebook as a more flexible, interactive approach to packaging analysis code. We're confident that you won't be surprised when both workflows and notebooks make an appearance later in this chapter.

However, reproducibility is only one facet of the kaleidoscope of open science. As we set out to tackle this case study, we made a deliberate decision to look at it through the lens of the FAIR framework. As we mentioned briefly in Chapter 1, *FAIR* is an acronym that stands for findability, accessibility, interoperability, and reusability. It refers to a framework for evaluating the openness of research assets and services based on those four characteristics. The underlying idea is simple enough: all four pillars are requirements that must be satisfied for an asset to be considered *open* (as in *open science*). For example, to consider that an analysis is open, it is not enough for the code to be technically reusable and interoperable with other computing tools; other researchers should also be able to find the code and obtain a complete working copy. The original publication of the FAIR principles (*https://oreil.ly/JyTlX*) highlighted the necessity of applying them to scientific data management purposes, but it also made clear that they could be applied to a variety of other types of assets or *digital research objects* such as code and tools.

Although the bulk of the case study is focused on reproducibility, which we take to be mostly synonymous with reusability as spelled out in the FAIR framework, the three other FAIR characteristics will be reflected in our various implementation decisions. We'll circle back to this when we go over the methodology that we chose to follow, then again at the conclusion of the chapter when we discuss the final outcomes of this work. For now, it's time to take a look at the original research study that we chose to showcase in this project.

Original Research Study and History of the Case Study

The original study was authored by Drs. Donna J. Page, Matthieu J. Miossec, et al., who set out to identify genetic components associated with a form of congenital heart disease called nonsyndromic tetralogy of Fallot (*https://oreil.ly/XBv-Q*) by analyzing exome sequencing data from 829 cases and 1,252 controls collected from multiple research centers.[2] We go over the analysis in more detail further down, but to summarize for now, they first applied variant discovery methods based on the GATK Best Practices to call variants across the full set of samples (including both cases and controls), and then they used functional effect prediction to identify probable deleterious variants. Finally, they ran a variant load analysis to identify genes more frequently affected by deleterious variants in case samples compared to the controls.

---

2 Page, et al. originally shared their manuscript with the research community as a preprint in bioRxiv (*https://oreil.ly/tztSW*) in April 2018, and ultimately published it in the peer-reviewed journal *Circulation Research* in November 2018: "Whole Exome Sequencing Reveals the Major Genetic Contributors to Nonsyndromic Tetralogy of Fallot," *https://doi.org/10.1161/CIRCRESAHA.118.313250*.

As a result of this analysis, the authors identified 49 deleterious variants within the *NOTCH1* gene that appeared to associate with the tetralogy of Fallot congenital heart disease. Others had previously identified *NOTCH1* variants in families with congenital heart defects, including tetralogy of Fallot, so this was not a wholly unexpected result. However, this work was the first to scale variant analysis of tetralogy of Fallot to a cohort of nearly a thousand case samples, and to show that *NOTCH1* is a significant contributor to tetralogy of Fallot risk. As of this writing, it is still the largest exome study of non-syndromic tetralogy of Fallot that we are aware of.

We connected with Dr. Miossec about a workshop on genomic analysis that we were developing for a joint conference of several societies in Latin America, ISCB-LA SOIBIO EMBnet, to be held in Chile in November in 2018, while the manuscript was still at the preprint stage. At time, our team had also committed to developing a case study on computational reproducibility for the ASHG 2018 workshop, mentioned earlier, which was scheduled for October. Serendipitously, Dr. Miossec's preprint was a great fit for both purposes because it was typical of what our intended audience would find relatable: (1) a classic use case for variant discovery and association methods, (2) performed in a cohort of participants that was large enough to pose some scaling challenges but small enough to be within the means of most research groups, and (3) using exome sequencing data as was most common at the time. Given the Goldilocks-like "just right" (*https://oreil.ly/XDwv5*) character of the study itself and Dr. Miossec's collaborative attitude during our initial discussions of the ISCB-LA SOIBIO EMBnet workshop development project, we approached him about building a case study from his preprint as basis for both workshops. We then worked together over the following months to develop the case study that we eventually delivered at the ASHG 2018 meeting and at the ISCB-LA SOIBIO EMBnet 2018 conference, respectively.

 To be clear, the selection of this study is not intended to demonstrate GATK Best Practices; technically, the authors' original implementation shows deviations from the canonical Best Practices. In addition, this does not constitute an endorsement of the biological validity of the study. In this case study, we are focused solely on the question of computational reproducibility of the analysis.

Assessing the Available Information and Key Challenges

Before we even had the Tetralogy of Fallot study selected, we had been reviewing the key challenges that we expected to face based on our experience with scientific publishing. As illustrated in Figure 14-2, a fundamental asymmetry exists between the information and means available to the author of a scientific paper compared to what is available to the reader. As an author, you generally have full access to an original dataset, which in the case of human genomic data is almost always locked down by data access and data-use restrictions. You have your computing environment, which might be customized with a variety of hardware and software components, and you have tools or code that you apply to the data within that environment to generate results.

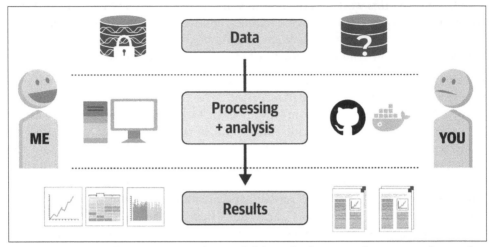

Figure 14-2. Typical asymmetry in the availability of information between author and reader.

On the other side, however, when you're a reader, you usually first see the preprint or published paper, which is in itself a highly processed, curated, and, to some extent, censored view of the author's original results. From the paper, you must walk backward up the stream to find a description of the methods that were applied, which are often incomplete, as in the dreaded "We called variants with GATK Best Practices as described in the online documentation." (No indication of version, specific tools, nada.) If you're lucky, an unformatted PDF buried in the supplemental materials includes a link to code or scripts in a GitHub repository, which might or might not be documented. If you're extremely lucky, you might even find a reference to a Docker container image that encapsulates the software environment requirements, or at least a Dockerfile to generate one. Finally, you might find that the data is subject to access restrictions that could prove unsurmountable. You might be able to cobble together the analysis based on the available information, code, and software resources in the paper—which, to be fair, can sometimes be quite well presented—but without the original data, you have no way to know whether it's running as you expect. So when you go to apply it to your own data, it can be difficult to know how much you can trust the results it produces.

Now that we've painted such a grim picture, how did our situation measure up? Well, the preprint that was available at the time for the Tetralogy of Fallot paper fell squarely within the norm. Figure 14-3 summarizes the methodological information it contained.

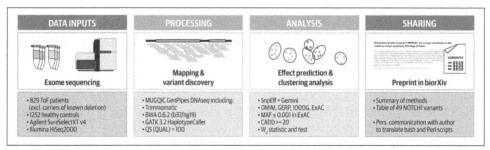

Figure 14-3. Summary of the information provided in the original preprint of the Tetralogy of Fallot paper.

The description of the methods used in the early stages of data processing (alignment up to variant calling) referenced a third-party pipeline operated at another institution, which was definitely too vague to be sufficient to reproduce the work exactly. On the bright side, it did reference key tools and their versions, such as `HaplotypeCaller` from GATK 3.2, and we knew enough about those tools to figure out their requirements by ourselves. For later parts of the analysis, the method descriptions were more detailed and pointed to scripts in Bash and Perl code, but these were still not entirely sufficient by themselves.

Spoiler alert: we benefited immensely from having a lead author in our corner with detailed knowledge of how the analysis was done; we would not have been successful without direct help from Dr. Miossec.

Where we hit the wall, however, was entirely predictable: we knew early on that the data was restricted-access, and as it turned out, we were not able to gain access to any of it at any point, even for development purposes. But hang on to your hat; that doesn't mean the project was dead. We get to solutions in a few minutes.

Designing a Reproducible Implementation

Given these initial conditions, we began thinking about how we were going to tackle the project in practice. We wanted to use a methodology that would allow us not only to reproduce the analysis ourselves, but also to package the resulting implementation for others to reproduce. To that end, we followed the guidance provided by Justin Kitzes et al. in *The Practice of Reproducible Research* (University of California Press, 2018). You can learn a lot from their book that we won't cover here, so we recommend that you check it out. One of the takeaways we quickly implemented was to break down the overall work based on the following questions:

1. What is the *input* data, and how are we going to make that accessible to others?
2. What (a) *processing* and (b) *analysis* methods need to be applied to the data, and how can we make them portable?
3. How can we bundle all the assets we produce and *share* them for others to use?

This was a simple but effective framework that reminded us of our priorities and guided our decisions. The distinction between processing and analysis methods was particularly useful after we agreed what those terms should mean. *Processing* would refer to all the up-front data transformations, cleaning up, format changes, and so on that are always the same and are typically automated to be run in bulk. This corresponds roughly to what is often described as *secondary analysis*. Meanwhile, *analysis* proper would refer to the more changing, context-dependent activities like modeling and clustering that are typically done interactively, on a case-by-case basis. This corresponds to *tertiary analysis*. There was some gray area regarding variant calling and effect prediction, but we ultimately decided that variant calling should fall into the processing bucket, whereas variant effect prediction would go into the analysis bucket.

Although those distinctions were somewhat arbitrary in nature, they were important to us because we had decided that they would guide our efforts regarding how closely we would seek to adhere to the original study. Obviously, we wanted to be as close to the original as possible, but we knew by then that some of the information was incomplete. It was clear that we would need to make compromises to strike a balance between achieving perfection and the amount of effort that we could afford to devote to the project, which was sadly not infinite. So we made a few assumptions. We posited that (1) basic data processing like alignment and variant calling should be fairly robust to adaptations as long as it is applied the same way to all of the data, avoiding batch effects, and that (2) the effect prediction and variant load analysis at the core of the study were more critical to reproduce exactly.

With these guiding principles in mind, we outlined the following implementation strategy:

1. *Data input*: To deal with the locked-down data, we decided to generate synthetic exome data that would emulate the properties of the original dataset, down to the variants of interest. The synthetic data would be completely open and therefore satisfy the *Accessibility* and *Reusability* requirements of the FAIR framework.
2. Data processing and analysis
   a. *Processing*: Because the synthetic data would be fully aligned already, as we discuss shortly, we would need to implement only variant calling in this phase.

Based on what we knew of the original pipeline, we decided to use a combination of existing and new workflows written in WDL. The open and portable nature of WDL would make the processing *accessible, interoperable,* and *reusable* from the perspective of FAIR.

b. *Analysis:* With Dr. Miossec's help, we decided to reimplement his original scripts in two parts: the prediction of variant effects as a workflow in WDL, and the variant load analysis as R code in a Jupyter notebook. As with WDL workflows, using Jupyter notebooks would make the analysis *accessible, interoperable,* and *reusable* from the perspective of FAIR.

3. *Sharing:* We planned to do all the work in a Terra workspace, which we could then share with the community to provide full access to data, WDL workflows, and Jupyter notebooks. This easy sharing in Terra would help us round out the FAIR sweep by making our case study *findable* and *accessible* from the perspective of FAIR.

That last point was a nice bonus of doing the original work for this case study in Terra: we knew that after we were done, we could easily clone the development workspace to make a public workspace (*https://oreil.ly/yj2ql*) that would contain all of the workflows, notebook, and references to data. As a result, we would have only minimal work to do to turn our private development environment into a fully reproducible methods supplement.

In the next two sections, we explain how we implemented all of this in practice. At each stage, we point you to the specific elements of code and data that we developed in the workspace so that you can see exactly what we're talking about. We're not going to have you run anything as part of this chapter, but feel free to clone the workspace and try out the various workflows and notebook anyway.

Now, are you ready to dive into the details? Saddle up and fasten your seatbelt, because we're going to start with what turned out to be the most challenging part: creating the synthetic dataset.

Generating a Synthetic Dataset as a Stand-In for the Private Data

The original analysis involved 829 cases and 1,252 controls, all produced using exome sequencing, and, unfortunately for us, all subject to access restrictions that we could not readily overcome. Indeed, even if we had been able to access the data, we would definitely not have been able to distribute it with the educational resources that we planned to develop. It would still have been useful to have the data available for testing, but in any case, we were going to need to resort to synthetic data to completely achieve our goals.

But wait...what does that mean exactly, you ask? In general, that can mean a lot of things, so let's clarify that here, we're specifically referring to *synthetic sequence data*: a dataset made entirely of read sequences generated by a computer program. The idea was to create such a synthetic dataset that would mimic the original data, as illustrated in Figure 14-4, including the presence of the variants of interest in specific proportions, so that we could swap it in and still achieve our goal of reproducing the analysis despite lacking access to the original data.

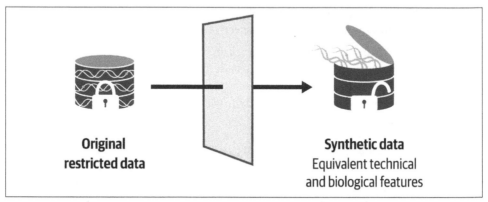

Original
restricted data

Synthetic data
Equivalent technical
and biological features

Figure 14-4. Replacing a real dataset that cannot be distributed with a synthetic dataset that mimics the original data's characteristics.

Let's talk about how that works in practice.

Overall Methodology

The idea of using synthetic genomic data is not new, far from it; people have been using synthetic data for some time. For example, the ICGC-TCGA DREAM Mutation challenges (*https://oreil.ly/DMj28*) were a series of recurring competitions in which organizers provided synthetic data that had been engineered to carry specific known variants, and participants were challenged to develop analysis methods that could identify those mutations with a high degree of accuracy and specificity.

Multiple programs (*https://oreil.ly/LjSJe*) can generate synthetic sequence data for purposes like this; in fact, several were developed in part to enable those challenges. The basic mechanism is to simulate reads based on the sequence of the reference genome and output these as standard FASTQ or BAM files. The tools typically accept a VCF of variant calls as secondary input, which modifies the data simulation algorithm in such a way that the resulting sequence data supports the variants contained in the input VCF. There are also programs that can introduce (or *spike in*) variants into sequence data that already exists. You provide a set of variants to the tool, which then modifies a subset of reads to make them support the desired variant calls.

In fact, in an early round of brainstorming, we had considered simply editing real samples from the 1000 Genomes Project to carry the variants of interest, and have the rest of the dataset stand in as controls. This would allow us to avoid having to generate actual synthetic data. However, our preliminary testing showed that the exome data from the Low Coverage dataset was not of good-enough quality to serve our purposes. At the time, the new High Coverage dataset introduced in Chapter 13 was not yet available, and because that data was generated using WGS, not exome sequencing, it would not have been appropriate for us to use anyway.

We therefore decided to create a set of synthetic exomes, but base them on the 1000 Genomes Phase 3 variant calls (one per project participant) so that they would have realistic variant profiles. This would essentially amount to reconstructing sequencing data for real people based on their previously identified variants. We would then mutate a subset of the synthetic exomes, assigned as cases, with the variants that Dr. Miossec et al. discovered in the Tetralogy of Fallot study, which were listed in the paper. Because there were more participants in the 1000 Genomes dataset than in the Tetralogy of Fallot cohort, that would give us enough elbow room to generate as many samples as we needed. Figure 14-5 presents the key steps in this process.

In theory, this approach seemed fairly straightforward but, as we rapidly learned, the reality of implementing this strategy (especially at the scale that we required) was not trivial. The packages we ended up using were very effective and for the most part quite well documented, but they were not very easy to use out of the box. In our observation, these tools tend to be used mostly by savvy tool developers for small-scale testing and benchmarking purposes, which led us to wonder how much of the high threshold of effort was a consequence of having an expert target audience versus the cause of their limited spread to less savvy users. In any case, we have not seen biomedical researchers use them individually for providing the kind of reproducible research supplements that we were envisioning, and in retrospect we are not surprised given the difficulties we had to overcome in our project. As we discuss later, this further motivated us to consider how we could capitalize on the results that we achieved to make it easier for others to adopt the model of providing synthetic data as a companion to a research study.

In the next section, we expose the gory details of how we implemented this part of the work, occasionally pausing to provide additional color around specific challenges that we feel can provide either valuable insight for others or, failing that, a touch of comic relief to keep things light.

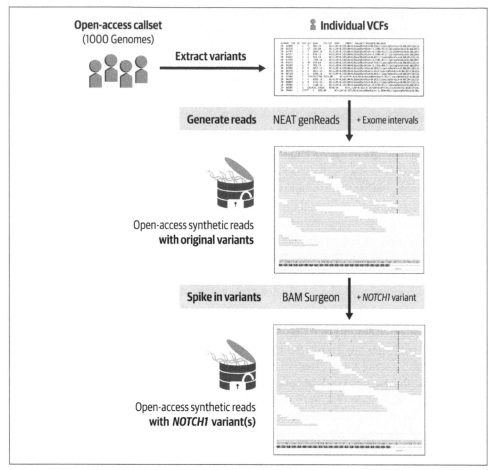

Figure 14-5. Overview of our implementation for generating appropriate synthetic data.

Retrieving the Variant Data from 1000 Genomes Participants

As we mentioned earlier, we decided to base the synthetic data simulation step on VCFs from participants from the 1000 Genomes Project. We chose that dataset because it was the largest available genomics dataset that was fully public, and a copy was freely available in GCS. The convenience stopped there, however. We had to jump through hoops from the get-go, first because the 1000 Genomes variant calls were provided at the time in the form of multisample VCFs containing variant calls from all project participants, split up by chromosome. For our purposes, we needed the opposite: a single-sample VCF containing all chromosomes' worth of data for each project participant.

So the first thing we implemented was a WDL workflow that, given a participant's identifier, would use GATK `SelectVariants` to extract the variant calls for the participant from each of the per-chromosome files, and then Picard `MergeGvcfs` to merge the data into a single VCF for that participant. This sounds simple enough, right? Well, yes, but this is bioinformatics, so of course things weren't that simple. We ran into weird errors due to malformed file headers when trying to process the original multisample VCFs, so we had to add in a call to Picard `FixVcfHeader` to patch them up. Then, we ran into sorting errors after the `SelectVariants` step, so we had to add a call to Picard `SortVcf` to address those issues. We crammed all three commands into a single WDL step to avoid having to do a lot of back-and-forth file copying. It wasn't pretty, but it worked. That became workflow 1_Collect-1000G-participants (*https://oreil.ly/wYw54*), which you can see in action in the case study workspace. For all tasks in this workflow, we used the standard GATK4 container image, which also contains Picard tools.

To run this in our workspace, we set up the participant table with minimal metadata from the 1000 Genomes Project, including participant identifiers and the cohort they originally belonged to, in case the latter came in handy down the line (it hasn't). Then, we configured the collector workflow to run per row in the table of participants and output the resulting VCF to the same row under a *sampleVcf* column.

In this workspace, we didn't need the distinction between participant and sample, so we decided to use a very simple and flat data structure that would use only one of the two. At the time, the participant table was mandatory, so we had no choice but to use it. If we were to rebuild this today, we might switch to using samples instead to make our tables more directly compatible with what we have in other workspaces.

The workflow suffers from a few major inefficiencies, so it took a while to run (4.5 hours on average, 12 hours for a set of 100 files run in parallel, which includes some preemptions) but it did produce the results we were looking for: a complete individual VCF for each participant. With that in hand, we could now move on to the next step: generating the synthetic data!

Creating Fake Exomes Based on Real People

We wanted to create synthetic exome data that would stand in for the original access-controlled dataset, so our first task here was to choose a toolkit with the appropriate features because, as we mentioned earlier, there are quite a few and they display a fairly wide range of capabilities. After surveying available options, we selected the NEAT-genReads toolkit (*https://oreil.ly/2GdMY*), an open source package developed by Zachary Stephens et al. It had a lot of features that we didn't really need, but it seemed to be well regarded and had the core capability we were looking for: the ability to generate either whole genome or exome data based on an input VCF of existing variant calls (in our case, the 1000 Genome participant VCFs).

How does it work? In a nutshell, the program simulates synthetic reads based on the reference genome sequence and then modifies the bases in a subset of reads at the positions where variants are present in the input VCF, in a proportion that matches the genotype in the VCF.

Figure 14-6 illustrates this process. In addition, the program is capable of simulating sequencing errors and other artifacts by modifying other bases in reads, either randomly or based on an error model. By default, the program outputs FASTQ files containing unaligned synthetic sequence data, a BAM file containing the same data aligned to the reference genome, and a VCF file that serves as a truth set containing the variants that the program introduced into the data. For more information on how this works and other options, see the GitHub repository referenced earlier as well as the original publication (*https://oreil.ly/GPUGj*) describing the toolkit in PLoS ONE.

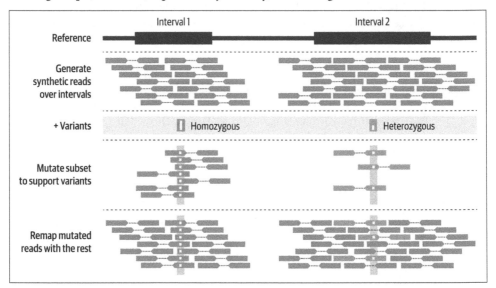

Figure 14-6. NEAT-genReads creates simulated read data based on a reference genome and list of variants.

In addition to this core functionality, NEAT-genReads accepts various parameters that control different aspects of the simulated data. For example, you can specify which regions should be covered or not using an intervals file, and at what target coverage they should be covered. You can also control read length, fragment length, and so on, as well as the mutation rate to use for errors and spontaneous mutations. We took advantage of all of these to tailor the resulting synthetic data to resemble the original study cohort as closely as possible based on the information available in the preprint.

To implement this, we developed the WDL workflow *2_Generate-synthetic-reads* (*https://oreil.ly/i42aY*), which takes a reference genome and a VCF of variants to use in generating the sequence read, and produces a BAM file as well as the truth VCF file. Again, you can see it in action in the workspace.

We'd love to say that this one went more smoothly than the first workflow, but no. We ran into even more problems that appeared to be due to either malformed data or the tools choking on elements that are allowed by the VCF format specification. For example, NEAT-genReads did not tolerate variant records that had multiple identifiers in the *rsID* field, so we had to apply an `awk` command to sanitize the VCFs by removing the contents of the *rsID* field from every record. Another fun one was that NEAT-genReads emitted incomplete VCF headers in the truth VCFs, so again we trotted out Picard `FixVcfHeader`. It also occasionally emitted malformed read names, so to fix that we deployed an unholy combination of `awk` and `samtools`. Finally, we also had to add `readgroup` information to the BAM files produced by NEAT-genReads, which clearly didn't consider that metadata important (GATK responds: blasphemy!).

Is your head spinning yet? Ours certainly was. Don't worry if you're not familiar with the more nitty-gritty details, though. Ultimately, the point is that after you get into a certain level of rolling your own analysis, this is the kind of thing you're likely to encounter, even with a great toolkit like NEAT-genReads (yes, we really like it, warts and all). What's worse is that this is the kind of thing that gets left out of methodology descriptions. Imagine that we had simply written, "We used NEAT-genReads with parameter *x*, *y*, and *z* to generate the synthetic data" and withheld the workflow code. If you tried to reproduce this work without our workflow (which includes comments about the issues), you would have to rediscover and troubleshoot all of that yourself.

Meanwhile, the other slight complication here was that we couldn't find a Docker container image of NEAT-genReads anywhere, so we had to create one ourselves. And this is where we realize with horror that we're 13-plus chapters into this book and we haven't even shown you how to create your own Docker image. Oops? Well, keep in mind that these days most commonly used tools are available on Docker images through projects like BioContainers (*https://oreil.ly/pzFEd*), which provide a great service to the community—they even take requests. Depending on what you do for a living, there's a good chance that you can get most, if not all, of your work done on the cloud without ever worrying about creating a Docker image. Yet we can't really let you finish this book without at least showing you the basics, so if you ever do need to package your own tool, you'll know what to expect. Let's take advantage of this NEAT-genReads exposé to do that in the accompanying sidebar.

Creating Your Own Docker Container Image

To be clear, this is not a hands-on exercise; it's just an explanation of the process. For a step-by-step example that you can follow along in practice, see this article (*https://oreil.ly/LKHQn*) in the Terra documentation.

Building your own Docker container image involves one single but important file: the Dockerfile. It's a plain-text file that contains the instructions for building your image, which we discuss in a minute. The key requirements are that the file must be located at the root of your project and it must be named exactly that: *Dockerfile*, without any file extension and with that exact capitalization.

Inside the Dockerfile, the nature of the instructions can vary wildly, but in general you can think of them as being similar to commands that you would run in your terminal if you were installing things on your local machine. Let's go over the Dockerfile we made for NEAT-genReads as an example:

```
FROM python:2

WORKDIR /usr

RUN pip install numpy==1.15.0
RUN git clone https://github.com/zstephens/neat-genreads
RUN cd ./neat-genreads && git checkout 73b0a5e0c452a3ca22765bd46212642eea2b75c2

CMD ["/bin/bash"]
```

First, you need to know that most Dockerfiles are built on a base image that already has a whole bunch of low-level stuff installed, and you're just adding a few things on top. This is what it looks like for a Python toolkit like NEAT-genReads: we specify FROM: `python:2` and, boom, that sets up a complete Python distribution for us. We can set the working directory with `WORKDIR`, which controls what location commands will be executed from by default inside the container. In our example, we set it to be in the `/usr` directory, which is there as part of the base image.

That sets the stage, so the next step is to add the software that's specific to the tool we're packaging, which typically involves using the `RUN` instruction to run command lines. In our example, we first use the classic Python installer `pip install` to add the `numpy` package, version 1.15.0 to the Python environment in the container. This is very typical of Dockerfiles for Python-based tools. To add the NEAT-genReads toolkit itself, we run `git clone` to copy the code from its repository in GitHub. We then set the version of NEAT-genReads to a specific snapshot by using the CD instruction to go into the code directory, then running a `git` command that sets up our preferred version of the code. Because NEAT-genReads is marked as being under active development, this allows us to point to a known stable version.

Finally, `CMD ["/bin/bash"]` specifies that all commands executed within the container will be run in the Bash shell.

To actually create the Docker image, you would run a command that looks like this, in the directory where your Dockerfile is located:

```
$ docker build -t <username>/<repo>:<tag> .
```

The *username/repo:tag* bit is the same as what you've already encountered in Chapter 4 and onward for the GATK image; for example, `us.gcr.io/broad-gatk/gatk: 4.3.1.0` or, in the case of the image referenced in our workspace, `vdauwera/neat-genreads:2-73b0a5e`.

Finally, you would push the image to a repository like Dockerhub, Quay.io, or GCR with a command like this:

```
$ docker push <username>/<repo>:<tag>
```

Depending on what you want to put on your image, the procedure might vary quite a bit. There are lots of other Dockerfile instructions that we didn't cover here; for example, you can copy local files to the container using CP, as is shown in the aforementioned documentation link. Our recommended approach to writing Dockerfiles is to look up another tool that has similar requirements and is available in a Docker image already, and use its Dockerfile as a template. Usually, if you're using an appropriate base image, you shouldn't need to install too much yourself.

We used the Dockerfile shown in the sidebar to create the container image and then we published it to a public repository on Docker Hub.

To run the workflow in our Terra workspace, we configured it to run on individual rows of the participant table, just like the first workflow. We wired up the *sampleVcf* column, populated by the output of the previous workflow, to serve as input for this second one. Finally, we configured the two resulting outputs, the BAM file and the VCF file, to be written to new columns in the table, *synthExomeBam* and *originalCallsVcf*, along with their index files.

Just like our first workflow, this one too suffers from a few inefficiencies due in part to the cleanup tasks that we had to add. It took about the same time as the first one to run, though within that amount of time it did quite a bit more work: it yielded a complete exome BAM file for each of the participants that we ran it on, with coverage and related metrics modeled after the original Tetralogy of Fallot cohort exomes reported in the preprint. We opened a few of the output BAM files in IGV, and were impressed to see that we could not readily distinguish a synthetic exome from a real exome with similar coverage properties.

This brings us to the final step in the process of generating the synthetic dataset: adding in the variants of interest.

Mutating the Fake Exomes

Walking into this, we were already familiar with an open source tool developed by Adam Ewing called BAMSurgeon (*https://oreil.ly/fFw0e*), which some of our colleagues had previously used successfully to create test cases for variant-calling development work. In a nutshell, BAMSurgeon is designed to modify read records in an existing BAM file to introduce variant alleles, as illustrated in Figure 14-7. The tool takes a BED file (a tab-delimited file format) listing the position, allele, and allele fraction of variants of interest, and will modify the corresponding fraction of reads accordingly. Because we had a list of the variants of interest identified in the Tetralogy of Fallot cohort, including their alleles and positions, we figured that we should be able to introduce them into a subset of our synthetic exomes to re-create the relevant characteristics of the case samples.

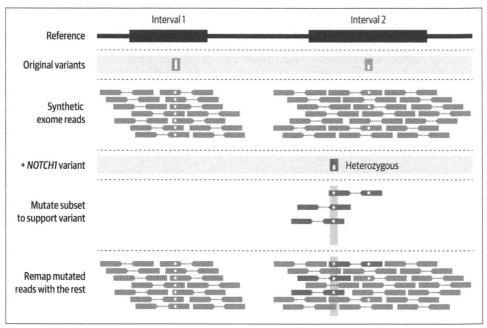

Figure 14-7. BAMSurgeon introduces mutations in read data.

We implemented this as the WDL workflow 3_Mutate-reads-with-BAMSurgeon (*https://oreil.ly/H43dN*), which takes the reference genome, a BAM file, and a BED file listing the variants to introduce, and produces the mutated BAM file. In the first iteration, we limited it to just adding SNPs, because BAMSurgeon uses different commands for different types of variants. In later work, contributors from the community added some logic to the workflow to also handle indels and copy-number variants using the relevant BAMSurgeon commands.

How much trouble did we have with this one? Hardly any trouble at all, if you can believe it. The only real problem we had was when we initially tested this on the 1000 Genomes Low Coverage data, because BAMSurgeon refused to add variants at positions where the sequence coverage was too low. However, when we switched to running on the lovely, plump data produced by NEAT-genReads (which we had given a target coverage of 50X), it worked like a charm.

As previously, in our Terra workspace, we configured the workflow to run on individual rows of the *participant* table. Then, we just needed to determine how we would label cases versus controls and how to set up which cases would be mutated with which variant.

That's when we realized that we had oversimplified the experimental design for this part. Our original plan was simply to introduce the variants listed in the paper into a subset of exomes that would serve as case samples, whereas the rest would be left untouched as control samples. Unfortunately, this reasoning had two flaws.

First, not all case samples had yielded variants of interest in the original study, so there were fewer variants than case samples. In addition, a number of variants in that list fell in genes for which the variant load results were less convincing, so we had decided early on to focus on just the *NOTCH1* variants for the case study. Because *NOTCH1* was the main focus of the paper itself, we felt that reproducing the *NOTCH1* result would be sufficient as a proof of concept. Yet we did believe that we should keep the proportion of cases and controls the same as in the paper. As a result, many of the synthetic exomes labeled as case samples would actually *not* receive a variant of interest.

Second, putting some data through different processing steps compared to the rest would expose us to batch effects. What if something in the differential processing caused an artifact that affected the analysis? We needed to eliminate this source of bias and ensure that all of the data would be processed the same way.

To address these flaws, we decided to introduce a neutral variant into any exome that would not receive a *NOTCH1* variant, whether it was labeled as a case sample or as a control sample. To that end, we designed a synonymous mutation that we predicted should not have any effect.

In the Terra workspace, we added a *role* column to the *participant* table, so that we could label participants as *neutral* or *case* based on the mutation they would receive. We used the term *neutral* rather than *control* because we wanted to give ourselves the flexibility to use participants with the neutral mutation either as actual control samples or as case samples that do not have a *NOTCH1* mutation. We put the list of participants through a random selection process to pick the ones that would become *NOTCH1* case samples and assigned them the *case* role in the participant table, labeling all other participants as neutral.

Finally, we created BED files for each of the *NOTCH1* variants reported in the paper as well as for the neutral variant we had designed. We randomly assigned individual *NOTCH1* variants to the participants labeled *case* and assigned the *neutral* variant to all others. For each participant, we linked the corresponding BED file in a new column of the participant table named *mutation*. We configured the BAMSurgeon workflow to take that field (`this.mutation`) as input for the variant to be introduced in each participant's exome BAM file.

Even though the preparation that we had to do to allocate variants appropriately was not trivial, the workflow was straightforward and ran quickly (0.5 hours on average, 2.5 hours for a set of 100 files run in parallel, again including some preemptions). As planned, it produced mutated exome BAMs for all the participants we ran it on. We checked a few of the output BAM files in IGV to verify that the desired variant had indeed been introduced, and were satisfied that the process had worked as expected.

Generating the Definitive Dataset

Having successfully harnessed existing data and tools into the aforementioned workflows, we were now in a position to generate a synthetic dataset that could adequately stand in for the original Tetralogy of Fallot cohort. However, we had to make one more compromise, which was to limit the number of exomes that we created to 500 instead of the full 2,081 that we had originally planned on to emulate the original cohort. The reason? It's so pedestrian that it hurts: we simply ran out of time before the conference. Sometimes, you just have to move forward with what you have.

So we generated 500 mutated synthetic exomes, which we further organized into two participant sets in the workspace, imaginatively named A100 and B500. The A100 set was a subset of 100 participants that included eight *NOTCH1* cases. The B500 set was the superset of all participants for which we created synthetic exomes, and included all 49 *NOTCH1* cases reported in the paper. You might notice that these compositions don't quite match the proportion of *NOTCH1* cases in the overall Tetralogy of Fallot cohort, but the two sets have similar proportions to each other. Later, we discuss how this affects the interpretation of the final results.

At this point, we're ready to move on to the next section, in which we attempt to reproduce the study methodology itself.

Re-Creating the Data Processing and Analysis Methodology

We wouldn't blame you if you had forgotten all about the original study by now, so let's go through a quick refresher. As illustrated in Figure 14-8, Dr. Miossec et al. started out with exome sequencing data from a cohort of 2,081 study participants, split almost evenly between cases; that is, patients suffering from the congenital heart disease tetralogy of Fallot (specifically, a type called *nonsyndromic*), and controls (people not affected by the disease). They applied fairly standard processing to the exome data using a pipeline based on the GATK Best Practices, consisting mainly of mapping to the reference genome and variant calling, which you are an expert in since Chapter 6. Then, they applied a custom analysis that involved predicting the effects of variants in order to focus on deleterious variants, before attempting to identify genes with a higher variant load than would be expected by chance.

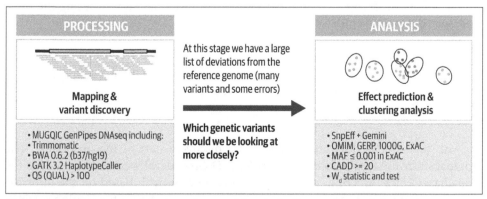

Figure 14-8. Summary of the two phases of the study: Processing and Analysis.

We go into more detail about the meaning and purpose of these procedures in a few moments. For now, the key concept we'd like to focus on is the difference between these two phases of the study. In the *Processing phase*, which can be highly standardized and automated, we're mainly trying to clean up the data and extract what we consider to be useful information out of the immense haystack of data we're presented. This produces variant calls across the entire sequenced territory, a long list of differences between an individual's genome, and a rather arbitrary reference. In itself, that list doesn't provide any real insights; it's still just data. That's where the *Analysis phase* comes in: we're going to start asking specific questions of the data and, if all goes well, produce insights about a particular aspect of the biological system that we're investigating.

We highlight this distinction because it was a key factor in our decision making about how closely we believed we needed to get to reproducing the exact methods used in the original study. Let's dive in, and you'll see what we mean.

Mapping and Variant Discovery

Right off the bat, we bump into an important deviation. The original study started from exome sequencing data in FASTQ format, which had to be mapped before anything else could happen. However, when we generated the synthetic exome dataset, we produced BAM files containing sequencing data that was already mapped and ready to go. We could have reverted the data to an unmapped state and started the pipeline from scratch, but we chose not to do that. Why? Time, cost, and convenience, not necessarily in that order. We made the deliberate decision to take a shortcut because we estimated that the effect on our results would be minimal. In our experience, as long as some key requirements are met and the data is of reasonably good quality, the exact implementation of the data processing part of the pipeline does not matter quite as much as whether it has been applied consistently in the same way to the entire dataset. We did, however, make sure that we were using the same reference genome, of course.

 Note the preconditions; we're not saying that *anything* goes. Sadly, a full discussion on that topic is beyond the scope of this book, but hit us up on Twitter if you want to talk.

So with that, we fast-forward to the variant discovery part of the processing. The original study used a pipeline called *GenPipes DNAseq* developed and operated at the Canadian Centre for Computational Genomics in Montreal. Based on the information we found in the preprint and in the online documentation for GenPipes DNAseq, the pipeline followed the main tenets of the GATK Best Practices for germline short variant discovery. This included performing single-sample variant calling with GATK `HaplotypeCaller` to generate a GVCF for each sample and then joint-calling all samples together. We did flag one deviation, which was that according to the preprint, the variant callset filtering step was done by hard filtering instead of using VQSR.

Accordingly, we repurposed some GATK workflows that we had on hand and customized them to emulate the processing described in the materials available to us. This produced the two WDL workflows *4_Call-single-sample-GVCF-GATK* (*https://oreil.ly/I8CO6*) (featuring your favorite `HaplotypeCaller`) and *5_Joint-call-and-hard-filter-GATK4* (*https://oreil.ly/UW5kj*) (featuring `GenotypeGVCFs` and friend), which you can find in the Terra workspace. We configured the first to run the synthetic exome BAM file of each participant in the table and output the GVCF to the corresponding row. In contrast, we configured the second to run at the sample set level, taking in the GVCFs from all participants in the set and producing a filtered multi-sample VCF for the set.

Here we allowed ourselves another notable deviation: instead of using version 3.2 of the GATK as the original study did, we used GATK 4.0.9.0 in order to take advantage of the substantial improvements in speed and scalability that GATK4 versions offer compared to older versions of GATK. We expected that given the high quality of our synthetic data, there would be few differences in the output of `HaplotypeCaller` between these versions. We would expect important differences to arise only on lower-quality data and in the low-confidence regions of the genome. Again, some caveats apply that we don't get into here, and we were careful to apply the same methods consistently across our entire dataset.

All in all, we made quick work of the Processing phase, in part thanks to those two deviations from the original methods. Since then, we've considered going back and reprocessing the data from scratch with a more accurate reimplementation of the original methods, to gauge exactly how much difference it would make. It's not exactly keeping us awake at night, but we're curious to see whether the results would support our judgment calls. If you end up doing it as a take-home exercise, don't hesitate to let us know how badly wrong we were.

For now, it's time to move on to the really interesting part of the analysis, in which we discover whether we're able to reproduce the main result of the original study.

Variant Effect Prediction, Prioritization, and Variant Load Analysis

Let's take a few minutes to recap the problem statement and go over the experimental design for this part given that, so far, we've mostly waved off the details. First, what do we have in hand, and what are we trying to achieve? Well, we're starting from a cohort-level VCF, which is a long inventory of everyone's variants, and we're hoping to identify genes that contribute to the risk of developing a form of congenital heart disease called tetralogy of Fallot. To be clear, we're not trying to find a *particular variant* that is associated with the disease; we're looking for a *gene* that is associated with the disease, with the understanding that different patients can carry different variants located within the same gene. What makes this difficult is that variants occur naturally in all genes, and if you look at enough variants across enough samples, you can easily find spurious associations that mean nothing.

The first step to tackling this problem is to narrow the list of variants to eliminate as many of them as you can based on how common they are, whether they're present in the control samples, and what kind of biological effect they are likely to have, if any. Indeed, the overwhelming majority of the variants in the callset are common, boring, and/or unlikely to have any biological effect. We want to prioritize; for instance, focus on rare variants that are found only in the case samples and are likely to have a deleterious effect on gene function.

With that much-reduced list of variants in hand, we can then perform a variant load analysis, as illustrated in Figure 14-9. This involves looking for genes that appear to be more frequently mutated than you would expect to observe by chance given their size.

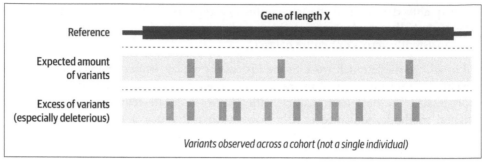

Figure 14-9. Comparing variant load in a gene across multiple samples.

The original study used a tool called `SnpEff` to predict variant effects, and then `vt` from GEMINI to prioritize variants based on their predicted effect, their frequency in population databases, and their presence in only the case samples. We implemented this as a two-step WDL workflow, *6_Predict-variant-effects-GEMINI* (*https://oreil.ly/ GBesI*), that used the same versions and commands for both tools, as well as the same population resources. In Terra, we configured the workflow to run at the participant set level and output the list of prioritized variants to the same row in the table. For this step, we had to add a pedigree file to each participant, specifying whether it should be counted as a case sample or as a control sample. For a given participant set, we selected one half of the participants, all carrying the neutral mutation, to be control samples. We then assigned the rest of the participants in the set, including all those carrying *NOTCH1* mutations, to be case samples.

Finally, Dr. Miossec helped us rewrite the original Perl scripts for the clustering analysis into R in a Jupyter notebook (*https://oreil.ly/cRROQ*), which you can also find in the workspace. The notebook is set up to pull in the output file produced by GEMINI. It then runs through a series of analysis steps, culminating in a clustering test that looks for an excess of rare, deleterious variants in the case samples. The analysis is documented with explanations for each step, so be sure to check it out if you'd like to learn more about this analysis.

In both the workflow and the notebook, we were careful to reproduce the original analysis to the letter, because we considered that these were the parts that would have the most influence on the final results.

Analytical Performance of the New Implementation

So the burning question is…did it work? Were we able to pull out *NOTCH1* from the haystack? The short answer is, yes, mostly; the long answer is rather more interesting.

As we mentioned earlier, we ended up being able to generate only 500 synthetic exomes in the time we had, so we defined a 100-participant set with eight *NOTCH1* cases and a 500-participant set with all 49 *NOTCH1* cases. Although these proportions of *NOTCH1* cases were higher than those in the original dataset, what mattered to us at that point was that our two participant sets were roughly proportional to each other. Because we could not test the method at full scale as originally intended, we would at least be able to gauge how the results changed proportionally to the dataset size, which was a point of interest from early on.

We ran the full set of processing and analysis workflows on both participant sets and then loaded the results of both into the notebook, where you can see the final clustering analysis repeated on both sets. In both cases, *NOTCH1* came up as a candidate gene, but with rather different levels of confidence, as shown in Figure 14-10. In the 100-participant set, *NOTCH1* was ranked only second in the table of candidate genes, failing to rise above the background noise. In contrast, in the 500-participant set, *NOTCH1* emerged as the uncontested top candidate, clearing the rest of the pack by a wide margin.

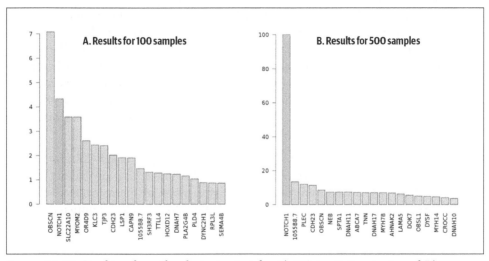

Figure 14-10. Ranking from the clustering test for A) 100-participant set, and B) 500-participant set.

We considered that this constituted a successful enough reproduction of the analysis because we were able to generate a result that matched our expectations, even though we had to make compromises in terms of fidelity.

We were particularly encouraged to see that this approach could be used to test the scaling of statistical power depending on the dataset size. We can easily imagine setting up additional experiments to test the scaling observed here to a finer granularity and a larger scale. We would also want to test how varying the proportion of *NOTCH1* cases relative to the overall cohort size would affect the results, as well, starting with the proportion that was actually observed in the original study.

So many possibilities, so little time. And speaking of which, it's getting late; we're almost at the end of...

The Long, Winding Road to FAIRness

Given what we just described, you should now be able to reproduce the Tetralogy of Fallot analysis in Terra as well as reuse the data and/or the methods for your own work if they are applicable. You should also be able to do so outside of Terra for the most part because most of the components are directly usable on other platforms. You can download the data to use on different platforms, and the Docker images and WDL workflows can be run just about anywhere. The notebook might take more effort to reuse given that the environment is not as cleanly bundled as the Docker images used for workflows. You can access all necessary information regarding the computing environment, but you would still need to independently set up an equivalent environment. We're hopeful that this is something that will improve further over time; for example, imagine if there were an option to emit a Dockerfile that could recreate the software environment of a particular notebook. We would also like to see a standardized way to obtain a digital object identifier (DOI) for a Terra workspace, to use in publications that include a companion workspace as supplemental material. This ties in with important work that others are doing to bundle research artifacts for easier archival, retrieval, and reuse. It's a complicated topic, however, as tends to be the case when developing standards.

So what can you do to achieve computational reproducibility and FAIRness in your own work? In the course of this case study, we pointed out several key factors that are typically under your direct control. We discussed many of them in the context in which they arose, but we thought it might be helpful to provide a summary of what we consider to be the most important guidelines:

Open source
> It's hard to overstate the importance of using open source tools and workflows. Using open source tooling helps keep your work reproducible in two ways: it ensures accessibility to all, and it increases the transparency of the methodology. If someone isn't able to run the exact same code for any reason, they still have a chance to read the code and reimplement the algorithm.

Version control
> It's imperative that you track the versions of tools, workflows, and notebooks you use as systematically as possible because changes from one version to the next can have a major effect on how tools work and what results they produce. If you are using other people's tools, record the versions you use and include that information when you publish your analysis. If you develop your own tools and workflows, make sure to use a version-control system like GitHub. In Terra, workflows are systematically versioned, either through GitHub and Dockstore or through

the internal Methods Repository. Notebooks are currently not versioned in Terra, so we recommend downloading your notebooks regularly and checking them into a version-control system like GitHub.

Automation and portability

When it comes to developing the part of your analysis that you will run many times on a lot of data, choose a method that will allow you to automate as much as possible and reduce your dependence on a specific environment. For example, choose a pipelining language like WDL or CWL, which can easily be run by others, rather than writing elaborate scripts in Bash, Python, or R that others will have a difficult time running even if you provide the code with your publications.

Built-in documentation

For analyses that involve a lot of interaction with the data and judgment calls in how to go from one step to the next, consider providing a Jupyter notebook that recapitulates your analysis with built-in commentary explaining what's happening at each step. Even if you prefer working in an environment that gives you more flexibility in your day-to-day work, like RStudio, for example, packaging the finished product as a Jupyter notebook will greatly enhance both its reproducibility and its intelligibility. Think about the last time you helped a new lab member or classmate get up to speed on a new analysis method; imagine being able to give them a notebook that explained what to do step by step rather than a loose collection of scripts and a *README* document that might or might not be up-to-date.

Open data

Finally, the elephant in the room is often going to be the data. There will be many valid reasons you might not be able to share the original data you worked with, especially if you're working with protected human data. However, you might be able to refer to open-access data that is sufficient to demonstrate how your analysis works. When that's not an option, as in the case study we just described, consider whether it might be possible to use synthetic data instead.

Does that last point mean you should go through what we did to generate the synthetic dataset? Well, hopefully not. You're most welcome to use the synthetic exomes we generated, which are freely accessible, or use the workflows we showed you to create your own. Our workflows are admittedly not very efficient in their present form and would benefit from some optimization to make them cheaper to run and more scalable, but they work. And, hey, they're open source, so feel free to play with them and propose some improvements!

Looking at the bigger picture, however, we believe there is an opportunity here to develop a community resource to minimize the amount of duplicative work that anyone needs to do in this space. Considering how standardized the generation of sequencing data has become (at least for the short read technologies), it should be possible to identify the data types that are most commonly used and generate a collection of generic synthetic datasets. These could then be hosted in the cloud and used off the shelf in combination with tools like BAMSurgeon by researchers who want to reproduce someone else's work, or to make their own work more readily reproducible, along the lines of what we described in this chapter.

Final Conclusions

Well, here we are at the end of the final chapter. How does it feel? You've covered a lot of ground in the cloud, so to speak. You started on a puny little Cloud Shell, and then moved quickly on to a beefier virtual machine, where you ran real genomic analyses with GATK; first manually, step by step, and then through simple WDL workflows with Cromwell. You applied your inner detective to deciphering mystery workflows, learning in the process how to approach new workflows methodically. Next, you graduated to using the Pipelines API, scaling up your workflow chops and tasting the freedom of launching a parallelized workflow into a wide-open space. From there, you jumped to Terra, where you peeled back layers of functionality—workflows, notebooks, data management—to finally find yourself on solid ground with our case study of a fully reproducible paper.

Based on what you learned in this book, you can now head down the path of using these tools in your own work. Take advantage of massive cloud-hosted datasets and a wealth of Dockerized tools that you don't need to figure out how to install. Develop your own analyses, execute them at scale, and share them with the world—or just your labmates. Whether you continue to use the specific tools we used here for those purposes, or with other tools within the growing cloud-based life sciences ecosystem, you can rest assured that similar principles and paradigms will apply.

Be sure to check back for updates to the book's GitHub repository (*https://oreil.ly/genomics-repo*), companion blog (*https://oreil.ly/genomics-blog*), and latest developments around GATK, WDL, Docker, and Terra.

Have fun, and keep it reproducible!

Glossary

ADC

Application Default Credentials—an authentication method

AI

artificial intelligence—a subset of machine learning

API

application programming interface

ASHG

American Society for Human Genetics—a professional society

ASIC

application-specific integrated circuit

AWS

Amazon Web Services

BAM

Binary Alignment Map—a sequence data format (binary compressed version of SAM)

BCL

Basecall—a sequence data format

BQSR

Base Quality Score Recalibration—a data processing technique that accounts for technical biases in sample preparation and sequencing processes

CIGAR

Concise Idiosyncratic Gapped Alignment Report—describes the alignment of a sequence read to the genome reference

CLI

command-line interface

CNA

copy-number alteration—a type of genomic variation that consists of a change in number of copies of the affected segment of DNA

CNN

convolutional neural network—a type of machine learning algorithm

contig

In the original context of genome assembly, a set of overlapping DNA segments that together represent a consensus region of DNA; by extension, can refer to a chromosome or any other free-standing representation of DNA such as a partially assembled segment of chromosome, a plasmid or a virus genome (from *contiguous*)

CPU

central processing unit

CRAM

a sequence data format (similar to BAM; another compressed version of SAM)

CWL

Common Workflow Language—a portable workflow language

DSL

domain-specific language—a programming language tailored for a specific application

FAIR

findable, accessible, interoperable and reusable

FP

false positive—a variant call that is not biologically real, resulting from error or data artifacts

FPGA

field-programmable gate array—a type of computer processor

GA4GH

Global Alliance for Genomics and Health—an organization promoting the development of standards and APIs

GATK

Genome Analysis Toolkit (see software list)

Gb

gigabases—one million nucleotide bases

GCE

Google Compute Engine—a GCP service for compute resources

GCP

Google Cloud Platform (see services list)

GCS

Google Cloud Storage—a GCP service for data storage

GiaB

Genome in a Bottle—(see datasets list)

GPU

graphical processing unit—a type of computer processor

GUI

graphical user interface

GWAS

genome-wide analysis studies

HDD

hard-disk drive—a type of data storage device

hg38

Human genome build 38—the current reference genome sequence

HGP

Human Genome Project—the first project to map the full human genome

HMM

hidden Markov model—an algorithm used to compute probabilities

HPC

high-performance computing

IDE

integrated development environment—an application designed to facilitate code development

IGV

Integrated Genome Viewer (see software list)

ISB

Institute for Systems Biology

JSON

JavaScript Object Notation—a structured format for providing key:value content

JVM

Java Virtual Machine—the application that runs Java programs

linkage disequilibrium

nonrandom association of alleles at two or more loci in a general population

locus (plural: loci)

a genomic location or feature; often used to refer to a gene, but can also refer to a single position or a short segment of sequence

MAF

Mutation Annotation Format—a file format for storing genomic variation data mostly used in cancer genomics

MCMC

Markov-Chain Monte Carlo—a type of algorithm

Mendelian

relating to the principles of genetic inheritance described by Gregor Mendel

Mendelian violation

genotype of a child that cannot occur through Mendelian inheritance given the parental genotypes

ML

machine learning—a discipline and type of algorithm

NCI

National Cancer Institutes

NHGRI

National Human Genome Research Institute

NHLBI

National Heart, Lung, and Blood Institute

NIH

National Institutes of Health

NIST

National Institute of Standards and Technology

OS

operating system

PAPI

Pipelines API—a GCP service

PCR

polymerase chain reaction—a technique for making many copies of a piece of DNA

PGM

Probabilistic Graphical Model—a type of machine learning algorithm

PoN

Panel of Normals—resource based on normal samples, used in somatic analysis to account for technical artifacts and germline background

SAM

Sequence Alignment Map—a sequence data format

SFTP

SSH File Transfer Protocol

SGE

Sun GridEngine—a type of job scheduler used in many high-performance computing environments

SNP

single-nucleotide polymorphism—a SNV in the context of population genetics

SNV

single-nucleotide variant—a form of genomic variation consisting of a single letter change

SSD

solid-state drive—a type of data storage device

SSH

Secure Shell

TCGA

The Cancer Genome Atlas—a cancer genome analysis project

TPU

tensor processing unit—a type of computer processor

TSV

tab-separated values—a tabular file format where values are separated by tab characters

UUID

universally unique identifier

VCF

Variant Call Format—a file format for storing genomic variation information

VM

virtual machine

VQSR

Variant Quality Score Recalibration—a variant filtering technique

WDL

Workflow Description Language—a portable workflow language

WES

Whole Exome Sequencing—a somewhat misguided name for exome sequencing, a technique that produces DNA sequence data covering a subset of the genome (sometimes also referred to as WEX)

WGS

Whole Genome Sequencing—a technique that produces DNA sequence data for (nearly) the entire genome

Index

Symbols

! (exclamation mark) preceding commands in Jupyter Notebook, 351, 359

"" (double quotes) enclosing strings, 231

(hash sign) in terminal prompt, 121, 128

$ (dollar sign)
    preceding variables, 128, 222
    VM prompt ending in, 85, 128

&& (logical AND) operator in GATK filtering commands, 143

+ (plus sign) concatenation operator in Python, 355

== (equality) operator, 252

? (question mark) indicating optional variable, 256

[] (square brackets), enclosing array values, 231

\ (backslash), ending lines in multiline commands, 123

{} (curly braces)
    enclosing code blocks in WDL, 217
    enclosing Python variable in Jupyter Notebook, 351

| (pipe symbol), piping data between utilities, 136

|| (logical OR) operator in GATK filtering expressions, 143

A

A, C, G, T (adenine, cytosine, guanine, thymine), 14

abstraction layer, 59

access token to view private data in Jupyter Notebook, 357

accessory files for GATK in WDL workflow, 228

add-on tools and services for Jupyter Notebook, 334

AggregatedBamQC, 259, 267

aliases in WDL
    import statement aliased to Utils, 264
    referring to content from imported files, 262, 266
    using to call task in different ways within a workflow, 267

alignment information for reads, 42

alleles, 15

allelic copy ratio analysis, 207

allelic ratio, alteration of, 27

alpha or beta evaluation stage, GATK tools, 192

alterations, 32

alternate haplotypes, 23

alternative contigs, 201

Amazon Web Services (AWS), 73
    AWS Batch job scheduler, 68
    Spot instances, 286

amino acid chain, synthesis of, 17

amplicon preparation, 38

amplification and deletion in segments, 206

annotating predicted functional effects with Funcotator, 195

API Library (GCP), 270

Application Default Credentials (ADC), 275

ApplyVQSR, 170

argument names (GATK), POSIX convention for, 118

ASIC (application-specific integrated circuit), 55

assemblies or builds (human genome), 23
authentication
    accessing data outside Terra in private GCS
        bucket, 374
    checking for custom VM in GCP, 102-103
    generating file of credentials that Cromwell
        gives to PAPI, 275
    Google account linked to Terra, 298
    viewing private data in embedded IGV
        browser in Jupyter Notebook, 357
automating analysis execution with workflows,
    209-244
    installing and setting up Cromwell, 212-216
    using scatter-gather parallelism, 236-244
    WDL and Cromwell, 210-212
    your first GATK workflow, Hello Haploty-
        peCaller, 226-236
    your first WDL script, Hello World, 216-225
autosomes, 24

B

b37 reference genome, 24
BAMs (Binary Alignment Maps)
    AggregatedBamQC task call, 259
    BAM files as input to HaplotypeCaller, 127
        loading into IGV, 129
    bamout file created by HaplotypeCaller,
        feeding to CNNScoreVariants, 178
    bamout generation to troubleshoot Haplo-
        typeCaller call, 132-134
    bamout produced by Mutect2 command,
        191
    converting CRAM to BAM, 248
    defined, 441
    entering GCS file path to view in IGV, 111
    file output by duplicate marking in germline
        short variant discovery, 152
    loading BAM files into embedded IGV
        browser in Jupyter Notebook without
        specifying index file, 357
    output from samtools view command, 250
    SAM format compressed into, 35
    unmapped read data stored in, production
        implementation of GATK Best Practices,
        150
    UnmappedBamToAlignedBam task, 258
BAMSurgeon, 430-432
BamToCram task, 259
BamToGvcf task, 259

Base Quality Score Recalibration (BQSR), 153
Basecall (BCL) data format, 34
basename function (WDL), 228, 239
    naming outputs based on name of input
        files, 252
    specifying file extension to trim off at end of
        input filename, 254
    using with sub function, 253
bases, 14
Bash shell
    installation scripts written in, 344
    running Docker container within Google
        Cloud Shell, 91
    scripts in WDL command blocks, 251
Bash, sed, and awk, 66
batch analysis using multiple VMs via cloud
    batch services, 76
Bayesian statistics, use in HaplotypeCaller, 167
Best Practices workflows (GATK), 143-145
    covered in this book, 145
    evolution and deviations, 144
    for germline short variant discovery,
        147-181
            data preprocessing, 147-155
    fully loaded workspaces in Terra, 320
    other major use cases, 145
beta evaluation stage, GATK tools, 192
billing account for GCP, connecting to Terra,
    299
billing project, 275
    creating for Terra, 298-301
    selecting for Terra workspace, 302
Binder project, 334
biochemical alterations in samples preserved
    with formalin and paraffin, 185
biochemical, physical, and software artifacts, 50
Boolean variables
    is_cram, 252
    use_gatk3_haplotype_caller, 263
boot disk, customizing for VM in GCP, 98
BQSR (Base Quality Score Recalibration), 153
Broad Institute of MIT and Harvard, 7, 9
Broad Methods Repository, 380
buckets (see storage buckets)
Budgets and alerts settings in GCP, 82
BWA mapper, 130
bwa mem mapper, 149

C

C++, effort to rewrite GATK in, 59
caching, 57
CalculateContamination tool, 192
CalculateGenotypePosteriors, 171
call caching (Cromwell), 318-320
call statement in WDL script, 218
CallCacheReading stage, 318
CallCopyRatioSegments tool, 206
callsets annotation, 137-140
The Cancer Genome Atlas (TCGA), 7
Cancer Genomics Cloud (CGC), 7
cancer genomics, challenges in, 183
cancer versus normal cell line, contrast in
    genomic composition, 197
cat command (gcloud), 85
cat utility, 351
Cell menu (Jupyter Notebook), 350
central processing units (see CPUs)
centromere, 15, 38
chromatids, 15
chromosomal recombination, 16
CIGAR (Concise Idiosyncratic Gapped Align-
    ment Report), 441
    soft clips, read data and, 130
    strings describing structure of read align-
        ment, 35
CLI (command-line interface), 441
cloud computing
    categories of research use cases for cloud
        services, 74-76
        batch analysis, using multiple VMs via
            batch services, 76
        framework analysis with multiple VMs
            via framework services, 76
        intermediate development and analysis,
            using single VM, 75
        lightweight development, Google Cloud
            Shell, 75
    cloud providers, workflow optimizations
        and, 286
    cloud to cloud federated analyses, 10
    cloud to on-premises analyses, 10
    cloud, defined, 56
    getting started, 79-113
        configuring integrated genome viewer to
            read data from GCS buckets, 109-113
        running basic commands in Google
            Cloud Shell, 84-94

setting up Google Cloud account and
    first project, 79-83
setting up your own custom VM, 94-108
introduction to, 72-74
    abstraction of infrastructure concerns,
        72
    cloud pros and cons, 73
    evolution of cloud infrastructure and
        services, 73
    toward an ecosystem for data sharing and
        analysis, 4-10
Cloud Storage JSON API, 270
clusters, 56
    access to Spark cluster in GATK tools, 120
CNNs (convolutional neural networks), 173,
    441
    applied to variant calling, 173
    applying 1D CNN to filter single-sample
        whole genome sequencing callset,
        176-177
    applying 2D CNN to include read data in
        the modeling, 178-180
    using for filtering germline short variants,
        overview, 175
CNNScoreVariants tool, 176
    --tensor-type read_tensor argument to
        switch on 2D model, 178
code blocks in WDL, 217
code cells, running in Jupyter Notebook, 332
code optimizations, 58
codons, 17
Colaboratory service for Jupyter Notebook
    (GCP), 334
collaboration and sharing, Jupyter Notebook in
    Terra, 339
CollectedReadCounts tool, 202
CollectRawWgsMetrics task, 259
CollectWgsMetrics task, 259, 267
CombineGVCFs tool, 164
command block in WDL code, 217
command line
    GATK command-line syntax, 117
    gcloud, connecting to VM by, 102
    (see also gcloud)
command-line tools
    Hello World in Jupyter Notebook, running
        using Python magic methods, 350
    in runtime environment for Jupyter Note-
        book in Terra, 342

long, winding road to FAIRness, 438-440
overview of the case study, 413-421
    assessing available information and key
        challenges, 417-419
    computational reproducibility and FAIR
        framework, 414-416
    designing a reproducible implementa-
        tion, 419-421
    original research study and history of
        case study, 416-417
re-creating data processing and analysis
    methodology, 432-438
    analytical performance of new imple-
        mentation, 436-438
    mapping and variant discovery, 433-435
    variant effect prediction, prioritization,
        and variant load analysis, 435
Funcotator tool, 195
functional annotations, resources for, 195
functional equivalence pipeline specification,
    51

G
GA4GH (see Global Alliance for Genomics and
    Health)
GA4GH Cloud Work Stream, 65
GATK (see Genome Analysis Toolkit)
gatk --java-options command, 118
GATK Best Practices workspace, creating new
    workspace from, 390-407
    cloning GATK Best Practices workspace,
        391
    copying data tables from 1000 Genomes
        workspace, 396
    examining data tables in GATK workspace,
        391-394
    learning about 1000 Genomes High Cover-
        age dataset, 394-396
    running joint-calling analysis on federated
        dataset, 400-407
    using TSV load files to import data from
        1000 Genomes workspace, 398-400
gatk wrapper script, 122
Gaussian mixture models, 167
Gaussian-kernel binary-segmentation algo-
    rithm, 204
GCE (see Google Compute Engine)
gcloud (Google Cloud SDK), 84
    alpha genomics pipelines run, 288

auth command, 275, 357
compute SSH command, 102
init command, 128
specifying Project ID, 84
GCP console, 80
    (see also Google Cloud Platform)
    APIs and Services section, enabling APIs
        and services, 270
    Billing account permissions page, 299
    billing notifications feature, 83
    Billing section, 81
    managing buckets through, 87
    stopping, starting, or deleting your VM
        instance, 108
    Terra interface, similarity to, 297
    using to download output files from Terra,
        328
    VM instance management page, 121
gene expression, 17
gene panels, 39
Gene Transfer Format (GTF), 195
genes, 14-16
    mutations in, 18
    relationship between DNA and, 19
genetics versus genomics, 20
Genome Analysis Toolkit (GATK), 4
    about, 115
    Best Practices for germline short variant
        discovery, 147-181, 268
        data preprocessing, 147-155
        joint discovery analysis, 155-172
        single-sample calling with CNN filtering,
            173-180
    Best Practices for somatic copy-number
        alterations, 197-208
        additional analysis options, 207-208
        tumor-only analysis workflow, 198-207
    Best Practices for somatic variant discovery,
        183-208
        challenges in cancer genomics, 183
        somatic short variants, SNVs and indels,
            185-196
    choice of Java programming language, rea-
        sons for, 59
    code modules optimizing PairHMM calcu-
        lations for different CPUs, 58
    command-line systax, 117
    development of DRAGEN-GATK pipelines,
        59

N

N + 1 problem, 158
    defeating by adding samples incrementally
        to GenomicsDB datastore, 164
n1-standard-2 machine type (in GCP), 98
NA12878 and NAA12878_small samples, 322
names
    bucket names in GCS, 86
    Docker containers, 92
    naming VM instance in GCP, 96
namespace in file paths, 221
nano text editor, 215
    WDL Hello World script in, 216
Nanopore sequencing, 32
National Cancer Institutes (NCI), 7
National Human Genome Research Institute
    (NHGRI), 20
National Institutes of Health (NIH)
    expansion in public cloud offerings, 8
    genomic data hosted by, 5
native code optimizations, 58
NeMo repository, 377
networking (cloud), 72
neural networks, 167
next-generation sequencing (NGS), 32
Nextflow, 67
nodes/machines, 56
nonribosomal protein synthesis, 17
normal versus cancer cell line, contrast in
    genomic composition, 197
notebook runtime, 336
    becoming blank slate after customization
        and regeneration, 337
    checking for newly added notebook in
        workspace, 389
    customizing, 337
    inspecting and customizing configuration in
        Terra, 341-346
        inspecting runtime configuration, 342
        startup script installing GATK and IGV
            on, 343
    local storage space in, 336
    stopping and saving its state, 337
nucleic acids, 14

O

"omics", 20
1000 Genomes High Coverage dataset, 377
    cloning the workspace, 407
    copying data tables from workspace, 396
    getting to know, 394-396
    using TSV load files to import data from
        1000 Genomes workspace, 398-400
1000 Genomes project, 190
    retrieving variant data from participants,
        424
one-off run execution mode, running Crom-
    well in, 218
openjdk-8-jre-headless option, installing, 212
OpenWDL, 210
optional variables, 256
order of operations, 241
os.environ command, 352
output block in WDL code, 218
outputs configuration panel for workflows in
    Terra, 316, 324
Oxford Nanopore sequencing, 32

P

PacBio DNA sequencing, 32
packages installed in default notebook runtime
    environment, 342
Page, Donna J. et al., 416
paired-end sequencing, 42
PairHMM, 58
Panel of Normals (PoN), 187, 322, 443
    creating Mutect2 PoN, 188
    creating somatic CNA PoN, 201
parallel computing, 60-64
    cloud workflows, parallelization of, 282
    many levels, from cores to clusters and
        clouds, 61-63
    parallelizing a simple analysis, 60-61
    read data distributed into multiple read
        groups, 150
    scatter-gather parallelism in GATK work-
        flow, 236-244
    trade-offs, speed, efficiency, and cost, 63
participant table (Terra), 322
passenger mutations, 183
PathSeq tool, 49
performance bottlenecks in computing, 56-60
    data storage and I/O operations, hard drive
        vs. solid state, 57
    memory, 57
    specialized hardware and code optimiza-
        tions, trade-offs, 57
Perl, 67

About the Authors

Dr. Geraldine A. Van der Auwera is the Director of Outreach and Communications for the Data Sciences Platform (DSP) at the Broad Institute of MIT and Harvard. As part of her outreach role, she serves as an educator and advocate for researchers who use DSP software and services including GATK, the Broad's industry-leading toolkit for variant discovery analysis; the Cromwell/WDL workflow management system; and Terra.bio, a cloud-based analysis platform that integrates computational resources, methods repository, and data management in a user-friendly environment. Van der Auwera was originally trained as a microbiologist, earning her Ph.D. in biological engineering from the Université catholique de Louvain in Belgium in 2007, then surviving a four-year postdoctoral stint at Harvard Medical School. She joined the Broad Institute in 2012 to become Benevolent Dictator for Life of the GATK user community, leaving behind the bench and pipette work forever.

Dr. Brian O'Connor is the Director of the Computational Genomics Platform at the University of California Santa Cruz (UCSC) Genomics Institute. There, he focuses on the development and deployment of large-scale, cloud-based systems for analyzing genomic data. These include the NHGRI AnVIL and NHLBI Bio Data Catalyst platforms as well as the Dockstore site for workflow and tool sharing. Brian is active in standards efforts and is the cochair of the Global Alliance for Genomics and Health Cloud Work Stream where he works on API standards for cloud interoperability. Brian joined UCSC from the Ontario Institute for Cancer Research where his previous projects included leading the technical implementation of worldwide, cloud-based analysis systems for the PanCancer Analysis of Whole Genomes project, creating the Dockstore, and managing a successful rebuild of the International Cancer Genome Consortium's Data Portal.

Colophon

The animal on the cover of *Genomics in the Cloud* is a milkspotted pufferfish (*Chelonodon patoca*), native to and present today in the estuaries, mangroves, coastal regions, and brackish waters of the Indo-Pacific. Pufferfish, also called blowfish, are known for their elastic bellies, which expand to ward off predators.

Like other species of pufferfish, the milkspotted variety has tetrodotoxin on its skin, a mucus layer that is poisonous to humans and other predators. Its scales are brownish gray with dark stripes and white spots. The underside is yellowy-white and the eye has a yellow ring. An adult milkspotted pufferfish can grow 10 inches long. These fish eat invertebrates such as mollusks and worms and plants like algae.

Many of the animals on O'Reilly's covers are endangered; all of them are important to the world.

The cover illustration is by Karen Montgomery, based on a black-and-white engraving from *Shaw's Zoology*. The cover fonts are Gilroy Semibold and Guardian Sans. The text font is Adobe Minion Pro; the heading font is Adobe Myriad Condensed; and the code font is Dalton Maag's Ubuntu Mono.

Milton Keynes UK
Ingram Content Group UK Ltd.
UKHW031816121223
434254UK00008B/668